Traffic Management

Research for Innovative Transports Set

coordinated by
Bernard Jacob

Volume 3

Traffic Management

Edited by
Simon Cohen
George Yannis

iSTE

WILEY

First published 2016 in Great Britain and the United States by ISTE Ltd and John Wiley & Sons, Inc.

ISTE Ltd
27-37 St George's Road
London SW19 4EU
UK

www.iste.co.uk

John Wiley & Sons, Inc.
111 River Street
Hoboken, NJ 07030
USA

www.wiley.com

Library of Congress Control Number: 2016936177

British Library Cataloguing-in-Publication Data
A CIP record for this book is available from the British Library
ISBN 978-1-78630-028-7

Contents

Acknowledgments

The European Commission, DG MOVE and RTD, the Conference of European Road Directors (CEDR), the European Road Transport Research Advisory Council (ERTRAC), the European Rail Research Advisory Council (ERRAC) the European technology platform WATERBORNE-TP are acknowledged for their support and active contribution to the Programme Committee of the TRA2014, in charge of reviewing and selecting the papers presented at the conference, which forms the main input of this volume.

The French Institute for science and technology for transport, development and network (Ifsttar) is aknowledged for the successful organisaation organization of the conference TRA2014, in which 600 high quality papers were presented.

Joëlle Labarrère, former secretary of the Programme Committee of TRA2014, and executive assitant of the department COSYS with Ifsttar, is aknowledged for her valuable help to the editors and for this volume making.

Preface

The transport sector is very much concerned about environmental adaptation and mitigation issues. Most of these are related to the objective of curbing GHG emission by 20% by 2020, alternative energy and energy savings, sustainable mobility and infrastructures, safety and security, etc. These objectives require the implementation of advanced research work to develop new policies, and to adjust education and industrial innovations.

The theme and slogan of the Transport Research Arena held in Paris (TRA2014) were respectively: "Transport Solutions: From Research to Deployment" and "Innovate Mobility, Mobilise Innovation". Top researchers and engineers, as well as private and public policy and decision-makers, were mobilized to identify and take the relevant steps to implement innovative solutions in transport. All surface modes were included, including walking and cycling, as well as cross modal aspects.

Policies, technologies and behaviors must be continually adapted to new constraints, such as climate change, the diminishing supply of fossil fuels, the economic crisis, the increased demand for mobility, safety and security, i.e. all the societal issues of the 21st Century. Transport infrastructures and materials, modal share, co-modality, urban planning, public transportation and mobility, safety and security, freight, logistics, ITS, energy and environment issues are the subject of extensive studies, research work and industrial innovations that are reported in this series of books.

This book is a part of a set of six volumes called the *Research for Innovative Transports* set. This collection presents an update of the latest academic and applied research, case studies, best practices and user perspectives on transport carried out in Europe and worldwide. The presentations made during TRA2014 reflect on them. The TRAs are supported by the European Commission (DG-MOVE and DG-RTD), the Conference of European Road Directors (CEDR) and the modal European platforms, ERRAC (rail), ERTRAC (road), WATERBORNE, and ALICE (freight), and also by the European Construction Technology Platform (ECTP) and the European Transport Research Alliance (ETRA).

The volumes are made up of a selection of the best papers presented at the TRA2014. All papers were peer reviewed before being accepted at the conference, and they were then selected by the editors for the purpose of the present collection. Each volume contains complementary academic and applied inputs provided by highly qualified researchers, experts and professionals from all around the world.

Each volume of the series covers a strategic theme of TRA2014.

Volume 1, *Energy and Environment*, presents recent research work around the triptych "transports, energy and environment" that demonstrate that vehicle technologies and fuels can still improve, but it is necessary to prepare their implementation (electromobility), think about new services and involve enterprises. Mitigation strategies and policies are examined under different prospective scenarios, to develop and promote alternative fuels and technologies, multi-modality and services, and optimized transport chains while preserving climate and the environment. Evaluation and certification methodologies are key elements for assessing air pollution, noise and vibration from road, rail and maritime transports, and their impacts on the environment. Different depollution technologies and mitigation strategies are also presented.

Volume 2, *Towards Innovative Freight and Logistics*, analyzes how to optimize freight movements and logistics; it introduces new vehicle concepts, points out the governance and organization issues, and proposes an assessment framework.

Volumes 3 and 4 are complementary books covering the topic of traffic management and safety.

Volume 3, *Traffic Management*, starts with a survey of data collection processes and policies and then shows how traffic modeling and simulation may resolve major problems. Traffic management, monitoring and routing tools and experience are reported and the role of traffic information is highlighted. Impact assessments are presented.

Volume 4, *Traffic Safety*, describes the main road safety policies, accident analysis and modeling. Special focus is placed on the safety of vulnerable road users. The roles of infrastructure and ITS in safety are analyzed. Finally railway safety is focused upon.

Volume 5, *Materials and Infrastructures*, is split into two sub-volumes, investigating geotechnical issues and pavement materials' characterization, innovative materials, technologies and processes and introducing new techniques and approaches for auscultation and monitoring. Solutions to increase the durability of infrastructures and to improve maintenance and repair are presented, for recycling as well as for ensuring the sustainability of the infrastructures. Specific railways and inland navigation issues are addressed. A focus is put on climate resilient roads.

Volume 6, *Urban Mobility and Public Transport*, highlights possible innovations in order to improve transports and the quality of life in urban areas. Buses and two-wheelers could be a viable alternative in cities if they are safe and reliable. New methodologies are needed to assess urban mobility through new survey protocols, a better knowledge of user behavior or taking into account the value of travel for public transport. The interactions between urban transport and land planning are a key issue. However, these interactions have to be better assessed in order to propose scenarios for new policies.

Bernard JACOB, Chair of the TRA2014 Programme Committee

Jean-Bernard KOVARIK, Chair of the TRA2014 Management Committee

March 2016

Introduction

Advances in telecommunications and information technologies are changing the practices used in both everyday life and in professional life. The transport world, sensitive to innovation, does not escape to this movement.

Our daily environment demonstrates successful mutations. New equipment is deployed along the roads or on board vehicles. Variable message signs display real-time travel times. Cameras detect incidents and trigger alerts. Information terminals provide service schedules and waiting times for buses or trains. Other technologies facilitate the management of daily travel, making it more reliable, safer and more comfortable.

These developments highlight various aspects of advanced traffic management as well as transport safety. Behind, there is transport research. Its role is to imagine, assess and support the emergence of new approaches and innovative systems. Multi-disciplinary by essence, transport research is well adapted to deal with these issues. This is the purpose of this volume, resulting from the international TRA2014 Conference, held in Paris in April 2014. The Conference was organized under the sign of the transition from research to deployment in transport solutions.

The topic of traffic is organized into two separate but complementary volumes, Volume 3 on *Traffic Management* and Volume 4 on *Traffic Safety*, both presenting a selection of chapters in the aforementioned fields. As a major event on transport in Europe, the Conference covered a broad range of issues linked to Traffic Management and Safety. Naturally, the shortlist presented here does not cover the wide spectrum of these area. It aims to highlight its diversity through a choice of updated papers from the Conference. Selection is primarily based on a quality

---

Introduction written by Simon COHEN and George YANNIS.

criterion, taking also into account the geographical diversity of papers in order to restore the originality and richness of the current research.

I.1. Main findings

The selected 23 chapters included in this volume on traffic management demonstrate how technological innovations as well as new methodologies applied to traffic management can modify usual practices, and offer efficient solutions to the ongoing challenges of increasing congestion, environmental issues and economical constraints. Both theoretical papers and practical case studies explore topics such as data collection, modelling, traffic operations, information and assessment.

The quality of traffic management is strongly dependent on the availability of reliable and accurate data. But what are good practices in data collection? Surveys can help to outline the best practices among existing systems and identify possible areas for improvement. How can managers prepare for the evolution towards a new generation of sensors while taking into account the contributions of cooperative systems? These questions concern both the nature of data as well as the system architecture. These are themes discussed during the conference and reported in this volume.

Many lessons are provided to traffic managers from practical case studies. In this regard, the recent concept of living laboratory (living lab), as a large-scale sensor network, is becoming more and more popular. In the last few years, living labs have been expanding. They enable *in situ* experimentation and stimulate activities in sensor technology, data collection, innovative services and products. In living labs, stakeholders such as public, private, research institutions, industry, and especially users, closely collaborate in an open-innovation process. A large amount of open data become available and enables a better insight into what happens on the network. The involvement of users is increased. Finally, living labs lead to the development of better adapted innovations at a lower cost, for all the different stakeholders.

Traffic modeling is at the heart of planning issues and projects assessment. Several of the innovations illustrated in some papers concern the traditional four steps process. In the field of modeling, the Conference reveals a more important use of comprehensive approaches. The objectives and criteria become multiple and simultaneously concern the issues of efficiency, safety and environmental aspects. Sustainability becomes a key theme. Multimodality issues are discussed frequently. Management becomes integrated and juxtaposes the stages of operation and decision. Several categories are taken into account, such as transport operators, users and institutions.

Different case studies in railway and road traffic management illustrate these trends.

Information tools play a fundamental role in network management. Before a journey, they influence the route choice, the departure times and, to a lesser extent, the modal choice. Traveler information systems are more and more becoming real time, interoperable, multimodal, regional and even transnational.

Assessment should provide answers to the new questions arising. For example, what are the impacts of cooperative systems? What are the comparative performances of classical and innovative systems? How to introduce the social costs of congestion of rail networks and therefore improve cost-benefit analysis? Impact assessment as described in several case studies facilitates the decision of the different stakeholders.

This third volume extracted from the TRA Conference 2014 will interest both the research community and higher education, professionals in the management of road and rail traffic, economic and institutional decision-makers increasingly require new forms of network management. They will find both state of the art of some key issues, chapters on some methods and illustrative case studies.

The presentation of *Traffic Management* is split into five parts. It begins with data collection, continues with modeling, describes some traffic operations and information systems and ends with evaluation.

Part 1 considers both classic and innovative data collection systems. The reader, especially the traffic manager, will find descriptions of best practices as well as the potential of some advanced technologies. Issues on cooperative ITS architectures are also discussed. The living lab in the Dutch City of Assen shows how it contributes to improve use cases.

In Part 2, researchers will be interested by theoretical innovations in the traditional four steps process: a new approach to trip generation, optimized intermodal roundtrips and an alternative assignment method.

Part 3 deals with traffic operations, congestion monitoring and routing. It will be of particular interest to professionals. Behavioral responses to daily traffic congestion are investigated, showing that marginal adjustments are preferred to alterations. Lane changing behavior is also analyzed under free flow and heavy traffic. Effects of variable speed limits are estimated while various dynamic route guidance algorithms are compared.

Part 4 lists first the influence of pre-trip information systems from a literature review. Concrete experiences of real time passenger information and multimodal traveler information are reported.

Finally, Part 5 focuses mainly on impact assessment. The impact of various cooperative systems on safety, traffic efficiency and environment are reported. The conclusion provides decision support for road authorities on future investments in the field. Performance of classic and innovative technologies for travel time calculation are compared in a pilot. The issue of the social cost of rail congestion is also addressed.

I.2. Conclusion

The chapters gathered in this volume provide an insight into research, best practices and transport policies with focus on state-of the-art advances in traffic management. They demonstrate the progress made in the various process of data collection, modelling, management, information and assessment, assisting academics, transport professionals, practitioners and decision makers to a better understanding of the current and future trends. The crucial and increasing role of ITS applications becomes evident and more frequently researchers and practitioners apply a universal approach and interdisciplinary methodologies to address transport related issues, including global approaches in modelling. Furthermore, special focus is given to the sustainability of presented solutions, with an emphasis onto optimized and sustainable traffic management, as well as new concepts such as living labs.

PART 1

Data Collection

1

A Review of Statewide Traffic Data Collection, Processing, Projection and Quality Control

Three of the most essential metrics of highway system operation are the volume, composition and weight of traffic using the roadway and street network. Agencies need timely and reliable traffic information to perform their varied duties in the areas of planning, design, construction, maintenance and operation of roads. If the collected data are not current and accurate, decisions made by the agency may be delayed or incorrect. This study identifies appropriate measures to ensure that high-quality traffic data are collected, processed, analyzed and reported in an optimal and cost-effective way by the New Mexico Department of Transportation. This is achieved through an in-depth review of the Department's current procedures, including interviews with individuals, both inside and outside of the agency, who have traffic data responsibilities. A survey of best practices in traffic data collection at the national level, as identified in the technical literature and an examination of programs in selected states was also undertaken.

1.1. Introduction

One of the most essential metrics of highway system operation is the volume of traffic using the roadway and street network. The New Mexico Department of Transportation (NMDOT) needs timely and reliable traffic volume information to perform its varied duties in the areas of planning, design, construction, maintenance and operation of roads. If volume data collected by the NMDOT and others are not current, the decisions made by the department may be delayed or incorrect.

Chapter written by Rafiqul TAREFDER and James BROGAN.

On the other hand, if the data are current but erroneous, then any decisions made on the basis of faulty data will certainly be wrong. Part of the challenge, however, is that the collection, processing and storage of traffic volume data is decidedly not glamorous and, like all traffic studies, is subject to cutbacks in financial support when department resources are tight.

There can easily be adverse financial consequences for not collecting and maintaining traffic volume data in a manner consistent with recommended practices. To improve the results of other applications of traffic data, the NMDOT foresees a need to: (1) identify inefficiencies, inaccuracies and redundancies in the department's current practices of data collection, analysis, and forecasting and (2) develop justifiable recommendations for enhancing data collection, quality control and data use.

1.2. Current traffic data collection in New Mexico

The State of New Mexico has approximately 68,000 centerline miles of roadway (20). Over 14,000 of these miles are on non-local roads which are monitored by the NMDOT with both short-term volume (coverage) counts and approximately 150 active permanent count locations. The permanent count sites include both Automatic Traffic Recorders (ATRs) recording volume, speed and classification data; 15 Automatic Weight and Classification (AWAC) sites collecting weigh-in-motion data in addition to volume and classification, and 30 ITS/camera sites, primarily in the Albuquerque area. A map of the permanent site locations is shown in Figure 1.1 and a complete listing of the permanent sites may be found in a separate document. The number of active sites may vary slightly due to maintenance and construction schedules as well as down time caused by incidents.

Equipment installed at a typical volume, classification and speed site includes both inductive loops and piezoelectric sensors; the weigh-in-motion sites have either bending plates, piezoelectric sensors or load cells along with inductive loops while the ITS sites have Smart Sensors (microwave) and cameras installed to provide both volume and speed data. For data polling from the ATR and AWAC sites, the NMDOT uses TDP (Peek) and TRADAS software developed by Chaparral Systems for data processing and analysis. Traffic count data is stored in an Oracle database.

Based on the Department's Consolidated Highway Data Base (CHDB – recently replaced by TIMS – Transportation Information Management System), a total of 14,853 short-count (coverage count) roadway sections have been identified; it appears that these sections were established based not only on ADT (the TMG suggests that homogeneous segments have traffic volumes that remain within ±10%) but also by the lengths of various construction projects, the location of political boundaries, and physical reference points such as interchange or intersection locations. Broken down by functional classification, these sites, along with their roadway mileages, as reported to FHWA for 2009, are shown in Table 1.1.

Counts at locations on these sections, except for the urban local system and minor rural collectors and local roads, are supposed to occur for 48 hours on a three-year cycle for the higher functional classes and on a six-year cycle for the lower functional classes. For example, in preparing the count program for the years 2012, 2013 and 2014, all of the high functional class sections counted in 2011 would be placed on the 2014 count program, 2010 sections would be counted again in 2013, and the remaining sections would be placed on the 2012 program. The traffic technician conducting the count may place the counter anywhere in the section where it is safe to do so. While the department has approximately 120 portable counters, only about 90 are currently being used because of staff shortages.

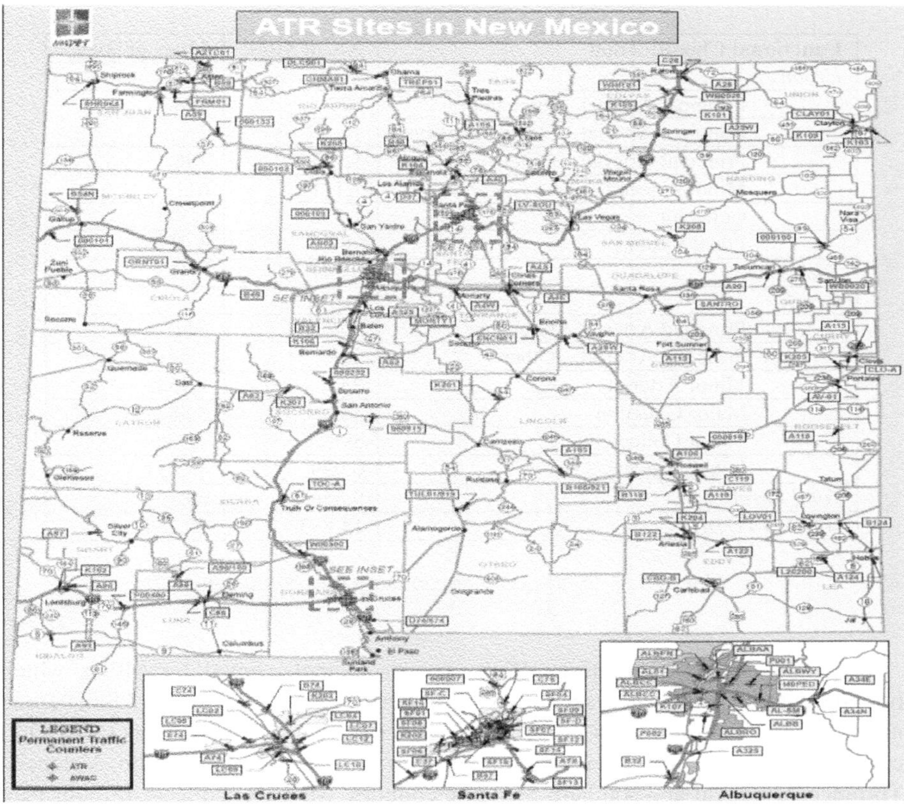

Figure 1.1. *ATR and AWAC sites in New Mexico. For a color version of this figure, please see www.iste.co.uk/jacob/traffic.zip*

While the current number of identified short-count roadway sections is adequate given the rural nature of the state, not enough sections in the lower functional classifications (minor arterials, collectors) are actually being counted because of staff and funding shortages. In fact, information from the Data Management Bureau indicates that a total of only 1,597 short-term counts from all agencies were conducted in 2009 and 1,690 in 2010. While these numbers may indicate adequate coverage of the Principal Arterials on a three-year cycle, they show that little coverage was provided to the lower functional classes.

Within the NMDOT, the counter shop at the General Office in Santa Fe conducts the counts statewide. District offices do not provide count data to the General Office although they may conduct specific counts (turning movements, speed, etc.) within their jurisdictions. Although there is no seasonal rule on when short counts are performed, the technicians try to avoid snow plows which tend to tear up the road tubes. Otherwise, counts are performed anytime the technician is in the area.

Functional Classification		No. of Sites	Miles of Rwy.
Urban			
	Principal Arterial – Interstate	529	156
	Principal Arterial – Other Freeways	1	5
	Principal Arterial – Other	1,117	706
	Minor Arterial	1,111	611
	Collector	1,552	1,503
	Local System	987	5,012
Total Urban		5,297	7,993
Rural			
	Principal Arterial – Interstate	966	844
	Principal Arterial – Other	298	1,841
	Minor Arterial	252	1,953
	Major Collector	513	3,882
	Minor Collector	598	3,150
	Local System	6,929	48,721
Total Rural		9,556	60,391
GRAND TOTAL		14,853	68,384

Table 1.1. *Short-count sections by functional class*

NMDOT traffic monitoring efforts are also supplemented by MPOs which provide data on many road sections within their jurisdictions. In the Albuquerque metropolitan area, for example, the Mid-Region Council of Governments (MRCOG) collects traffic data for all major state and non-state roads in Bernalillo, Valencia, Torrance, Sandoval and southern Santa Fe counties. MRCOG collects 48-hour data at a location every three years, usually on a Monday or Tuesday. Growth factors are applied to the counts during off-years and classification data from MRCOG is also available. Among the products produced by the MRCOG are annual traffic flow maps. Currently, traffic monitoring activities have not been contracted to any consultants by either the state or the MRCOG.

1.2.1. *Data collection at the NMDOT ITS Bureau*

The ITS Bureau maintains a number of camera and sensor locations in the Albuquerque metropolitan area, primarily along Interstates I-25 and I-40. XML data feeds from sensor locations provide lane by lane count, speed and occupancy information by one-minute intervals. Average speeds and volumes are also computed and a four-bin length-based classification system is collected. The data collected is used primarily for traffic management and emergency response applications and is being archived and shared with MRCOG for Federal reporting and other purposes.

1.2.2. *Other agency programs*

The New Mexico Department of Public Safety (NMDPS) Smart Roadside program uses electronic screening to improve its commercial vehicle enforcement operations. It employs imaging systems for automatic USDOT number and license plate recognition and provides alerts to roadside inspectors for high risk vehicles. Real-time safety information, as well as pass/fail indications for compliance with weight/distance tax requirements and various registration requirements, are gained. Three fixed sites (at the ports of entry at San Jon, Gallup and Anthony) and one mobile reader (in the Albuquerque area) are operational, with an additional seven fixed and two mobile sites planned.

Short-term counts for volume and classification (because of unreliable data, no short-term (portable) WIM data is collected) take place over a 48-hour period of time, while speed data, when required, is ordinarily obtained over a 24-hour period. Turning movement data, typically used by the Districts or by consultants for traffic impact analyses, is collected for a total of nine hours, focused around the AM, Mid-Day and PM peaks. AADTs and AAWDTs, however, are never calculated from turning movement counts.

Equipment failure prior to the completion of the indicated data collection time requires the entire count to be retaken for the entire 48-hour period. Missing values from permanent counters are never estimated; that day's data is left blank. Seasonal correction factors are calculated from similar functionally classified routes and are applied to all short-term counts. Axle correction factors are calculated from the ATR classification sites and are also applied to all short-term counts.

Vehicle occupancy data is required by 23 CFR 500 Part B. 500.202(e) further states that this data is to be collected on the average number of persons per automobile, light two-axle truck and bus, as appropriate to support the data uses identified in 500.203(a). One of those uses is in transportation management systems, such as those at the MRCOG. Similarly, while speed data can be collected at ATR sites, it is not clear whether/how this information is reported/used.

1.2.3. *Current WIM technology in NM*

Currently, the type of sensors that NMDOT are using for their 15 WIM stations are: piezo sensors (Mikros Raktel 8000, all except US 550) and bending plates (IRD 1058, three locations on US 550). In 2011, the bending plates at the three US 550 locations (San Ysidro, Cuba and Bloomfield) will be replaced by PAT plates. Also, the three counters will be replaced at the same time at these three WIM sites. Table 1.2 shows the name, code, location and type of technology of each of the WIM sites.

Site Name	Site Code	County	Road Name	Milepost	Technology
Hatchita	4	Grant	I-10	50.05	Piezo
Logan	100	Quay	US-54	328	Piezo
Gallup	111	McKinley	I-40	10.7	Piezo
Hobbs	202	Lea	US-62/180	84	Piezo
Lemitar	252	Socorro	I-25	158.8	Piezo
Rincon	300	Dona Ana	I-25	37.2	Piezo
Tucumcari	B20	Quay	I-40	340.9	Piezo
Raton	B28	Colfax	I-25	445	Piezo
Roswell	916	Roosevelt	US-70	354.3	Piezo
Vado	74	Dona Ana	I-10	155.6	Piezo
Tularosa	919	Otero	US-70	231.65	Piezo
San Antonio	915	Socorro	US-380	15.7	Piezo
San Ysidro	103	Sandoval	US-550	24.738	Bending Plate
Cuba	102	Sandoval	US-550	71.051	Bending Plate
Bloomfield	155	San Juan	US-550	121.5	Bending Plate

Table 1.2. *Location and type of WIM technology in NM*

The performance and accuracy of the bending plate sensors is much better than that of the piezo sensors, but they are much more expensive and difficult to install. However, the reliability and accuracy of piezo sensors is good regardless of the surface of the road if calibration is performed often.

The main reasons for inaccuracy in collected WIM data appear to be lack of calibration and the influence of temperature. Changes in temperature produce a bias in the weight measured by the sensor. If the temperature gets lower, the weight measured decreases, and vice-versa. Temperature sensors at all piezo WIM sites could correct the error due to temperature; currently, these stations do not have such a sensor.

Bending plate sensors are calibrated twice a year. Piezo sensors have not been calibrated since 2008, although it is recommended they be calibrated at least once a year. Therefore, some inaccuracy in the weight data collected by piezo WIM sites is not surprising. Calibration is not being carried out more frequently due to budget constraints. There are two new WIM sites that are planned to be installed soon, one on I-25 and one on I-40.

The Long Term Pavement Performance (LTPP) program also has two WIM sites located in New Mexico for specific pavement studies (SPS). These two sections are: 350110, located on I-25 North at M.P. 36.1, and 350500, located on I-10 East at M.P. 50.2. The data at both sites is processed and the corresponding axle load spectra are available in the LTPP database.

1.3. NMDOT data processing and reporting

In addition to the monthly data submitted to FHWA for truck weight studies and volume trends, Highway Performance Monitoring System (HPMS) data for the previous calendar year is required to be submitted annually to FHWA by June 15. This data is used not only by the DOT for pavement design but is also input at the Federal level for apportionment of highway funding, the development of performance measures, such as crash rates, and summary reports to Congress.

The traffic survey data collected by the NMDOT is broken down into classification, volume and weight categories. Classification data is further divided into annual class summaries and percentages, both overall and by day of the week. Class percentages of monthly average daily traffic (MADT) at all continuous count sites are also provided, as is the overall percentage of traffic statewide at the permanent sites by functional classification. Typical examples for 2009 are shown in a separate document.

Volume information is also broken down into several categories. In addition to annual volume summaries by site which compare AADT, AAWDT and AAWET totals to the previous year, annual day of week, the 500 highest hours and hourly day of week tables are provided. Tables listing day-of-week percentages, a commercial AADT summary and the highest hours by direction are also reported. Typical examples are again provided separately.

WIM data, by lane, direction and for the entire roadway, are provided from each AWAC site for each of the 13 FHWA classifications. "Off Scale" and "Unclassified" data columns are also listed. The tables list number of vehicles, EASLs for both flexible and rigid pavements (calculated by equations provided in the table) and gross vehicle weight.

Tables providing growth factors, axle factors and daily/seasonal factors, all by both site and functional class, are also provided as part of the annual report. Tables of daily vehicle miles of travel (DVMT), by county, NMDOT district and functional class are also provided. Examples of all are included in attachments.

1.3.1. *Summary of New Mexico practice*

The documents describing New Mexico's traffic monitoring program appear to be in compliance with both Federal Regulations and the several guidelines and standards available at the national level; in actual practice, however, the state is not. For example, while the number of counts on those roads classified as urban or rural Principal Arterials appears to be adequate, this is not the case on roads of lower functional classification. This is somewhat surprising since the State has been a leader in the development and enhancement of traffic monitoring activities since the late 1980s.

However, like many agencies currently, the traffic monitoring program suffers from a lack of resources, both personnel and equipment, necessary to increase and improve data collection efforts on minor roadways in both urban and rural areas. Additional resources are also necessary to improve data collection activities at weigh-in-motion sites. This critical item could be provided either in-house or through contract personnel.

1.4. Traffic data projection and quality control

In order to identify current NMDOT traffic data procedures, policies, practices and qualities, interviews and written surveys were conducted with NMDOT employees and selected individuals from other agencies, both public and private, who collect, process, store or utilize traffic data.

The project technical panel members were asked to recommend a set of individuals who should be interviewed. The recommendations included 26 individuals, including persons from the planning bureau, traffic engineers at headquarters and in the districts, pavement engineers, ITS experts and individuals from three consulting firms, FHWA and MRCOG. Additional interviews were conducted via an email survey of individuals not previously contacted in person. A second round of surveys with a more detailed questionnaire was conducted. The following sections summarize the input the researchers received from all of these efforts.

1.4.1. *In-person interviews*

Question 1 asked if the individual or his/her office collected traffic data. Not surprisingly, all except one reported that they did. Some actually did collect data, others processed the data and still others supervised the data collection.

Question 2 inquired about the types of data collected. The emphasis of the planning bureau was on traffic volume, vehicle classification and weigh-in-motion (WIM). According to the interviewees, the department maintains about 122 permanent count stations, and conducts shorter-term counts with portable counters, principally in the southern part of the state during the late fall, winter and early spring and in the northern part of the state during the remainder of the year. NMDOT has 120 portable traffic volume counters, but is currently using only about 90 due to staffing shortages. The department has 15 permanent weigh-in-motion sites; formerly, the department employed portable WIMs on a 3-year cycle at about 95 sites. The bureau is able to collect speed data, in bins, and believes that it may be required to do so in the future.

The district traffic engineers collect a more diverse traffic data set; in addition to daily traffic volume, they routinely collect manual traffic volume counts, spot speed data and vehicle delay for traffic signal warrants, citizen complaints, speed zoning and lane blockage/lane rental in construction zones. None of the individuals interviewed reported that they conducted travel time studies.

Question 2a asked why the individuals/offices collected the traffic data. Individuals at the planning bureau gave two primary reasons for the data collection: Federal reporting requirements, both monthly and annual, and in support of engineering purposes. The traffic engineers had more varied reasons for their data collection. In addition to concerns expressed by citizens, these engineers must conduct studies to document the need for traffic control devices, as specified in the *Manual on Uniform Traffic Control Devices (22)*. They reported that data for turning movement volume counts at intersections and short-term counts at other locations are not available from Santa Fe, although these data were more commonly available in the past.

Question 2b addressed the issue of processing the traffic data. Planning indicated that they use the TRAffic DAta System (TRADAS) from Chaparral Systems

Corporation for collecting, editing, summarizing and reporting traffic data. The software meets the data processing requirements of AASHTO's *Guidelines for Traffic Data Programs* and FHWA's *Traffic Monitoring Guide*. Because of the diversity of traffic data collected by the district traffic engineers, their data processing was more varied. For example, the data collected by consultants for traffic impact analyses (TIAs) is processed in accord with the state's *Access Management Manual (23)*. These studies also make use of ITE's *Trip Generation Manual (24)*. Except for special circumstances, traffic volume data in the districts are only collected on Tuesdays, Wednesdays and Thursdays. None of the state's counts are done using traffic cameras. The pavement engineer reported the need for data processing to address future MEPDG requirements.

Question 2c asked who the data are reported to. The planning bureau indicated that a primary use of the data was for making reports to FHWA, although they also respond to requests for data from others within the department as well as consultants and the public. The reporting by the district traffic engineers appears to vary among the districts. For the most part, the data are used by the districts for the purposes for which it was collected, but are rarely, if ever, shared with Planning. As a result, there is no central database that contains all the traffic data collected by the NMDOT, other agencies or consultants.

Question 3 inquired about documents prepared by the interviewees or their offices related to traffic data collection, policies or practices. The planning bureau relies on the state's traffic monitoring standards, which are essentially an updated version of the standards developed by the Planning/Research Bureau in 1989–90. None of the districts have developed their own documents; they reportedly rely to some extent on guidance from ITE's latest *Manual of Transportation Engineering Studies (25)*.

This question also elicited the response that there is no communication among those in different districts collecting traffic data. The researchers believe that information sharing among the districts and between the districts and the General Office could be beneficial.

Question 4 asked about the use of documents on traffic data collection prepared by others, either inside or outside of the department. Planning reportedly uses AASHTO's *Guidelines*. The districts do use ITE's *Manual*, but individuals suggested that there is a need for a department manual on setting up equipment so that data are consistent among the districts and to allow input to a common database. Engineers involved with pavement design rely on AASHTO documents regarding traffic volume, vehicle classification, weight and related issues.

Question 5 sought input on the use of traffic data. Planning primarily collects the data, reports it to FHWA and upper management and shares it with others inside or outside the department who request it. The districts clearly use the traffic data for making decisions regarding design and operations. District 3 has some interaction with MRCOG regarding traffic data.

Question 6, regarding the types of data used and the purposes for using the data, was essentially addressed in the responses to other questions. However, the pavement engineer mentioned the specific needs for ESALs, vehicle volume and classification and traffic growth rates as essential parameters for the work of his office.

Question 7 asked the individuals how they accessed the information and the format in which it is stored. Planning indicates that they respond to requests from individuals within and outside the department. Based on the researchers' experience on previous projects, an email request to the bureau for volume information is typically accommodated in two days or less. The department is working to get the information available online for all to use. In Santa Fe, the data are stored in TIMS, which is the new version of the former consolidated highway database.

Question 8 inquired about the completeness, reliability, accuracy and timeliness of the data. With respect to completeness, some concerns were expressed about the WIM devices. The devices used in New Mexico include bending plates and piezoelectric sensors. The specific problems mentioned include power failures and the tendency of both devices to lose their calibration, sometimes after just a short period. Portable counters are placed for a 48-hour count and are deployed on a three-year cycle. The traffic engineers commented on the lack of turning movement counts and the lack of "k" factors, the ratio of the thirtieth highest hourly volume to the average daily traffic. The pavement engineer felt that the data were complete for AASHTO's current design policy, but incomplete/insufficient for the MEPDG.

With respect to reliability, planning reports that the traffic volume counts are adjusted to AADT using computed daily and seasonal factors. Procedures exist to promptly identify ATRs that are producing suspicious or erroneous data, and technicians are dispatched to resolve the problems. The interviewees note, however, that the ATRs are aging and that some need replacement. A concern was expressed about the quality control for traffic data.

Regarding the accuracy of the traffic data, there was a sense that the data meet current needs but are not perfect. Planning is considering the use of manual or portable counters near permanent ATRs to verify accuracy. There were no comments on the accuracy of vehicular speed or traffic delay data.

The respondents seemed to be pleased with the timeliness of traffic data. The WIM data are reported monthly and the ATRs are autopolled on a daily basis. The traffic engineers were satisfied with the timeliness of data that their offices collected, but somewhat less with data collected by others. The proposed move to making traffic data available online was welcomed by all as a means of enhancing both timeliness and accessibility.

The final question sought input on current or projected data needs that are not being met by the existing system. The pavement engineer expressed a need for the inputs required for MEPDG, including axle load spectra, seasonal adjustment factors and speed data. The limited number of WIM sites was also a concern; the need for the use of portable WIM sites to provide data for MEPDG was mentioned. Traffic engineers felt there was a greater need for TIA data for proposed developments. Some concern was also expressed about the reliability of traffic projections.

1.4.2. *Additional written interviews*

In addition to the interviews described in the previous section, inputs were obtained from 13 additional individuals. These individuals represented not only NMDOT personnel from the General Office, the ITS bureau, MRCOG and District Traffic Engineering staffs, but also several private consultants. Particularly helpful was a very detailed response provided by the New Mexico Division of FHWA. Additional input was also provided by a representative of the NM Department of Health with an interest in traffic data from an epidemiological standpoint. The responses to the survey form are summarized below.

Question 1: All but one of the additional respondents indicated that they either collect, process or use traffic data. One of the engineering consultants indicated that they do not collect the data themselves but rather hire sub-consultants who specialize in data collection. Only one response, from the software firm that developed TRADAS, indicated that while they do not collect data, they develop software related to data collection.

Question 2: Planning groups, especially at the NMDOT and, to a lesser extent, at the MRCOG, report collecting volume, speed, classification and WIM data. Consultants, in addition, collect turning movements and also some crash and pedestrian volume data. One consultant reported collecting origin-destination data through license plate surveys as well as GPS-based travel time studies. Another consultant expressed a need for gap data. The NM Department of Health routinely collects data on EMS (ambulance) patient data.

Question 2a: Data is used internally for project planning and design, as well as for Federal reporting requirements.

Question 2b: The NMDOT Planning Bureau processes the data through TRADAS, their traffic data software. One consultant uses video data collection software from Miovision Technology, while the NMDOT ITS bureau uses a third party who configures and manages the data from their Traffic Management Center in an SEQUEL environment.

Question 2c: Data from the ITS Bureau is shared with MRCOG and with contractors who are posting travel times on NMROADS. Consultant data is for internal use or is shared with clients. NMDOT district counts are used internally and not usually shared with the General Office. MRCOG data goes into its database and is shared with NMDOT Planning.

Question 3: Some consultants routinely detail data collection and processing procedures in reports to their clients.

Question 4: The AASHTO *Guidelines* and the FHWA *Traffic Monitoring Guide* were mentioned by several respondents.

Question 5: Traffic data is used for design decisions at the NMDOT. It is also used by Traffic and Planning groups within consulting firms. The NMDOT ITS Bureau uses data to configure its Dynamic Message Signs.

Question 6: Consultants use data for their own studies including classification and WIM data for LTPP studies. They also report using volume data for marketing studies as well as engineering analyses.

Question 7: Data is typically stored on agency/company servers in Excel, pdf or csv formats. Backup is by hard copy in project folders. GIS shape files and dbt tables are also used for storage. NMDOT ITS has data accessible in a web environment; its contractor stores in a SEQUEL database.

Question 8: Most responses indicate completeness, accuracy, reliability and timeliness as "sufficient for our needs", particularly if the data were collected in-house. There was some concern relating to WIM sensors going out of calibration and the need for better monitoring. Some consultants also expressed concern with the currency of the MRCOG database and the need for more frequent volume counts. MRCOG, for its part, laments the "significant lack of ATR or continuous count locations", and NMDOT's inability to preserve the ATRs that it does have.

Question 9: NMDOT ITS still has some gaps in Interstate coverage, particularly on flyovers; it expects to have these completed soon. Some respondents would like to see turning movement counts to be a "standard inventory requirement", (stored in a central database, along with weights, classifications, etc.) as well as the establishment of a central data collection tool. The MRCOG would like link speeds and travel times on non-interstate facilities to aid in their congestion management process.

1.4.3. *NMDOT project panel responses*

A series of advanced questions, based on Transportation Research Circular E-C120, Traffic Monitoring Data and Successful Strategies in Collection and Analysis [EC 07], was distributed to the project's Technical Panel. This section is an attempt to summarize the four completed questionnaires that were returned, three from NMDOT personnel and one from a representative of the Federal Highway Administration. A summary of the Technical Panel's responses with those detailed in the Circular is described below.

NMDOT Responses – The problems facing the NMDOT in regard to traffic monitoring revolve around money, politics and the resulting inability to ensure that the data collection meets all of the Federal requirements. A longstanding lack of funding has affected both the ability to collect the data that is required by Federal agencies through insufficient staff as well as through inadequate installation, calibration and repair of data collection equipment. Funding constraints have also resulted in little or inadequate staff training involving proper data collection procedures as well as inadequate database maintenance and reporting, with only static annual reports available on the web. WIM data is downloaded weekly, although no automated quality control checks are performed. Manually flagged data are investigated and a technician is sent to review conditions at the site.

Other traffic data from continuous monitoring are downloaded and reviewed daily; automated data quality checks are applied, and flagged or suspect data is investigated, including sending a technician to examine the site. Unexplained large data variations are purged and recounts scheduled.

NM FHWA Response – There is concern at the FHWA as to whether some of the Federal reporting requirements are being satisfactorily addressed; this may be attributed to the department's less-than-complete understanding of the *Traffic Monitoring Guide* (TMG). For example, the TMG calls for a three-year counting cycle on the National Highway System (NHS), on Principal Arterials, and on HPMS sample sections. Every major system section should be monitored to provide truck counts. The Traffic Monitoring System (*TMS/H, 23 CFR 500 Subpart B*) also calls for vehicle occupancy monitoring and a testing program for field equipment. These requirements are not currently being met. Finally, while realizing that keeping counters working is always a challenge, operational counters used to determine the annual growth rates need enhancement and additional counters are needed to provide statistical validity [BRO 11].

1.5. Conclusions

The survey of traffic data collection professionals in the state was used to both identify current problems and describe possible opportunities, and to provide a summary of both the technical literature and the best data collection and analysis practices employed at the national level. Several conclusions associated with both data collection and analyses are presented and recommendations for improving the processes are given below:

New Mexico's involvement with issues associated with traffic monitoring dates back to the late 1980s when the NMDOT uncovered significant quality and reliability problems in its data collection and analysis processes. Rather than conducting a recount when missing or suspicious data readings were observed, for example, staff often used "engineering judgment" to replace the missing or questionable values. This revelation led to a national effort, in many ways spearheaded by New Mexico, to standardize data collection processes and improve data quality.

Since that time, the state has continued to make progress in its traffic monitoring activities and, in most instances, its documents appear to be in compliance with both Federal requirements and the several published national guidelines and standards. That is not to say, however, that areas of concern have not been identified through interactions with traffic data specialists in the state. Like many public enterprises nationwide, the department's traffic monitoring program is in need of additional funding to acquire both additional staff to oversee program activities and to conduct additional counts as well as to purchase additional and upgraded equipment.

The traffic data needs associated with new pavement design procedures as well as the possibility of additional data requirements related to the speed and vehicle occupancy data also point to the importance of collecting timely and accurate data.

Interviews with department staff and others also pointed out the need to address additional data requirements. District personnel, in particular, expressed a need to collect and, equally importantly, store intersection turning movement counts and other manually collected data. Archiving of these data, which are typically collected as part of a traffic impact study, are also critical for other engineering applications.

1.6. Acknowledgments

This study was funded by NMDOT.

1.7. Bibliography

[BRO 11] BROGAN J., TAREFDER R.A., RODRIGUEZ-RUIZ I.J. *et al.*, "Statewide traffic data collection, processing, projection and quality control", Final Report, ORA 456-310, Research Bureau, New Mexico Department of Transportation, pp. 1–175, 2011.

[EC 07] E-C120, Traffic Monitoring Data: Successful Strategies in Collection and Analysis Transportation Research Circular E-C120, 2007.

2

SYNCRO – An Innovative Public Procurement of an Advanced Data Gathering System for Interurban Roads Based on its Technologies

The SYNCRO project emerges as an original public procurement of an interurban road data gathering and processing system by two cross-border road operators. The goal is to improve information for road users and dynamic traffic management. The objective is to investigate the advanced technologies, test them, and then deploy an innovative ITS solution. The SYNCRO system is expected to collect traffic count data, travel times, and possibly any other relevant data (such as weather road data). Technological innovation is expected to enable a massive deployment of the SYNCRO system across an interurban road network and prepare the emergence of cooperative intelligent transport systems, in particular vehicle-infrastructure communication. A legal framework was elaborated to support this cross-border public procurement of innovation. A market notice was published on 6 December 2013 in order to select three candidates for a competitive dialogue. Finally one tenderer will be awarded with supplying a pilot system then possibly an operational system for roll up on the road network.

2.1. Introduction

In terms of mobility issues, the current environmental, energy and economic context requires us to change our individual and collective behavior. Public road

---

Chapter written by Jean-Christophe MAISONOBE, Jean Daniel DEMOND, Giannicola MARENGO, Dolores ADAMSKI, Diego ALBESANO and Olivier LATOUILLE.

operators must revolutionize the services they provide to road users by offering them real-time, personalized, high-quality information. The data-gathering technologies currently in use are inadequate to fulfill this new ambition: they are too costly, too complex to implement and sometimes inappropriate or even obsolete.

Because most road operators in Member States face the same issues, there is an obvious need to build partnerships among them in order to pool resources and to design and deploy innovative solutions, with the aim of achieving efficient smart road systems.

Therefore, SYNCRO emerges as the first cross-border public low energy road data gathering and processing system. This project is granted by the European Commission, by the Competitiveness and Innovation Framework Program (CIP). It brings together a consortium of four partners from two Member States (two public procurers, two SME support organizations) to establish the conditions for a technological breakthrough in Europe or even internationally. The major goal of SYNCRO is to provide road data for traffic management, including road user information and road network exploitation. SYNCRO has to be a road data gathering and processing system which efficiently provides data to the current road management center.

The SYNCRO road data gathering system is expected to collect road data for use with three main issues. First, the SYNCRO system has to collect near real-time road data necessary for operating the road network. As an illustration, collected road data are expected to be used by the decision-making tool for traffic management and for road exploitation in the case of critical meteorological winter conditions. Second, the SYNCRO system has to collect near real-time data useful or necessary for better informing road users of traffic conditions. In the longer term, the supply of advanced mobility services to the road user is targeted. To provide real-time traffic information services is besides one of the priority actions driven by the EU directives about intelligent transport systems. Third, the SYNCRO system has to provide near exhaustive traffic data in postponed mode to elaborate statistics. These statistics are used for informing public policies and investment strategies. They could be also used to elaborate traffic management plans, which can be dedicated to different types of events: cultural or sporting events, road works, natural and technological disasters, etc.

The global issue is concretely ensuring safer and more comfortable mobility for the road user, and more efficient and rationalized road network management and operation.

2.2. Elaboration of the SYNCRO technical vision: the SYNCRO functional program

2.2.1. *Innovation for more complete data collection by a single system*

Collecting different types of data from a single system is one of the major challenges of the SYNCRO project.

The SYNCRO system must collect and transmit, at least, the types of road data identified as strictly required:

– conventional traffic data, identified as obligatory to supply efficient decision-making support for the operation of the road network and to supply count statistics that are sufficiently accurate, reliable and exhaustive to inform public transport policies;

– travel time, identified as obligatory to provide users with high quality information for their mobility;

– the technology providers are free and even encouraged to propose the collection of additional road data, such as road weather data or any other relevant type of data achieving the above objectives.

Conventional traffic data include vehicle count, vehicle speed, vehicle flow rate, lane occupancy rate and vehicle categorization (including light vehicles, heavy goods vehicles, coaches and caravans). Any technology could be investigated provided that the functional need is met. For example, vehicle categorization could be achieved from any measurement type: shape, length or weight for instance. The traffic management center must be supplied with adequate data to monitor the traffic conditions (congestion, load traffic, dense traffic and light traffic conditions). These data are also necessary to elaborate the different statistics.

The SYNCRO system must supply a first evaluation or calculation of travel times. At least partial data as measurement of individual travel times are expected. Other evaluation methodologies of travel times could be used, provided that they are adapted to a very meshed and diversified road network. The objective is to provide users with high quality information and even more elaborate advanced mobility services.

Depending on the technologies to be implemented, and subject to compliance with the economic and operational constraints, additional data could be collected. The collection of road weather data, such as data on temperature, type and quantity of rain, or condition of the road surface, may be interesting. Sensors could be deployed to a reduced number of "strategic" points of the road network or more systematically across the road network. The collection of road weather data must meet two main objectives: providing decision-making support for the operation of the road network and better informing users of the traffic conditions on the road network.

Figure 2.1 outlines the targeted multiple data collecting system.

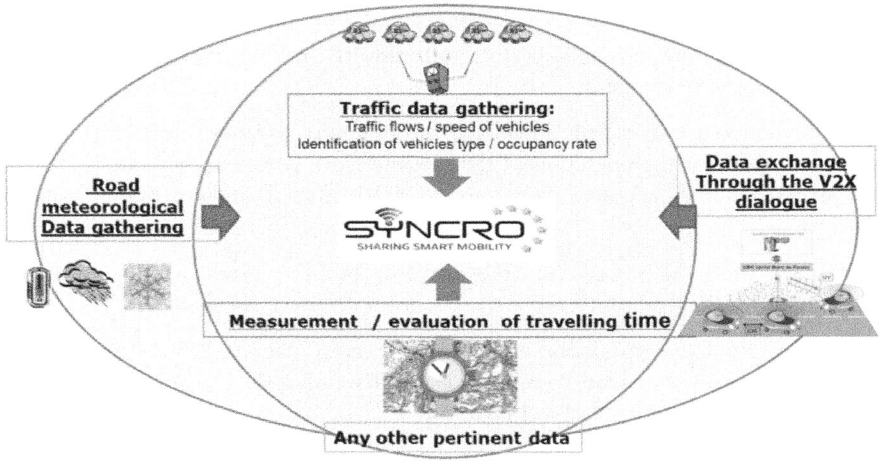

Figure 2.1. *Multiple road data collecting system. For a color version of this figure, please see www.iste.co.uk/jacob/traffic.zip*

2.2.2. *Innovation for enabling massive deployment across an interurban road network*

Innovative technologies are expected to enable the deployment of the data gathering system across the entire interurban road network of the Departement of l'Isère and the Province of Turin. The technological solution must provide solutions costing significantly less than the solutions already in place in order to enable the massive deployment of the systems across the entire area. The SYNCRO system is expected to be rolled out at a significantly larger scale than the current one, at acceptable economical and operational conditions. These networks are extensive (5,100 km of roads for Isère, 3,500 km for the Province of Turin) including interurban roads, peri-urban roads, rural roads and roads serving ski resorts. As an example, the road network managed by the council (*Conseil general*) of Isère serves 36 ski resorts, which involve significant traffic issues.

The reduction of global cost is also required: it obviously includes the costs of purchasing the system and its components, the installation cost, the operating, maintenance and renewal costs, and the dismantling and recycling costs. The level of quality and robustness of the current road data system must at least be maintained or even improved to ensure cheap maintenance and long lifecycle. Both energy cost and telecommunication cost have to be very significantly reduced.

The technological innovations must facilitate the installation of the system and its integration into the various road environments. It requires minimizing civil engineering works, and self-reliance of electrical or wired telecommunication networks.

Miniaturized devices, low energy or even self-sustained technologies, and flexible communication solutions (included wired and wireless) are technological bricks likely to make possible a roll-out at a larger scale.

Low energy technologies reduce operating costs and environmental impacts. Energy harvesting and energy autonomous systems could be particularly attractive, in particular to equip sites far from the electrical grid. Miniaturized technologies often enable us to manufacture cheaper and use less energy. Miniaturized systems also facilitate the physical installation on the public domain and their societal acceptance. Miniaturization reduces volume on the road edge, facilitates aesthetic integration into the landscape and reduces certain risks of vandalism. Sensors, power supply and communication components need to be miniaturized.

2.2.3. *Transition toward cooperative systems*

Currently, road data are collected by sensors integrated into the infrastructure such as inductive loops or cameras. But a new concept is emerging: cooperative road systems. The vehicles are equipped with on board ITS technologies which enable them to communicate with each other and with the infrastructure, if the latter is also equipped with the adequate technologies. Data gathering and information exchange will be based on the real-time transfer of data or even information from vehicle to vehicle, or vehicle to infrastructure via radio interface. In this configuration, road vehicles could become data sources.

The SYNCRO system is expected to support two paradigms: first, to be operational at short and middle term, the SYNCRO system has to gather data from sensors integrated into or on the edge of the infrastructure. The objective is to detect vehicles and more largely to collect data, even if the vehicles are not yet equipped with any on board ITS technologies (which is the current situation that will probably last for some time). Second, the same SYNCRO system may also be able to support data and information exchanges with the first vehicles equipped with on board ITS technologies. The objective is to ensure the transition toward smart cooperative roads by supporting V2X dialogues: the SYNCRO system could accompany the first deployment of smart vehicles, to progressively develop new data gathering practices. The challenge is to have an efficient investment strategy aimed at preparing the compatibility of the infrastructure with the V2X technologies before they are massively deployed and used. The SYNCRO system could be compatible with the on board technologies used on the vehicles and must comply with the V2X European ITS standards. According to the technological opportunities, the V2X dialogue supported by the SYNCRO system could at least be demonstrated on some basic but significant use cases. The SYNCRO system is expected to be able to support both paradigms (data collecting from infrastructure sensors and V2X dialogue) as shown in Figure 2.2.

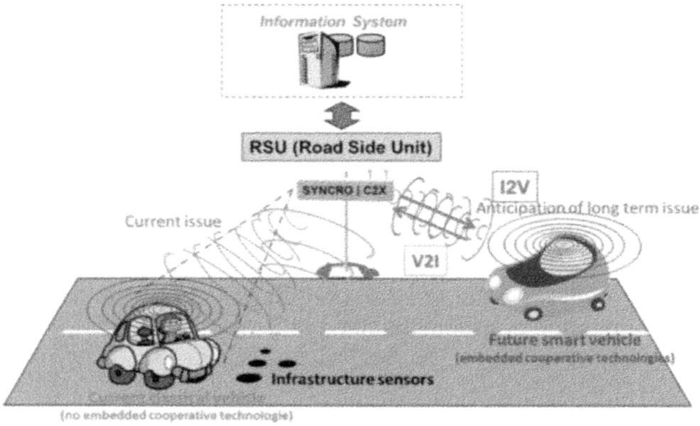

Figure 2.2. *Double functional requirements of the SYNCRO system For a color version of this figure, please see www.iste.co.uk/jacob/traffic.zip*

2.3. A system to gather road data and to provide the current operational road management center with data

The SYNCRO system must share an interface with the current traffic management center of the *Conseil général* of Isère and the traffic management center of the Province of Turin, in order to provide them with the required road data. Short range communication technologies are expected to support communication with cars in a smart cooperative road approach. SYNCRO contributes to a cleaner, safer and more efficient transport system.

Itinisère, the current road management center of the CG38, monitors a 5,100-km interurban road network: 108 counting stations based on inductive loop technologies are deployed on the road network. Around 40 sites are equipped with video cameras which enable operators to view images in real-time situations and in strategic places, for instance for views of congestion or snow covering the road, as shown in Figure 2.3. The operators control around 40 variable message signs. Older equipment communicate with the road manager center by Public Switch Telephone Network (PSTN) but more and more equipment communicate using a wired network (by xDSL, ADSL or SDSL) or a wireless network (by GPRS, 3G).

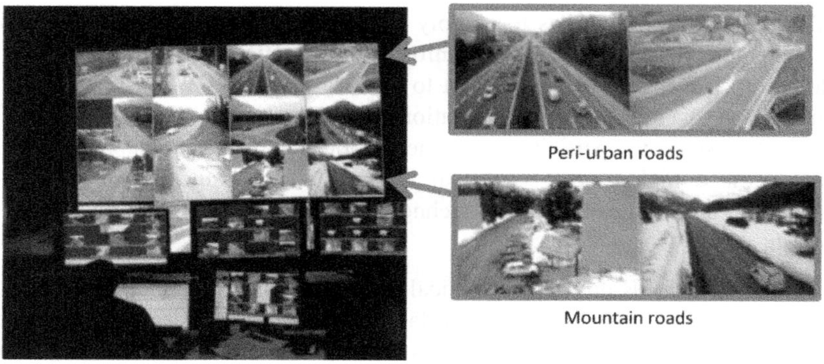

Figure 2.3. *Road management center of the Conseil général of Isère. For a color version of this figure, please see www.iste.co.uk/jacob/traffic.zip*

Cameras for automatic number plate recognition measure individual travel time on the road to "L'Alpe d'Huez" ski resort (TEMPO project). Embedded numerical notebooks enable snow removal operators to send traffic information and events in real-time to the traffic manager center for the winter service vehicle.

Information about traffic condition is delivered to the road end-user through different technological supports: interactive voice response, website, web applications which are available from smart phones, table computers and laptop computers. Figure 2.4 gives examples of advanced services delivered by Itinisère to the road users.

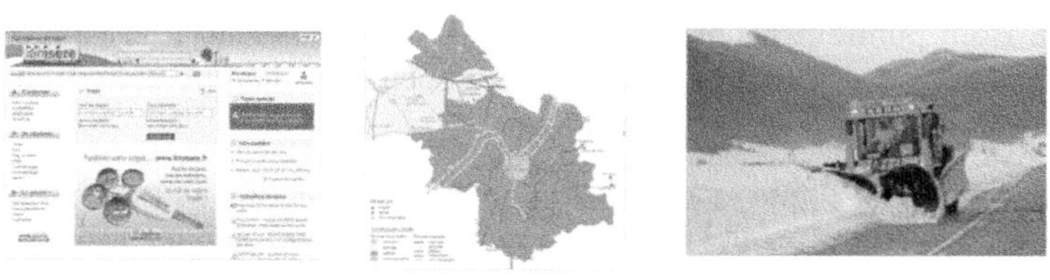

Website www.itinisere.fr delivering qualified information about traffic condition in Isère to interurban road users

Mapping condition in winter critical conditions delivered on the web site www. Itinisere.fr

Snow removal on a moutains road of Isere (Vercors)

Figure 2.4. *Three examples of advanced services to road users currently delivered by the Conseil général of Isère. For a color version of this figure, please see www.iste.co.uk/jacob/traffic.zip*

2.4. Impact and potential of the SYNCRO system

2.4.1. *Possible impact of SYNCRO project*

The SYNCRO system has to comply with the well-defined targets of the local road manager concerning road user information, traffic management support and exploitation. But it could be important to analyze how the SYNCRO system could be integrated at different scales (national or European). The capacity of the SYNCRO system to participate in the traffic data management coordination initiatives at a higher scale (national or European) should be evaluated. The transposition and replication of the technology across another road network are an important factor.

– Functional needs and later technical specifications could specify compliances with strategic standards. First, the interface between the gathering system and the traffic management center could ensure a road data format for exchanging at local, national and European levels. The format as DATEXII could be specified. The data gathering system could provide future mobility platforms which support traffic data or information exchange with other mobility operators, possibly at local, national or European levels. Second, the possible V2X technologies have to comply with V2X ITS European standards developed by the ETSI or by other standardization organizations.

– More generally, the participation in collaboration and dissemination could facilitate the coherence between the SYNCRO project and the traffic data initiatives at national or European levels.

– The SYNCRO project involves two distinct cross-border road network managers, one French (the *Conseil général* of Isère) and one Italian (the Province of Turin). Both public procurers will participate in a competitive dialogue with the technological providers, and will also be involved in the tender's analysis and selection. Both public procurers will be involved together in a test and validation phase of a pilot system. It could favor a more open and interoperable solution, to be implemented across two different countries.

– SYNCRO is a public procurement of innovation; the targeted technologies should be innovative while having a minimum level of maturity, especially concerning their technological elementary bricks. Competitiveness between the technological operators may facilitate the emergence of a solution which complies with standards, interoperability requirements and conditions for coherence with traffic data issues at national and European levels. Economic operators gain in proposing solutions which could be easily used on different road networks and countries.

2.4.2. *Exploration of a diversified technological offering*

A large diversity of technologies could be considered, with different degrees of innovation: the technological offering includes intrusive sensor technologies such as inductive loop, magnetometer, piezoceramic sensors, etc. Road edge technologies could also be considered as radar technologies, laser technologies, video camera combination technologies, etc. Other technologies could be considered, such as floating car data, Bluetooth detection and technologies as RFID (Radio Frequency Identification) or DSRC (Dedicated Short Range Communication). Numerous innovative sensors, energy management and communication technologies could be used to improve the system, without being necessarily identified for the road transport sector. Experimentations were carried out by the *Conseil général* of Isère before the SYNCRO project started. For example in the TPV project, the magnetometer was tested to measure traffic and to measure individual travel time by matching magnetic signals. These tests and feedbacks highlight the interest and need to further explore this very complex technology. One major conclusion was that the purchase of innovative technologies requires iterative discussions with the actors of innovation, based on a functional program. Preliminary investigations are of the utmost importance.

Following the recently published market notice, a selection of three candidates should enable us to compare these technologies. The first results of the competitive dialogue will enable us to define the awarded technologies. Beyond the technological opportunity in compliance with the economical and operational requirements, the perimeter of the system will be clarified, in particular in term of sets of collected road data. This public procurement is a public procurement of innovation, and not a precommercial procurement; it addresses innovative solutions which have both a minimum level of innovation and of maturity. It mainly concerns innovative solutions which are not already implemented on the ground but which don't require much research and development work. The system was planned to be operational in 2015.

2.5. An innovative legal framework to implement three phases of the SYNCRO project

An innovative legal framework is currently implemented: a market notice was published on 6 December 2013.

In practical terms, the SYNCRO project is composed of three fundamental steps:

– a competitive dialogue procedure which will implement a dialogue with technological providers based on a functional program. The functional program describes functional needs and targets without specifying the technological mean. This step enables the emergence of an innovative technological solution and the elaboration of technological specifications. An economic operator (one entity or a group of entities) will be awarded the contract at the end of the competitive dialogue;

– the procurement of a pilot system that will be tested: this pilot system allows us to experiment, to qualify the system for validation. This step enables us to take (or not) the possible decision for the procurement and scale deployment of the technology;

– the procurement of a final system: this step corresponds to the final public procurement and to the large scale deployment of the technology. The fully operational system will be implemented and interfaced with the existing road manager centers.

As described above, the purchase of a road data gathering system able to support efficiently high level road management and advanced services provided to end-users requires three major criteria:

– exploring a wide technological offering and purchasing an innovation which is impossible to specify at this stage of the project;

– co-elaborating a complex system by using existing technological bricks from technology suppliers;

– undertaking a transnational public procurement to efficiently meet an ITS issue, which increasingly arises at the European scale.

To meet such technical challenges requires us to elaborate an innovative legal framework. Therefore, the SYNCRO partners are carrying out a legal study, supported by external experts (an expert team from the legal office DS Avocat managed by Frédérique Olivier). Figure 2.5 describes the three steps of the SYNCRO public procurement.

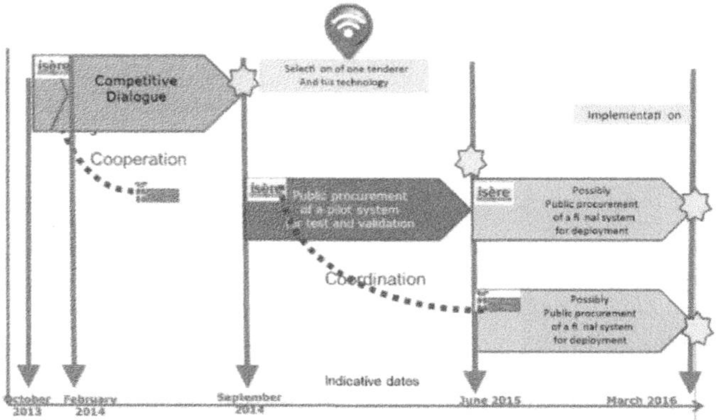

Figure 2.5. *Organization of the cross-border public procurement of ITS System in the SYNCRO project*

2.5.1. *Purchasing innovation thanks to a competitive dialogue*

The legal procedure is expected to offer a dialogue which is as flexible and complete as possible, while complying with the rules of transparency, equity and equality of treatment of the economic operators. The competitive dialogue procedure will be used. Under French and Italian law, a competitive dialogue procedure may be used in the frame of public procurement (governed by the *Code des Marchés Publics* (Code of Public Procurement) in France). The call for tender was launched. The competitive dialogue allows the public procurers to discuss with the enterprises, in order to explore the technological offers and to develop or to adjust technological solutions thanks to iterative exchanges. Collaborative and iterative cross-analysis of the technological offer and of the road operators' current and future needs is essential to elaborate the innovative systems by assembling different technological bricks. They should facilitate technology comparison and combination, while strictly ensuring the confidentiality of exchanges with technology providers to protect their intellectual and industrial property rights.

The procedure has to allow efficient and suitable remuneration of the non-selected enterprises which participate in the dialogue. It is strictly required to promote and to make possible the participation of innovative SMEs (small and medium-sized enterprises) thanks to a proportional sharing of risk. This competitive dialogue should enable the award of a two-phase procurement to an economic operator.

2.5.2. *A two-phase procurement: a pilot system and a final system*

The competitive dialogue will lead to awarding a tenderer, to contract a single awarded framework agreement, for differing subsequent contracts related to differing phases of the project.

In the first phase, a pilot system will be purchased for test and validation. Ideally this pilot system will be tested in real conditions, both on the road network of Isère and on the road network of the Province of Turin. A pilot system test is required by both public procurers to take a decision for the final public procurement. These tests could also be an opportunity to adjust or upgrade the technological system.

In the second phase, each procurer will possibly purchase a final system for larger scale deployment on their own road networks. The size of the roll out will depend on the economical, technological and operational performances of the proposed system, and on the internal strategy of each road manager. Deployments beyond the project could also be possible in the case of a very attractive successful system. The award of a framework agreement will enable the *Conseil général* of Isère to select a single awardee for both the pilot system and the possible final systems. Such a solution may open the competition to the maximum number of economic operators, by securing the investments made during the pilot system phase by actors such as innovative SMEs. It may favor balanced risk sharing between public procurers and innovation providers, and secure intellectual properties rights of each actor. But such a legal framework could probably be fully implemented for French procurement, due to problematic cross-border public procurement.

2.5.3. *A coordinated transnational procurement*

A legal framework is required to support joint procurement between two public procurers from different Member States, a French local authority (the *Conseil général* of Isère) and an Italian local authority (the Province of Turin). The legal framework must comply with European, French and Italian laws. The whole procurement framework has to be secured, including intellectual and industrial property issues (both between public procurer and economic operators and between each public procurer). Different legal tools have been investigated, such as outline agreements, statutes and contracts, between local authorities and proposed by the Outline European Convention on Transfrontier Co-operation between Territorial Communities or Authorities and European Economic Interest Grouping (EEIG) or European Grouping of Territorial Cooperation (EGTC).

Due to the lack of adequate legal tools, the procurement cannot really be a joint procurement, but it will be a coordinated procurement. The Province of Turin will participate in the competitive dialogue, for the selection of the tenderers and the pilot system experimentation led by the *Conseil général* of Isère according to French public procurement legislation. Each public procurer will take his decision to purchase or not purchase a final system. The Province of Turin will probably have to launch its own tender but on the basis of the results of competitive dialogue and the pilot system test.

2.6. Conclusion

Two local authorities are experimenting with an innovative public procurement of road ITS solutions for traffic management, thanks to the support of the European Commission through PPI projects funded by the Competitiveness and Innovation Framework Program. A technological vision has been achieved, concretized by a functional program. A legal framework has been elaborated to support a public procurement preliminary highlighted by the investigation and evaluation of the technological opportunities. The call for tender is launched with the publication on 6 December 2013 of a market notice in the Official Journal of the European Union. A competitive dialogue with three technological suppliers (or grouping of technological operators) is planned to begin from February 2014.

Several key points are going to be adjusted. The perimeter of the data gathering could be adjusted taking into account the technological opportunity presented during the competitive dialogue. The integration of the SYNCRO system in larger scale road data ecosystems and the capacity of the SYNCRO system to be duplicated and diffused should be investigated. The conditions of the pilot test implementation, including test methodology and protocol, will be achieved.

First, this project is an exceptional opportunity to explore large and diversified technological opportunities to improve traffic data gathering while anticipating the emergence of the smart road. The innovative system is expected to both deliver advanced services to road end-users and to improve the management and exploitation of the road network. Diversified technological opportunities will be considered: sensor technologies, miniaturized technologies, energy autonomous technologies, information and communication technologies and C2X technologies. This innovation initiative should be as coherent as possible with the traffic data and ITS issues at a larger scale, at national and European scales, and should lead to a sustainable solution from an economical, operational and societal point of view. Second, such an experiment contributes to developing the public procurement of innovation. An improved knowledge of legal framework and innovative procurement practices is developing the ability of the public procurer (a local authority in this case) to purchase innovation, by reducing technical and cultural barriers. In the case of SYNCRO, purchasing innovative ITS improves management of the road network and service supplied to the end-user, while contributing to the emergence of certain technologies (such as advanced sensors and road side unit technologies). Similar approaches could be implemented for other societal issues: ensuring high-quality affordable healthcare for aging or disabled people, improving energy efficiency, or ensuring higher quality and better access to education, for example.

In addition to the improvement of the quality and effectiveness of public services, innovative public procurement can stimulate innovation from the demand side and contribute to help innovative European enterprises to take international leadership in new markets.

The next step, the competitive dialogue, will define the technological choice made jointly by a French and an Italian road network manager. In the short term, the technologies compared during the competitive dialogue will be clarified.

2.7. Acknowledgments

This project is supported and funded by the European Commission, in the frame of the PPI project (Public Procurement of Innovative Solutions) of the CIP (Competitiveness and Innovation Framework Program). This funding is gratefully acknowledged.

We are grateful to SYNCRO's research team for all their contributions to the execution of our project.

We are grateful to Maître Frederique Olivier and her team from the legal office DS Avocat, for their contribution concerning the legal framework of the SYNCRO project.

3

Tailoring a Reference Model for C-ITS Architectures and Using a DATEX II Profile to Communicate Traffic Signal Information

There are well-accepted structures for the emerging Cooperative ITS (C-ITS) architectures which become clear in several R&D and FOT projects. The first release of standards for C-ITS in response to the European Commission Mandate M/453 tightens these structures. The big challenge is to merge these architectures with existing urban systems focusing on connecting legacy, i.e. to fit them into existing urban architectures and constraints, bringing forward urban seamlessness.

Road operators should be aware of the availability of more than one possibility to publish their data, which is a simultaneous opportunity and duty. This paper shows one example of how to tailor a generic C-ITS architecture developed in the SEAMLESS project to the needs of the use case "Traffic Light Phase Assistant" and gives a profile at hand for a DATEX II communication.

3.1. Introduction

There are well-accepted structures for the emerging Cooperative ITS (C-ITS) architectures which become clear in several research and development (R&D) and field operational test (FOT) projects. The first release of standards for C-ITS in response to the European Commission Mandate M/453 tightens these structures. The big challenge is to merge these architectures with existing urban systems, focusing

---

Chapter written by Jörg FREUDENSTEIN and Ian CORNWELL.

on connecting legacy. In the SEAMLESS project [SEA 13], it was shown with a generic approach that appropriate urban standards, like UTMC (Urban Traffic Management & Control) and OTS (Open Traffic Systems), can be mapped within the Cooperative ITS architecture. Considering the opposite situation, C-ITS architectures can be fitted into existing urban architectures and constraints, bringing forward urban seamlessness.

Nevertheless, there are challenges which remain. The greatest lack of clarity seems to apply to the role of the Roadside ITS Station (R-ITS-S) – the number of possible R-ITS-S scenarios is as huge as the number of different tasks for the ITS communication. A key question is where (and in which systems) the intelligence of the R-ITS-S is best allocated. A similar lack of uniformity can be seen in the use case for the Traffic Light Phase Assistant. Because of its overall nature, this use case covers a lot of topics, which are solved very differently in a variety of projects.

The progress as well as the problems will be shown in this paper in the example of an urban context, but they exist exactly the same way in interurban scenarios. For travel times or in-vehicle signage along motorways, other standards apply. But mapping in the generic architecture can be done in a similar manner. Also the question of R-ITS-S forming is the same in an interurban context.

The SEAMLESS[1] project did research on these issues. As a result of the project, road operators should be aware of having more than one opportunity to publish their data. This has at once possibilities and drawbacks. They can reach a wider audience and increase the amount of information and its reliability. But by handling this difficult task the wrong way, they can even reduce the reliability of the data and their own reputation. So careful consideration is needed to ensure the necessary quality for the task.

This paper shows the usage of a generic architecture developed in the SEAMLESS project for the use case "Traffic Light Phase Assistant" and gives a DATEX II[2] profile at hand for the communication of traffic signal information.

3.2. Architecture of intelligent transport systems

Due to the European Commission's ITS aspirations and ITS activities (ITS Action Plan and ITS Directive), a broad discussion on the design and concept of transport and transportation systems was initiated. This discussion is highly relevant for all who deal with the implementation and use of such systems. It is driven primarily by the efforts of

1 Seamless traffic data dissemination across urban and inter-urban networks – initiated by ERA-NET ROAD II – see http://www.eranetroad.org/. SEAMLESS is part of the program "Mobility – Getting the most out of Intelligent Infrastructure", which is a transnational joint research program funded across the border by the Road Authorities of Belgium, Germany, the Netherlands, Norway, Switzerland and the United Kingdom.

2 A European standard for traffic and travel related information. See www.datex2.eu.

the European Commission to renew the traditional understanding of transport and transportation by using the term "mobility". Therefore, less attention should be paid to the realization of systems than on the design of the so-called ITS services[3], which as "mobility services" are supposed to represent an entire service segment to enable and provide additional value from the perspective of territorial and cross-border applications for both the implementer and provider as well as for the user.

Considering that those discussions are not only about ITS systems (system architectures) but also about extensive ITS services, the understanding of architecture in the transport sector must be formed in the sense that whenever the concept of architecture is used, it must be made clear which aspects of an ITS or mobility service are being addressed.

3.2.1. *The ITS pyramid*

As a suitable meta-model and methodological tool for manageable and comprehensible representation and description of ITS services, we suggested the "ITS pyramid", described in FGSV [FGS 12] (see Figure 3.1):

The ITS pyramid:

– consists of five layers, which together span the field of potential observation and display areas of ITS services;

– represents the structural design of ITS services to better identify the characteristics of ITS, classify them and relate them to each other;

– provides necessary semantics for the description of ITS and its business models.

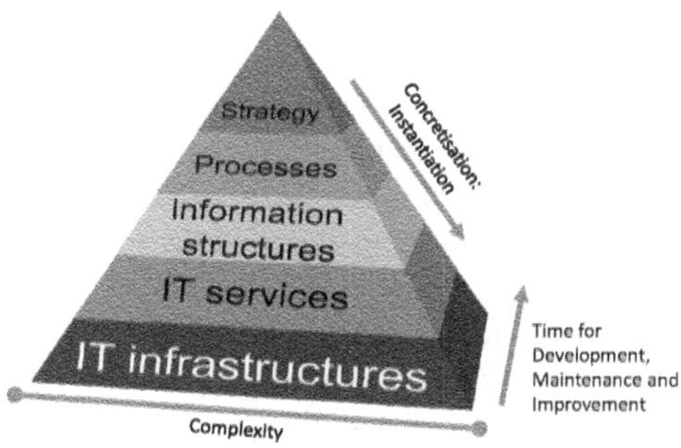

Figure 3.1. *The ITS pyramid. For a color version of this figure, please see www.iste.co.uk/jacob/traffic.zip*

3 Not to be confused with services in the traditional sense. ITS services are services for road users and travelers provided through the use of ITS – Intelligent Transport Systems.

The basic structure of the five layers from top to bottom are described as strategy, processes, information structures, IT services and IT infrastructure:

– the strategy layer describes the long-term (*visions*) and the medium-term (*missions*) objectives of an ITS service, i.e. the benefits of the ITS service. The strategy is closely related to the business model of an ITS service;

– the process layer describes the actions of operators within their business processes and their day-to-day behavior. Role models enable transferability by providing an abstraction from concrete organizations. Typical ITS service roles include: content owner, content provider, service provider, network provider. Typical role characteristics connected with a specific behavior are for example public, private, and external stakeholders. The tailored architecture shown later will serve as an example for this layer;

– the information structure layer deals with information that is generated by or processed in (business) processes. Information logistics, i.e. the collection and presentation of information and its distribution to where it is relevant and applicable can only be specified if appropriate information structures are mutually agreed upon by all stakeholders. The DATEX II profile shown later will serve as an example for this layer;

– the two IT layers (not to be confused with ITS) describe the IT services and IT systems. They are not the focus of this chapter.

3.3. A generic C-ITS architecture

In SEAMLESS Deliverable No. 3 [SEA 13] and in Freudenstein [FRE 12], the following generic architecture for C-ITS was introduced.

Figure 3.2 shows the aforementioned generic architecture for ITS. It covers all relevant C-ITS components and communication channels. Starting with this overall view, explicit and detailed architectures can be tailored from it for specific situations and systems (an example of this will be included in the next chapter).

The generic model is divided into components mostly driven by service providers, and components mostly driven by road operators. For Central ITS stations (C-ITS-S) as well as for Roadside ITS stations (R-ITS-S) both parties might operate instances of these components in suitable system environments.

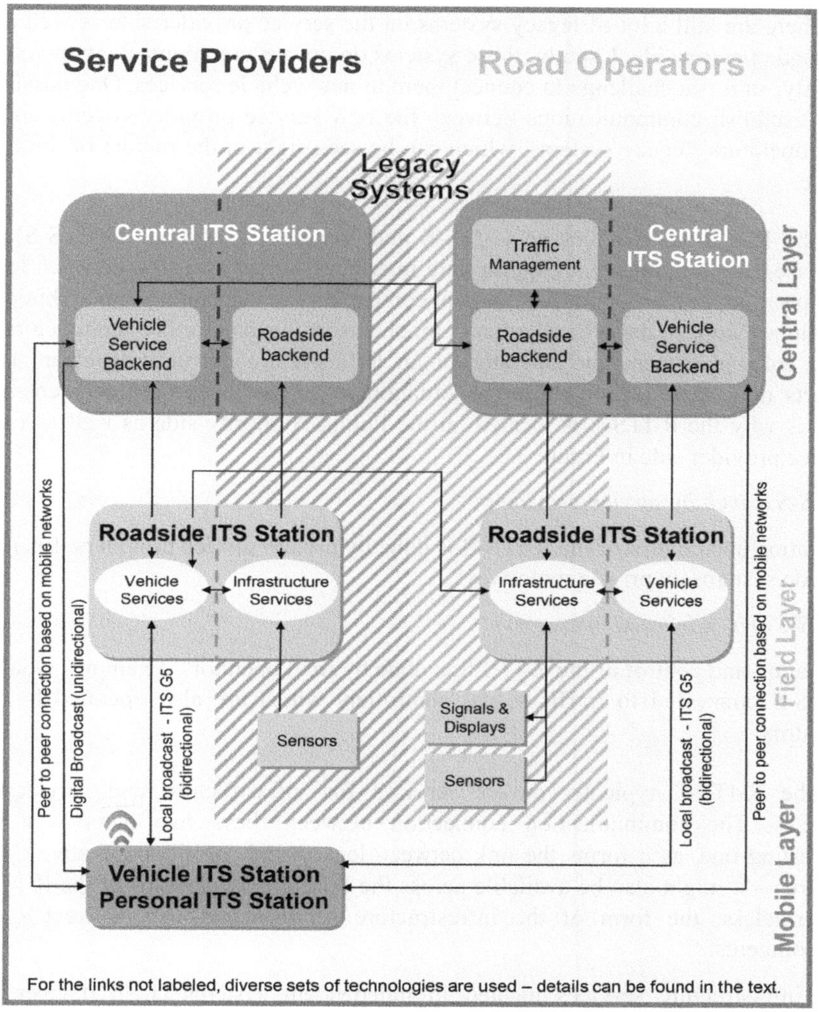

Figure 3.2. *A generic ITS model. For a color version of this figure, please see www.iste.co.uk/jacob/traffic.zip*

According to their main task and behavior, all components are clustered in layers named Central, Field and Mobile. Each of the components can be viewed as an independent system including different subsystems and back ends. Usually, these subsystems contain different services or back ends, which can be seen as black boxes. They need to provide several services or functionalities, but their inner structure or even implementation is not part of this generic architecture.

There are still a lot of legacy systems on the service provider side as well as on the road operator side. Usually, those systems do not yet communicate with vehicles directly, so it is a challenge to connect them to new vehicle services. One possibility is to establish communications between the new service provider systems and the road operators' legacy systems, which can be seen right in the middle of the figure above.

The R-ITS-S has a link to C-ITS-S and Vehicle-ITS-stations (V-ITS-S). But likewise, it is possible (with appropriate interfaces) to act as a link between legacy systems and new cooperative system components – depending on architectural decisions. More so than for all other components of the generic ITS architecture, the exact role, placement and operator of an R-ITS-S are not entirely clear. Those aspects may vary a lot in different situations, circumstances or implementations. That is why the R-ITS-S is located on the Public Authority side as well as on the service provider side in Figure 3.2:

R-ITS-S driven by service providers;

Operation and control of the R-ITS-S is done by private service providers. Examples include systems on private motorways.

R-ITS-S driven by road operators;

Operation and control of the R-ITS-S is done by road operators. Examples would be R-ITS-S connected to traffic signal controllers which are also operated by road operators.

The R-ITS-S typically covers services for infrastructure and services for vehicles. The communication connection between these two parts is a very interesting one, as it forms the link between legacy and "new" cooperative system features – it might also be available across the different ownerships of the R-ITS-S. Nevertheless, the form of the infrastructure services can vary in real system environments:

– infrastructure services located in independent systems (field devices, e.g. traffic light controller);

– without any link to the vehicle services (i.e. the legacy system cannot communicate with the Mobile Layer directly, but only with the Central Layer);

– with an external link to the vehicle services (i.e. the legacy system can use this connection to communicate with the Mobile Layer);

– it is also conceivable to combine infrastructure services and vehicle services in one system with an internal interface between them.

3.4. A tailored architecture for the use case "Traffic Light Phase Assistant"

The term Traffic Light Phase Assistant is used collectively for different objectives and applications (including some not yet known or developed). The common characteristic is their need for information describing the near-future behavior of traffic signals. Usually this should cover:

– operating state of controller (e.g. if working in regular or in an exceptional mode like "switched off");

– current signal program and/or sequence of the next signal programs (and also the specification of the signal programs as part of static information);

– current signal state including information which allows to relate it to the targeted traffic streams;

– time to green and time to red for controlled traffic streams, both for the current situation ("real-time") and often also for the future (short term prediction, "next changes" or schedule).

But even more detailed control information can be required:

– traffic adaptive or fixed time control;

– cycle-time (if fixed time control applies);

– special operation control (for emergency vehicles);

– public transport priority active or;

– manual control;

– queue length of the waiting vehicles.

This dynamic data might need to be enriched by static information about the controller (type and control method, signal groups, quantity and design of signals, etc.). In most scenarios of this use case, detailed geographical information about the whole junction (topology information) and especially its stop lines need to be present. At least an assignment of the information to the affected traffic streams must be enabled somehow. The source of information can vary between the controller itself or some central system (when the central system has knowledge about the signals and knows their states).

3.4.1. *Scenario via service provider*

In the following scenario, the traffic light information is gathered from the central component of a road operator and transmitted to a service provider. Note that there are other scenarios possible which are not shown here, for example communication via the field layer or a public scenario via Central ITS Station.

Usually, the road operator system has a direct connection to and from its field devices, but the central system might also already store all relevant information itself. The customer for the information is the service provider. The link between these two C-ITS stations might be established by a DATEX II connection; an example of this interface will be described in the next chapter.

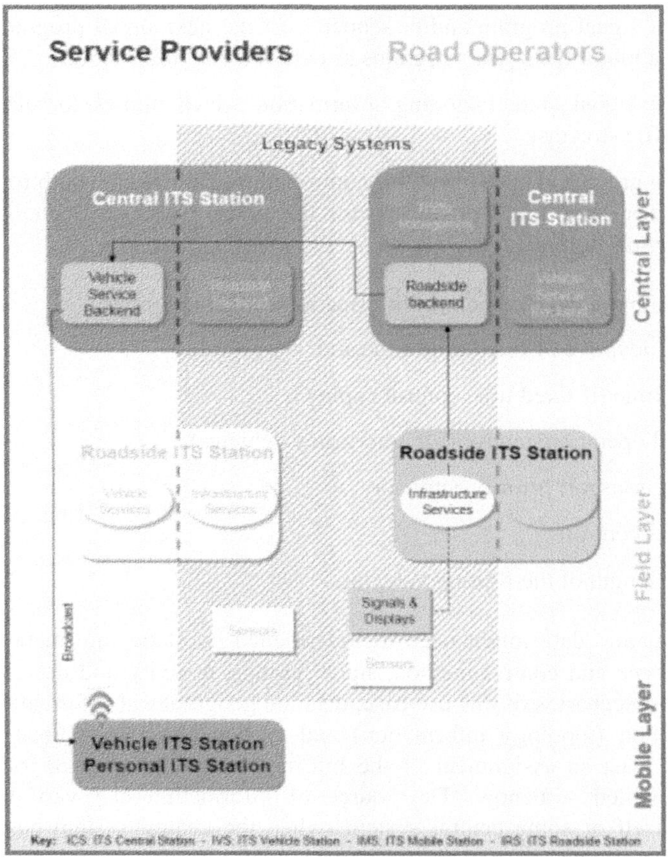

Figure 3.3. *Traffic light phase assistant via service provider. For a color version of this figure, please see www.iste.co.uk/jacob/traffic.zip*

The service provider will prepare the information and broadcast it to the vehicles. This type of communication chain should not be used for real-time data.

Useful information would be the next sequences of signal programs, content of signal programs or current changes in pre-calculated sequences. It is also possible to transfer the geospatial information of the intersection (see next section).

3.5. A DATEX II profile to communicate traffic light information

DATEX II (see www.datex2.eu) is the European open, interoperable interface specification for machine to machine communication of dynamic transport and travel time data between centrals and with third parties. It is based upon well-accepted IT standards (UML[4], XML[5], http, etc.) and is offered by the DATEX community (mostly road operators) free of charge and without license restrictions. Its content is standardized in CEN/TS 16157.

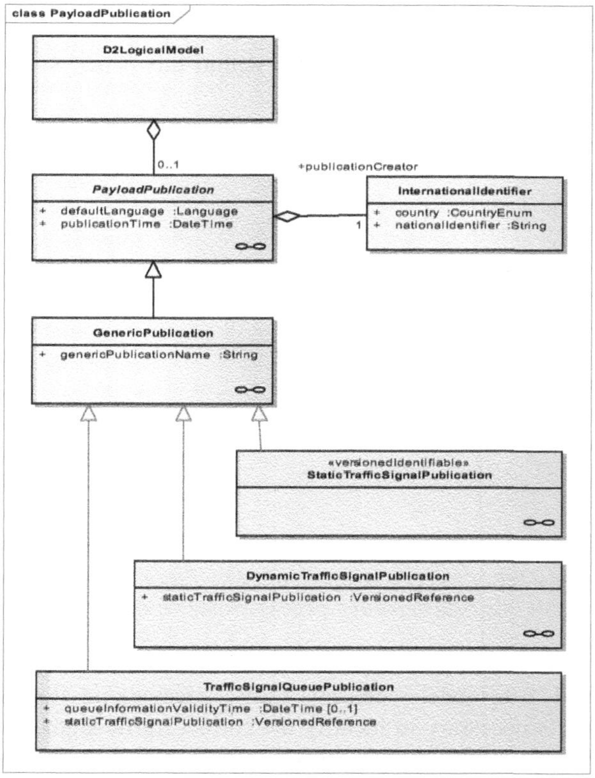

Figure 3.4. *Three publications for the traffic signal information*

4 Unified Modeling Language.

5 Extended Markup Language.

For the communication channel presented above, a DATEX II profile [DAT 13] has been developed and published to exchange traffic signal data between a (public) central and a (private) service provider. As some of the elements needed for this profile are not yet part of the DATEX standard, they have been added using an extension mechanism of DATEX II ("Level B").

The profile (which is available for free as XML schema including a documentation file, see DATEX [DAT 13]) defines three publications:

– a static publication to define traffic streams and stop line points;

– a dynamic publication to transport next signal states, optionally using a time vector;

– a queue information publication, which allows informing about queuing times on the intersections.

It is not possible to describe all details within this paper, so only some basic ideas are mentioned here.

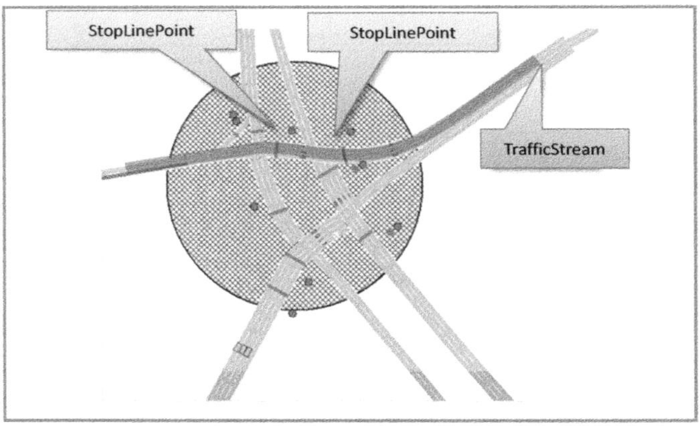

Figure 3.5. *Stop line points and traffic streams. For a color version of this figure, please see www.iste.co.uk/jacob/traffic.zip*

An elaborated data model represents traffic streams and stop line points (see Figure 3.5). This is necessary to identify the correct traffic signal that the dynamic data will refer to. In terms of georeferencing, a stop line is represented either by coordinates or by coordinates offset to a traffic stream. A traffic stream itself is identified by a linear object (for example using ALERT-C).

In the dynamic part of the model, the traffic signal times can be described with a number of different time values. Especially for traffic adapted signalization (where the exact time of change is unknown), additional probability values are necessary (see Figure 3.6).

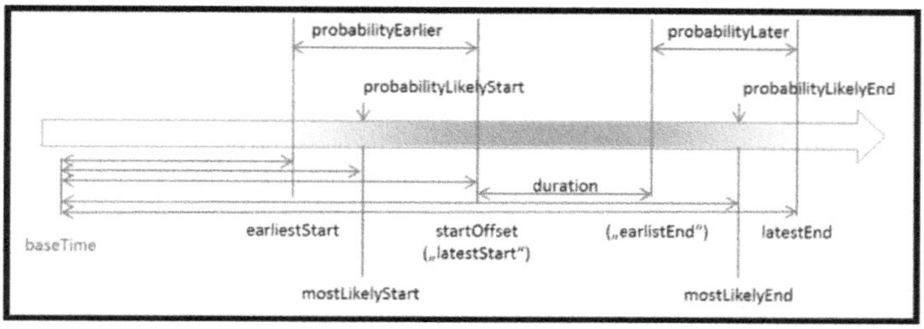

Figure 3.6. *Prognosis information. For a color version of this figure, please see www.iste.co.uk/jacob/traffic.zip*

Using common methods of XML data transfer, this data profile enables the transfer of any amount of traffic signal information to a service provider, who can broadcast the information to its customers.

3.6. Summary

Starting from a generic ITS architecture, this chapter details how to find a tailored architecture suitable for the needs of a special use case. For the traffic signal information, a DATEX II profile was presented, which is publicly available.

The SEAMLESS project also showed that there is a lack of clarity on the role of the R-ITS-S – the number of possible R-ITS-S scenarios is as huge as the number of different tasks for the ITS communication. A key question is where (and in which systems) the intelligence of the R-ITS-S is best allocated. A similar lack of uniformity can be seen in the use case for the Traffic Light Phase Assistant. Because of its nature, this use case covers a lot of topics, which are solved very differently in a variety of projects.

The progress as well as the problems are shown in this chapter in an urban context as an example, but they exist exactly the same way in interurban scenarios. For travel times or in-vehicle signage along motorways other standards apply, but mapping in the generic architecture can be done in a similar manner. The question of the architectural role of the R-ITS-S is also the same in an interurban context.

The work done in SEAMLESS brought together the standardization work currently running on European level and the standards which are already established in the Member States of the European Commission. The activities in Europe are driven by ETSI and CEN/ISO based on a mandate of the European commission.

They are essential in order to bring infrastructure information into vehicles because these kinds of services must be harmonized (at least in Europe). However, the results produced by the two organizations touch the needs of legacy systems only marginally. Helping to ensure the deployment of innovative services such as traffic signal assistance, mapping with a generic architecture and a DATEX II profile have been introduced.

3.7. Bibliography

[CVI 06] CVIS, D.CVIS.3.1, CVIS Reference Architecture, CVIS, 2006.

[DAT 13] DATEX II, DATEX Profiles Directory, "Traffic Light Information", available at: http://www.datex2.eu/d2-profile/2013/08/27/426, 2013.

[EC 12] EUROPEAN COMMISSION, 2nd Progress report to EC Mandate M/45, February 2012.

[FGS 12] FGSV, "Hinweise zur Strukturierung einer Rahmenarchitektur für Intelligente Verkehrs-systeme (IVS) in Deutschland", German Road and Transportation Research Association (FGSV), working group "Verkehrsmanagement", working committee "Telematik", study group "ITS Systemarchitekturen", 2012.

[FRE 12] FREUDENSTEIN J. *et al.*, Seamless traffic data dissemination across urban and inter-urban networks, ITS World Congress, Vienna, October 2012.

[HER 13] HERB, Teresina, Organisational architecture for Cooperative Systems, Bundesanstalt für Straßenwesen, FGSV Straßenverkehrstechnik, 2013.

[ISO 14] ISO TS 17425, Intelligent transport systems – Co-operative systems – Data exchange specification for in-vehicle presentation of external road and traffic related data, in development since 2014.

[ITS 11] ITS Action Plan, The frame architecture and the ITS Action Plan, booklet of the E-FRAME Project, June 2011.

[LUD 12] LUDOVIC S., "SCORE@F – Ready to deliver the promise, French Ministry in charge of transports", *5th EasyWay Annual Forum London*, 2012.

[MEN 11] MENIG C., "Field trials with coop. systems in Europe", *Audi AG, 8th ITS European Congress*, Lyon 2011.

[OPE 09] OPEN TRAFFIC CITY ASSOCIATION (OCA), OTS-Leitfaden, 2009.

[SAE 09] SAE J2735, Dedicated short range communications message set dictionary, SAE 2009.

[SCH 10] SCHADE H J., Convenor CEN TC278/WG16 and ISO TC204/WG18, 2nd ETSI TC ITS Workshop, 2010.

[SEA 13] SEAMLESS, Seamless traffic data dissemination across urban and inter-urban networks, initiated by ERA-NET ROAD II, Deliverables 1-4, available at: http://www.eranetroad.org/, 2013.

4

Sensor City Mobility: The City of Assen as a "Living Lab" for Smart Mobility Solutions Using Sensor Data

In the Dutch city of Assen, a large-scale urban sensor network is realized. The purpose of the urban sensor network is to enable several useful practical applications ("use cases") to stimulate the development of economic activities in the field of these applications and sensor technology. In the research project Sensor City Mobility, this network is used for the first time to develop and test use cases in the field of travel information and traffic management. In this paper, the need for this network and a real-life experiment is explained. The following use cases are discussed Multimodal Travel Information and Guidance, Dynamic Traffic Management in-car, Parking Guidance Service, Driving Style Monitor and TravelAlert. The project and its complete evaluation was finished in February 2014. Hence, the paper concludes with preliminary technical and organizational lessons learned from the Sensor City Mobility project.

4.1. Introduction

4.1.1. *What is necessary for travel information and traffic management?*

Traffic management and travel information are still largely based on static information, and travel information services for travelers and traffic managers are often not personalized. The value of travel information increases when a reliable prediction of the traffic situation can be generated so that real-time travel

Chapter written by Jan BURGMEIJER, Janiek DE KRUIJFF, Ernst Jan VAN ARK, Gerdien KLUNDER and Diana VONK NOORDEGRAAF.

information can be given to the travelers. Effectively influencing the behavior of travelers requires:

– services based on real-time data for traveling, for both road traffic and public transport;

– modeling to predict road traffic situations in the near future, say 30 min ahead;

– travel services which are personalized and aligned with the intentions and preferences of the traveler, offering maximum comfort and ease;

– cost-effective services as a future alternative for (part of) the current expensive static road side traffic management and radio broadcast traffic information.

4.1.2. *Why is a real-life test necessary?*

Many projects have tried to design, build and implement new travel information and traffic management services and to test these services in a small scale demonstration environment. The Sensor City Mobility project was started in 2010 as one of the first projects testing new mobility services in real life with a large number of voluntary participants. Real-life testing is more demanding than a technical test or a small scale demonstration. Therefore, the technology should be more robust, because voluntary participants will stop using a service if it does not work properly or is not user friendly.

The benefits of a real-life test are:

– a large number of participants enables the evaluation of the use of the services in addition to a technical evaluation; do these services affect travel behavior? What improvements could be made to make these services more effective and increase its value for the users?

– a real-life test is a step closer towards a commercial introduction of services by the project partners, who may introduce new prototypes of products and services which are already close to market introduction.

4.1.3. *What is meant by Sensor City?*

Sensor City is an ambitious project in which an infrastructure with a high number of sensor nodes is being implemented in the city of Assen. Each node bears several sensors, which are connected using a broadband communication network, with a large city database. The database is connected to a back office where real-time data processing is done, generating information to be used in real-time services. The aim of the Sensor City project is to:

– become a test facility for service providers from different societal domains;

– become a test facility for businesses and developers active in sensor technologies;

– test and showcase new societal services in a living lab with real users;

– get local business and societal parties in Assen involved in defining new and relevant use cases, and making them early users of innovative technological solutions.

In short, Sensor City is an ambitious project where a large-scale measurement network will be implemented, creating a living lab which allows the development of a variety of practical applications that use complex sensor systems. As a testing ground and showcase for sensor system applications, it is a unique facility throughout the world.

4.1.4. *Constructing the consortium and selecting the partners*

The consortium of Sensor City Mobility was organized in such a way to achieve a good mixture of small and large companies with backgrounds in different and relevant disciplines. The consortium was formed with partners including traffic system providers, mobile service providers, travel information service providers, hardware providers and software developers. TNO organized the consortium and the project as an "innovation catalyst", building the innovation ecosystem and facilitating cooperation between these partners in a complex and multidisciplinary environment. The municipality of Assen and the province of Drenthe started a foundation Sensor City for formulating and implementing the policies and goals of Sensor City. The foundation was also responsible for the implementation of the sensor network infrastructure and for contracting the consortium partners and the sensor network construction partners.

The Sensor City Mobility project is a cooperation between a consortium of 14 governmental and private organizations, being: DySI, Elevation Concepts, Municipality of Assen, Goudappel Coffeng, Magicview, Mobuy, NXP, 9292, Parkingware, Peek Traffic, Quest Traffic Consultancy, TNO and TomTom.

The next section will describe the general functional architecture, the sensor network and the technologies which have been used, followed by the description of use cases which are implemented in more detail. A section is further devoted to the real-time predictive modeling used to support the use case services. Finally, the preliminary results are discussed. The evaluation of the project, final conclusions and lessons learned will be useful for future pilots and implementations.

4.2. Architecture, sensor network and technologies used

In order to perform a real life experiment regarding travel information and traffic management, several conditions have to be fulfilled. This chapter describes the data obtaining sensor infrastructure and the technologies to bring the information to the participants.

4.2.1. *General architecture*

The general architecture is displayed in Figure 4.1. It consists of three parts:

1) on the left side the sensor networks are shown, which generate all sensor data which may be useful for travel information and traffic management use cases. The technologies of the sensor data may differ (inductive loops, Bluetooth etc.) and differentiation between fixed sensors and floating data centers (GPS, Mobile) is possible. In section 4.2.2, the sensor network is discussed in more detail;

2) in the middle of the figure, the modeling of the data is shown. Models are necessary to get useful information about the huge amount of data generated ("big data"). Models are described in more detail in section 4.4. Also positioned in the middle is the back office, which is necessary to support and facilitate the services. An internet portal (website) for the participants of the experiment is included. Back offices and portals are standard for mobile services and are not explained further;

3) on the right side, the use cases are listed. Use cases often require special devices and software at the client site (e.g. on board units, tablets, smartphones, navigation software, etc.) or hardware/software at the road side (e.g. traffic management systems).

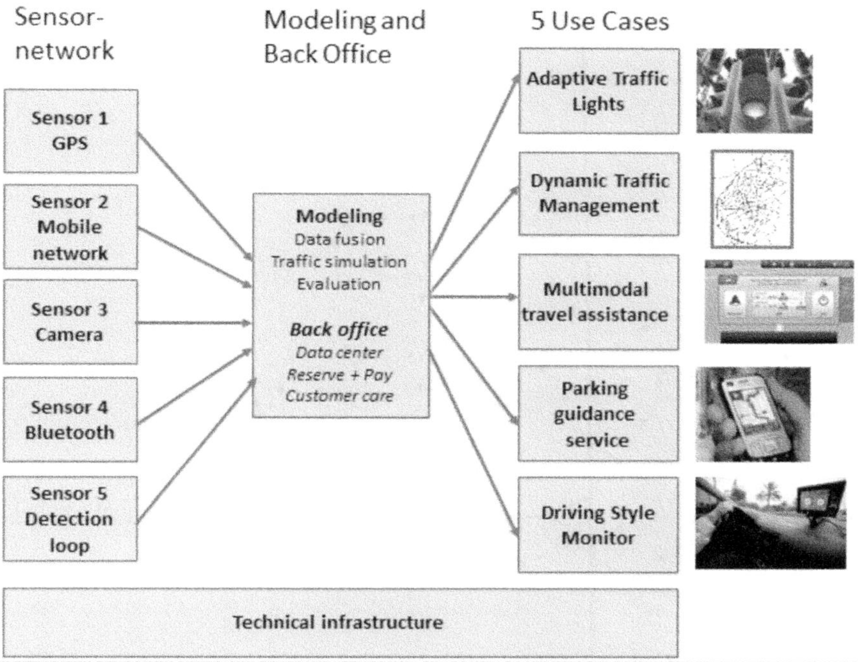

Figure 4.1. *Sensor City mobility architecture*

4.2.2. *Sensor network in Assen*

A large traffic monitoring network is being constructed, consisting of measurement points throughout the city of Assen, which will be equipped with various sensors. Traffic data from the road sensor network will consist of measurements at fixed locations from inductive loop detectors, cameras, radar and Bluetooth. Furthermore, floating car data will be provided by the detection of a large group of test participants with GPS and mobile phone tracking using a smartphone application (see section 3.5). Currently, 43 Bluetooth sensors have been installed and around 13 more were planned near future (see Figure 4.1). The count data of loop detectors at traffic signal controlled intersections are also available in real-time. The data are further complemented with data about the status of the Dutch road network from the National Data Warehouse (NDW) for traffic information. The NDW database contains several types of data, such as speeds from loop detectors and travel times, measured by cameras with automatic number plate recognition capability.

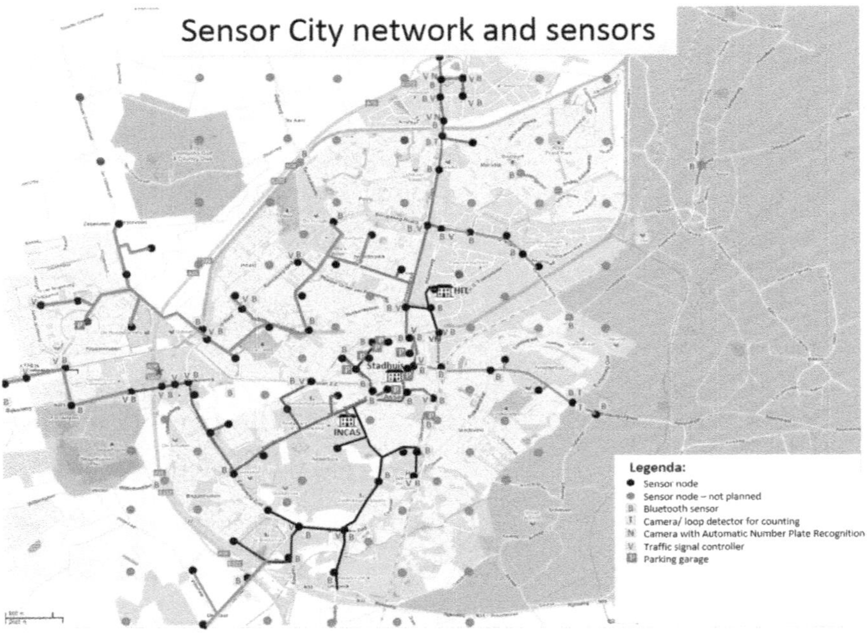

Figure 4.2. *Sensor City network and sensors. For a color version of this figure, please see www.iste.co.uk/jacob/traffic.zip*

4.2.3. *Technologies used*

There are several methods to provide travel information to the participants. Participants can get an on board unit (OBU) and a tablet installed in their car. The OBU is capable of transmitting GPS locations and data from the accelerometer from the car to the back office. The GPS data are used to calculate for example the speed of the vehicle, and the accelerometer is used to calculate the acceleration and deceleration. The OBU is also used to receive information from the back office, such as the maximum allowed speed on the traversed road sections. This information can be displayed on the OBU itself or passed through to the tablet. The tablet is used as an intelligent navigation system, which will be further explained in the next chapter.

Apart from the in-car devices, participants could also use their smartphone to test applications. Participants use their own smartphones. Hence, the threshold in participating in the experiment is lower than in the case of having to install in-car equipment. Smartphones are capable of sending and receiving almost the same information as the in-car devices, but can add information, for example about upcoming appointments in agendas.

The challenge lies in the combination of obtaining the traffic information, the processing of this data and sending the travel information to the participants. The received data are filtered depending on the personal profile of the participant. Different systems must be linked to the back office to execute the required proceedings. This is necessary for testing the use cases, which are the subject of the next chapter.

4.3. Use cases for mobility

In this chapter a description of the mobility use cases which are being developed and will be implemented and evaluated in a real-life experiment are discussed (see Figure 4.1). Adaptive traffic lights are not discussed in this chapter.

4.3.1. *Multimodal travel information and guidance*

The in-car system contains a turn-by-turn navigation system on the tablet. Where traditional navigation systems mainly focus on single-modality trips, the software applied within the Sensor City Mobility project is enhanced to support multi- and intermodal trips. During traveling the user is supported in making optimal choices for departure times (car or public transport) and modal switches (from car to public transport). The application also provides the user with the most accurate information about departure and arrival times of the different modalities of public transport.

4.3.2. *Dynamic traffic management in-car*

The objective of the use case dynamic traffic management is to manage the traffic flow using dynamic routing advice in-car. This requires accurate and real-time data of the traffic flow. In this use case, this data is collected mainly via loop detectors in the fixed sensor network. Using this data in dynamic traffic models, a prediction 30 min ahead in time of the traffic flow is generated. With this prediction, congestion can be forecasted and alternative traffic management scenarios can be generated. Car drivers get smart routing advice on their in-car navigation system directing them around the congested area. The more accurate the traffic flow can be monitored and predicted, the more effective the traffic management routing advice can be.

4.3.3. *Parking guidance services*

Parking in cities causes problems for both municipalities and car drivers. Municipalities want to reduce the number of cars driving around searching for a parking place, to increase utilization of the available parking spaces, reduce congestion in the city center and to increase income from parking lots and garages. An in-car parking guidance and information (PGI) system that guides drivers through the city to available parking garages, and furthermore offers the possibility to reserve a parking space, would help achieve these goals. For drivers, the added value of such a system is also clear.

This use case within the Sensor City Mobility project aims to develop such an in-car PGI. This system allows the drivers to:

– be informed about the available parking capacity in the parking lots;

– automatically reserve and pay for the parking location;

– be informed about the current traffic situation (i.e. congestion) and get routing advice to the parking space taking into account the current traffic situation;

– adjust their preferences for willingness to walk, money to pay and preferred parking lots.

4.3.4. *Driving style monitor for safe and eco-friendly driving*

Traffic safety is an important social issue. The number of casualties on the road is a substantial drawback of the modern motorized society. Most of the applied measures focus on both the infrastructure and the vehicle. The third factor, the user, has proved to be more difficult to influence. Although enforcement, information and education are applied integrally in the existing traffic policies, it is still difficult to intrinsically motivate road users to drive in a safe and eco-friendly fashion.

Opposed to enforcement methods that mainly punish undesired behavior, the continuous monitoring of driving behavior also enables the system to emphasize and perhaps reward desired behavior. The on-board unit applied in the Sensor City Mobility project can monitor and evaluate the driving style of the participants in three perspectives; speed, acceleration (and deceleration) and cornering behavior. This information is gathered by a combination of accelerometers and GPS data, and translated into a single performance indicator which provides feedback concerning the driving style; the main reference point for this performance indicator is the driving style of other users (peer feedback).

4.3.5. *The "TravelAlert" smartphone application*

The last discussed use case of the Sensor City Mobility experiment is the "TravelAlert" application (in Dutch "ReisAlarm"). The main aim of this application is to complement the on-trip multimodal traffic information and guidance by supporting the user in his decision-making process for departure and travel modality. The main functionalities of the "TravelAlert" application can be approximated from two perspectives; first, the functionalities of the application toward the user; and second, the functionality toward the provider/scientific community.

From the users' perspective, the main components of the TravelAlert application consists of a link to the digital calendar of the user and an intelligent multimodal travel time calculation model. Based on this calendar, the application deducts the origin and destination and predicts the travel time for the various available travel modalities. This travel time prediction not only includes the historically observed traffic situation on the route, but also takes the current traffic situations into account. The application monitors weather conditions, incidents or road maintenance works and subsequently uses this information to adjust the route advice. Based on the available information, the application suggests to the user to depart early, to take an alternative route or to choose an alternative mode of travel. When, due to critical incidents, no feasible arrangements or alternatives exist, it is envisioned that the application advises the user to cancel the trip and moreover supplies the contact details of the participants concerned in the calendar entry.

Individual travel information has been of significant interest to researchers. This is based on the fact that trips can place great strain on the transportation networks. While past research has investigated travel distance, travel times, model and associated route choices, less attention has been paid to investigating the day-to-day dynamics of travel behavior. While repetitive trips (such as commuting) are often assumed to be static and therefore highly predictable, research results indicate that users often change departure times, routes and also often use trip chaining that significantly affect the departure time and route choice behavior [HAI 04]. One major component of the TravelAlert application is the positional monitoring algorithm. This algorithm detects whether the phone is moving, and moreover periodically determines the exact position of the smartphone. This location tracking service is integrated within the background services of the operating system of the phone and automatically starts when the phone is switched on. Based on these location traces and a sophisticated data processing algorithm, the travel behavior of the user is deducted and stored on a server. This dynamic travel data can subsequently be applied for research purposes.

The preliminary results of the use cases will be further discussed in section 4.5.

4.4. Modeling

In order to design and evaluate the applications that are part of the use cases, offline testing is used in a "simulation laboratory" and through operational tests. Laboratory-based evaluation has many advantages since it can take place in a controlled environment and identify issues before the system under evaluation is actually implemented.

4.4.1. *Online modeling*

For real-time applications, a traffic macro-simulation model called Streamline has been developed by Goudappel Coffeng. The purpose of the macroscopic model of the Assen region is to develop a short-term online traffic forecasting model that describes the actual situation, including the upcoming half to one hour. Because of calculation times every 5 min, the half hour forecast will be updated at this interval; and at every 10 min for the one hour forecast. In this short term forecast, the chosen traffic management regime will be taken into account Ben-Akiva *et al.* [BEN 02]. Results of the short term forecast are expected travel times and information on the expected traffic conditions in and around the city of Assen. This kind of information has to be translated in useful travel advices for travelers in the Assen region.

The state estimation of the traffic situation by the online traffic prediction model will be based on historical and real-time traffic data together with OD patterns derived from MON [MOB 12], the Dutch national survey of personal travel patterns and the existing static model of Assen for an average working day. This forms a base set of OD matrices for each hour and day of the week. By calibrating the OD matrices on the available sensor data the model will be fine-tuned for the actual traffic situation in Assen. The base set of OD matrices will be updated continously in time by learning from the calibration effects. The short term prediction is then determined by applying calibration factors of the actual situation on the base OD matrix for the next prediction period. Currently, loop detector data are used for real-time calibration of the OD matrix. The predicted OD will be applied for the upcoming prediction period.

4.4.2. *Offline modeling*

The effects of the use cases on mobility and air quality will be investigated with an offline traffic micro simulation model. A traffic micro simulation model has been developed for offline evaluation, the ITS modeler [VER 05]. Impacts can be estimated in terms of travel time, delay, speeds, vehicle kilometers, fuel usage, etc. Furthermore, for the traffic management use case, a more detailed modeling of the traffic signal control is modeled in Vissim and optimized by an agent-based control method.

The network of the offline model is a part of the larger macroscopic model. The same data has also been used for calibration and OD matrices have been aggregated for the smaller network. As such, OD matrices and link results can be compared and exchanged easily.

Since the project is situated in a city environment, it concerns the simulation and prediction of urban networks [MAH 04]. This entails challenges for simulation and prediction concerning specific urban characteristics such as intersections, cues and traffic signal modeling and more route choice possibilities. This complicates the calibration and validation process and required new developments, such as detailed modeling of controlled intersections. Improvements concerning these subjects are in progress for the microscopic model.

4.4.2.1. *Modeling of parking guidance services*

One of the use cases to be evaluated with offline modeling is the parking guidance service. As explained above, this functionality provides real-time in-car parking and routing advice to the drivers, including a check for available capacity in the parking spot beforehand. This should prevent queues near parking garages and reduce traffic searching for parking.

In order to determine the effects of the parking system by simulation in Assen, the ITS modeler implementation of the network of Assen will be used. A Saturday will be simulated because that is the busiest day for shopping in the city center. An extensive parking model is being developed and calibrated with a survey among 500 participants. A multinomial logit model is developed with attributes such as price, distance from destination, travel time, parking type (off street/ on street) and probability of finding vacant parking spot.

The modeling will be done approximately as follows: a fraction of all vehicles with destinations in the city center are assumed to have access to the parking guidance and information service (PGI). Vehicles with parking applications will be assigned to a parking lot based on available space and a cost function of the mentioned attributes. They are dynamically assigned to the parking lots. The remaining vehicles in the center zones are assumed to park along the curbs or at private parking spaces. Each driver can either use the in-car parking application (including congestion information) or have no information. The size of the fraction of vehicles assigned to parking lots can be adjusted, as well as the penetration rate of the in-car PGI. Vehicles without the parking application do not have knowledge on the available parking spaces and may therefore drive to an already occupied parking lot, where they will wait or move to another parking lot, generating more travel time and vehicle kilometers.

4.4.3. *Smart Routing*

One of the other functionalities to be tested in the Sensor City Mobility experiment is Smart Routing. This functionality will be used to give individual routing advice to the participants using the in-car system, while distributing these drivers over a set of suitable routes in case of congestion (instead of the normally preferred route) in order to better spread the traffic over the available infrastructure. The travel times of this set of routes are of the same magnitude (with differences of several minutes) in order to minimize the impact of this algorithm on the individual traveler's travel time. The Smart Routing advice is determined based on a real-time prediction of the travel times by the online model. The 5-min predictions (up to half an hour forecast) are communicated to the online Smart Routing algorithm, which will in turn supply route advice to the in-car devices.

As a second input for the Smart Routing algorithm, the Bluetooth sensor measurements are used directly. For each 5-min interval, a dataset of actual sensor measurements is stored in a database. This database contains (among others) the average measured travel times between pairs of Bluetooth sensors. With an additional processing module, this travel time is further filtered and translated into a travel time per link. These travel times can be considered as a reliable monitor of the current situation and are also directly passed to the Smart Routing algorithm. Further investigation needs to reveal which input (online modeling or Bluetooth monitoring and processing) provides the best input for the Smart Routing functionality.

4.4.4. *Calibration and validation*

All modeling environments will make use of the different types of measurements (but the same sources) in Assen to calibrate and validate the models. Usually, calibration and validation is done manually by comparing measured and simulated traffic variables (flows, speeds) and adjusting model parameters such as the Origin–Destination matrix [CHE 08], or for microscopic simulation models behavioral parameters such as desired speed and following distance. For online application, manual calibration is not possible since it should be performed in the loop within a very tight time schedule. Therefore, automatic calibration and validation procedures are developed.

New types of data also offer new opportunities for better calibration. For example, floating car data offer another type of information to point data such as loop detectors. While floating car data give very little information on flows, since the penetration rate of the measuring device is usually unknown, it does provide other useful information, namely on individual vehicle routing and travel times. This information can be used for e.g. calibrating (and validating) route choice models, the assignment model and for improving the Origin–Destination estimation.

4.5. Preliminary results and evaluation of the experiment

As of September 2013, the quantitative results of the in-car experiments were not yet available. Almost 150 participants are driving with on board units and tablets in their cars. The systems and services that are required for the use cases are working stand-alone. The combination between the different systems also demonstrate the required functionalities. The main challenge is to combine the systems to provide the travel information in real-time to the participants.

Although the in-car equipment has been distributed recently, the TravelAlert smartphone application has already been deployed for several months. The application was installed on 101 unique devices. In total 11,490 trips were successfully deducted; the total travel distance was registered is 277,021. These preliminary results demonstrate that it is technically possible to gather data which provides extensive and detailed information concerning the travel behavior of the participants. Moreover, it appears that it is possible to automatically deduct and process the important factors of each trip such as route and mode choice. Based on a preliminary user survey, no significant impediments or issues have been found in relation to the user experience.

Given the largely technical focus of the Sensor City Mobility project on using and combining sensor data, the project team faced many technical challenges, in addition challenges relating to the cooperation between the 14 consortium partners. Furthermore, the institutional project structure in which the consortium partners were bringing in additional investments in the project (making them collectively the client) also added to the project management complexity. Also, the timely real-life implementation of the sensor network proved challenging. The time required for acquiring the necessary permits and contracts was underestimated. A last example of a challenge is the protection of the privacy of the participants. Some of the obtained information of the participants may possibly lead to privacy issues, so keeping track of the availability of this data for partners and making sure the data is safe and secure are important aspects of the Sensor City Mobility experiment.

Figure 4.3. *Location plot of travel alert users. For a color version of this figure, please see www.iste.co.uk/jacob/traffic.zip*

The services that were developed and tested as part of the use cases have contributed to the state-of-the-art knowledge on applying sensor technology, and provide opportunities for future implementations of innovative traffic information and traffic management services. The next step is to bring the services to the participants of the Sensor City Mobility project.

4.6 Acknowledgments

The Sensor City Mobility experiment is being sponsored by the European Union, the European Fund for Regional Development, the Ministry of Economic Affairs in the Netherlands and the organization "Noord- Nederland, Koers Noord".

4.7. Bibliography

[BEN 02] BEN-AKIVA M., BIERLAIRE M., KOUTSOPOULOS H.N., "Real-time simulation of traffic demand-supply interactions within DynaMIT", in GENDREAU M., MARCOTTE P. (eds), *Transportation and Network Analysis: Current Trends*, Miscellenea in honor of Michael Florian, Kluwer, pp. 19–36, 2002.

[CHE 08] CHEN Y.S., ZUYLEN H.J., ASMUNDSDOTTIR R., "Estimation of dynamic origin-destination matrices using floating car data", *Proceedings of the 10th TRAIL Congress*, Delft, 2008.

[FRI 11] FRISO K., REITSMA J., *Sensor City Assen: Onderweg Met Op Maat Gesneden Verkeersinformatie*. Bijdrage aan het Colloquium Vervoersplanologisch Speurwerk 24 en 25 November 2011, Antwerpen (in Dutch), 2011.

[KLU 11] KLUNDER G.A., CHEN Y.S., ZANTEMA K., "Consistent traffic modelling and calibration at different resolution levels based on various sensor data for real-time traffic management", *2nd International Conference of Models and Technologies for Intelligences Transportation Systems*, Leuven, 2011.

[MAH 04] MAHMASSANI H.S., "Application of a real-time dynamic traffic assignment-simulation model system for intelligent network traffic management", *Proceedings SuRe Seminar Beijing*, 2004.

[MON 12] MON, Mobiliteitsonderzoek Nederland, Last modified: October 12, 2012 available at: http://www.rijkswaterstaat.nl/zakelijk/innovatie/innovatieprogrammas/mobiliitsonderzoek/index.aspx, 2012.

[VER 05] VERSTEEGT E. *et al.*, "Modelling cooperative roadside and in-vehicle intelligent transport systems using the ITS modeler – an integrated umbrella for existing microscopic traffic models", *ITS World Congress*, 2005.

[ARK 13] VAN ARK E.J., VAN DEN HAAK P., DUIJNISVELD M., *ReisAlarm: Was u vanochtend op tijd op het CVS?* Bijdrage aan het Colloquium Vervoersplanologisch Speurwerk 21 en 22 november 2013, Rotterdam (in Dutch) available at: www.sensorcitymobility.nl, 2013.

PART 2

Traffic Modeling and Simulation

5

Forecasting Capabilities of a Micro-Simulation Method for Trip Generation

Traditional trip generation methods (i.e. cross-classification and regression) struggle to take into account a big part of the heterogeneity of urban mobility. The variable to be explained is the number of trips (usually home-based segmented by purpose). Therefore, the chaining of activities is lost and, in the case of journeys containing non-home-based trips, the main purpose of the journey is usually lost too. This paper presents an alternative approach to trip generation based on micro-simulation. In this approach, the mobility behavior of agents (variable to be explained) is characterized by mobility profiles (number and type of tours). The modeling approach uses the algorithm to partition the population. Using data from three household travel surveys in the Lyon conurbation, we test our approach and its forecasting capabilities over a period of 20 years (1985–2006). Preliminary results show that our approach yields results comparable to those of more traditional techniques, while keeping track of activity-chaining patterns, which could be used to refine other components of trip models.

5.1. Introduction

Trip generation is a crucial step of the classic urban transport modeling system (UTMS) not least because it is the first one. Errors that might be introduced during this step can propagate and result in poor forecasts [ZHA 02]. Because of the still widespread use of the UTMS [TRB 07], it is important to look for ways to improve its predictive accuracy. This paper presents the preliminary results of research carried on the forecasting capability of a micro-simulation method for trip generation.

Strambi and van de Bilt [STR 98] note that although challenged by developments from the activity-based analysis of travel patterns, conventional trip generation modeling techniques have evolved very little as cross-classification and

Chapter written by Jorge CABRERA DELGADO and Patrick BONNEL.

linear regression remain predominant. And yet, the specification of a trip generation model is not always straightforward. According to Strambi and van de Bilt, three elements must be taken into account when specifying a trip generation model: (1) selection of pertinent explanatory variables, (2) choice of categorization for each variable and (3) the treatment of potential interactions between explanatory variables. The two traditional approaches show limitations at different stages of the specification process. Variable selection and categorization in cross-classification are more the result of a trial and error process based on theory, experience, statistical analysis and expedience. Despite some efforts toward systematizing the process [STO 83, SUP 83], there is still no mention of any kind of "optimal" procedure. The strength of cross-classification is the possibility to consider complex interactions between explanatory variables. Regression methods benefit from stepwise procedures to help with variable selection. But they are limited in the interactions between explanatory variables that can be modeled (linear form, autocorrelation issues).

Moreover, cross-classification and regression struggle to take into account a big part of the heterogeneity of urban mobility. The variable to be explained is the number of trips, usually home-based (HB) segmented by purpose. Non-home-based (NHB) trips are mostly modeled in a very simplistic fashion, without recurring any kind of segmentation. Therefore, as the proportion of NHB trips increases, a well-documented tendency [BON 04], the explanatory power of the whole approach decreases. The increase in NHB trips means that individuals tend to do "more things" (participate in more activities) each time they go out. If activities which typically have a strong relation with an explanatory variable (for example, work trips and number of workers in a household) are not at an end of an HB trip then traditional models will not "use" the strong relation and instead will replace it with a weaker one (that of the NHB trips).

Furthermore, as models use more disaggregated data (going from zones to households or persons), the interdependence of trips becomes more problematic. Stopher *et al.* [STO 05] found that the use of the number of trips for each purpose as the variable to be explained had some limitations. The disaggregation in simulation reveals potential inconsistencies in the data. For instance, the independence in trips simulated for each purpose leads to an underestimation of the number of households and individuals making no trips. In order to take into account the interdependence between trips and trip characteristics, Stopher *et al.* [STO 05] proposed to use the tour [ADL 79, AXH 00] as the unit of measure of mobility.

The developments in activity-based modeling, mentioned earlier, have been possible in part because of the increase in computing power availability, which makes the use of micro-simulation methods possible [VAU 97, KIT 00, STO 05]. It also fosters the use of algorithmic methods, such as decision trees or computational intelligence, for the modeling of the traditional steps of the UTMS. For example, Strambi and van de Bilt [STR 98] applied the CHAID algorithm to model trip generation, using the number of trips per purpose as the variable to be explained. Badoe and Miller [BAD 98] applied an AID-like procedure to model mode choice.

The purpose of the research presented here is to analyze the forecasting capabilities of micro-simulation procedure for trip generation. This approach is similar to the work of Stopher *et al.* [STO 05] in its principle of using a decision tree method to create population segments homogeneous in terms of mobility behavior. It differs from it though in its definition of what constitutes mobility behavior. Instead of using just a number of tours, we constituted *mobility profiles* characterized by the number and type of tours an individual makes during one day. The other novelty of this research is that we use data from the last three household surveys in the Lyon conurbation (1985–1986, 1994–1995 and 2005–2006) to test the forecasting capability of the approach in a set similar to that used for testing such characteristics for traditional models [FOX 10, DEL 13].

The organization of this chapters is as follows. The first section presents the methodology modeling and testing the forecasting capabilities of the model and the data sources. The second section presents the results. A third section presents the conclusions.

5.2. Methodology

5.2.1. *Data*

For this study we used three data sets coming from three household travel surveys made in the Lyon conurbation between 1985 and 2006: the EMDs (for French Enquêtes ménages déplacements). The EMDs are conventional face-to-face home interview surveys. The three surveys took place during the following periods: 1985–1986, 1994–1995 and 2005–2006. They were carried out following a similar protocol [CER 08] for each period, which allows for comparisons of the results across surveys. In what follows, EMDs will be referred to as 1985, 1995 and 2006. The scope of the EMD Lyon has been considerably extended over time. In 1985, 1995 and 2006, the survey covered 71, 99 and 453 municipalities, respectively. The perimeter of the 1985 survey is close to that of the Grand Lyon. The perimeter of 1995 fully contains the perimeter of 1985 to which is added a part of the Ain department in the northeast. The perimeter of 2006, much broader than that of 1995, includes in particular a large part of the Beaujolais in the northwest of the conurbation. With the expansion of the surveyed perimeter, there has also been an increase in the number of households surveyed. For 1985, the sample contained 4,691 households (11,449 individuals). For 1995, there were 5,851 households surveyed in the perimeter of the 1985 survey and 6,001 on the perimeter of the 1995 survey (13,579 and 13,997 individuals, respectively). For 2006, there were 6,169 households in the 1985 perimeter, 6,544 in the 1995 perimeter and 11,229 in the 2006 perimeter (again 13,586, 14,523 and 25,656 individuals, respectively). Sampling is stratified by location and survey data is provided with expansion factors.

Only data for the perimeter of the 1985 survey is used in this study. This means that only households residing in this perimeter have been kept and, importantly, that only tours made in the perimeter have been kept. This was necessary to ensure the comparability of the results over time.

5.2.2. *Micro-simulation procedure*

The micro-simulation procedure has two steps. The first step could be seen as an equivalent of model calibration in the traditional trip generation modeling framework. It consists of applying a decision tree algorithm which partitions the population (travel survey data) into segments with similar travel behavior. The second step consists of defining a synthetic population for the period for which forecasts are needed. Then, for each member of the synthetic population, a "donor" is randomly selected from the surveyed sample. This "donor" has the same characteristics as the member of the synthetic population (i.e. belonging to the same population segment, or *seed*, as defined in step 1). The travel behavior of the "donor" (defined as the number of tours for each type of tour, see section 5.2.2.1) is then imputed to the member of the synthetic population. Different methods for the generation of synthetic populations may be applied, ranging from simple trend extrapolation to complex demographic models. These go beyond the scope of the present research. The reader may refer to Farooq *et al.* [FAR 13], among others.

5.2.2.1. *Trips, tours and mobility profiles: choosing the variable to be explained*

As noted in the introduction, using the trip as the unit of analysis of mobility presents some limitations mainly because of increased importance of the chaining of activities. Another important aspect of choosing tours instead of trips relates to the particular need for comparable/compatible data from different surveys for temporal validation. The definition of what constitutes a trip has remained stable for Lyon EMDs. However, as survey staff and quality control may change over time, it is possible that the application of the definition might change slightly from one survey to another, especially for short trips, like buying bread or gas. The choice of the tour as the unit of analysis minimizes the impact of the risk of inconsistency in the data.

In this study, the variable to be explained is not numerical (e.g. a number of trips or a number of tours), it is categorical: the *mobility profile*. Based on eight trip purposes (Work, School, Highschool, College, ShopServ (shopping for goods and services), Escort, Leisure and Other), we define the type of tour by the sequence of trips that compose it. Because using these tours directly for constructing mobility profiles would lead to an extremely high number of mobility profiles to be accounted for, especially taking into account data constraints, we simplified the description of a tour. Instead of being defined by all the purposes of the trips that compose it, we distinguished only 16 types of tours. These are defined respectively to what could be considered the "main" or more constrained (less flexible, see [KOS 84]) purpose of the tour according to the following hierarchy:

Work > Education (three purposes) > ShopServ > Escort > Leisure > Other.

For each main purpose, two types of tours are defined: simple (home – main purpose – home) and complex (other combinations including at least the main purpose and another one).

Using this segmentation of tours, a mobility profile is then a vector containing the number of tours of each type made by the individual during the day. Only the 48 more frequent profiles in 1985 were used, the 49th profile being "the rest".

5.2.2.2. *Creating a typology of the population according to mobility profiles (selecting explanatory variables)*

Loh [LOH 08] summarizes the decision tree approach, in mathematical terms, as follows. Let's assume that we want to explain/predict the value taken by a discrete variable y (mobility profile) using a vector of predictor variables x (socioeconomic characteristics). This problem can be interpreted as that of finding a function $d(x)$, that maps each point in X to a point in Y, where X and Y are respectively the domains of x and y. To construct $d(x)$, a training sample of n observations $L = \{(x1, y1),\ldots, (xn, yn)\}$ is used (i.e. the survey data). Decision tree methods determine $d(x)$ in a recursive manner. Each method corresponds to a particular algorithm that partitions the x-space one predictor variable at a time, thus giving $d(x)$ its tree-like structure.

For this study, we use the well-known CHAID (Chi-squared Automatic Interaction Detector, procedure) [KAS 80]. We chose this algorithm over other alternatives, like CART, because of CHAID's multiple-category splits which are easier to interpret (CART's splits are binary). To determine a split, the algorithm looks for the predictor variable that if split, better explains/predicts the category response variable. The explanatory/predictive ability of a particular split is determined via a hypothesis test regarding dependence of the split variable on the categorical response (using the Chi-square test for independence). The algorithm performs the splits until the test shows that the split variable and the responses are independent, using a pre-specified significance level.

In this study, we use the R statistical Software [DEV 11] and the implementation of the CHAID algorithm present in the package of the same name. As this article reports ongoing research, the CHAID algorithm was used with its default settings (5% significance level and no constraint on the minimal number of observations per node).

As shown in Table 5.1, changes observed between 1985 and 2006, at the sociodemographic level, in the study area correspond to those that have been observed in other cities in developed countries. We find the inexorable aging of the population that leads to an increase in the number of retirees. At the same time, there is a generalization of women employment. In the case of Lyon, the part of workers has stayed relatively stable between 1985 and 2006 due to the relative importance of these two phenomena. The increase in the number of pensioners has been almost completely offset by the increased participation of women in the labor market (which, in turn, results in a sharp drop in the share of people staying at home). Household size decreases as the number of one-person households increases and the number of households with children decreases. The number of women with a driver's license has increased sharply between 1985 and 2006.

Household and person characteristics	EMD 1985	EMD 1995	EMD 2006
Households			
Total (thousands)	431.4	515.8	541.7
Number of persons (%)			
1	27.2	32.6	34.8
2	30.4	31.1	32.7
3	17.8	15.1	13.1
4	16.7	13.5	11.8
5+	7.9	7.7	7.6
Number of workers (%)			
0	31.5	35.9	38.4
1	36.8	34.5	33.1
2+	31.7	29.6	28.6
Lifecycle (%) (see cycleDeVie)			
1	11.7	16.9	15.1
2	6.1	6.4	6.6
3	12.9	12.4	11.9
4	8.8	8.5	7.6
5	23.1	18.9	16.2

6	19.9	18.0	20.9
7	17.5	18.8	21.6
Number of vehicles (%)			
0	24.7	22.9	21.7
1	49.9	46.5	47.5
2+	25.5	30.6	30.8
Persons			
Total (thousands)	1,088.3	1,220.0	1,231.4
Surveyed persons (5 years)			
Total (thousands)	1,017.9	1,138.2	1,146.5
Age (%)			
5–10	7.3	7.9	7.5
11–14	5.9	5.8	5.5
15–17	5.3	4.4	4.4
18–25	14.5	14.0	11.5
26–35	14.6	16.7	14.4
36–45	15.2	14.6	15.3
45–60	21.2	19.3	19.9
61–75	11.8	13.5	14.9
75+	4.2	3.7	6.6
Occupation (%)			
Worker	44.0	43.9	43.9
School	7.2	7.9	7.5
Highschool	12.5	11.7	11.4
College	4.5	6.9	5.5
Retired	15.1	16.1	22.3
Stay home	11.1	7.6	4.1
Other	5.5	6.0	5.3
Persons over 18 years old			
Total (thousands)	830.3	932.3	947.1
Men (%)	47.1	46.6	46.6
With driver license (%)	87.9	89.5	91.6
Women (%)	52.9	53.4	53.4
With driver license (%)	57.8	69.4	77.8

Table 5.1. *Sociodemographic evolution of the inhabitants in the perimeter of the 1985 survey between 1985 and 2006 (expansion coefficients taken into account)*

The explanatory variables selected to perform the CHAID algorithm are:

sexe	sex. 1=male, 2=female
posMen	Position in the household. 1=head (HH), 2=spouse, 3=child, 4=other
permis	Driving license. 1=yes, 2=no
Statag	Occupation status. 1=worker; 2=child in school, age 5–10; 3=child in school, age 11–18; 4=college student (over 18 in school); 5=retired; 6=stay at home; 7=other (essentially unemployed)
cycleDeVie	Lifecycle stage of the household. 1=age household head <45 and no marital relation between household members (no spouse as defined by posMen); 2=age household head <45, couple without children; 3–5=presence of children where the younger child is, respectively, less than 5, between 5 and 14 and over 15; 6=couples without children where household head is over 45; 7=household head is over 45 and no spouse or children
spouse	1=there is a person posMen=2 in the household; 0=there isn't
distCentreComKmF	Distance from residence to city centerein km (by class)
ageF	age (by class)
accesVp	Certain=person has a driving license and the number of cars in the household>=number of drivers; probable=person has a driving license but number of cars household<number of drivers; impossible=person has no driving license or no cars in the household

It must be noted that all the explanatory variables are categorical. It must also be noted that trials including many more potential explanatory variables were made, but the results were very similar to those reported here.

5.2.2.3. *Evaluation of the forecasting capability*

In order to evaluate the forecasting capability of the approach, we use a methodology that makes it possible to compare the results from the micro-simulation procedure with those of a more traditional trip generation approach.

To assess the forecasting capability of the approach, we do not use a synthetic population per se but the "observed" populations in 1995 and 2006. These observed populations are obtained by replicating the individual in each sample according to the expansion factors. We assign the mobility profile of a donor to each person of the two 1995 and 2006 populations. The donor is randomly selected from 1985 individuals having the same mobility profile, taking into account the expansion coefficients. For each person of the 1995 and 2006 samples, we repeat the procedure 50 times in order to have an idea of the variability of the approach.

The micro-simulation procedure results in populations of 1995 and 2006 with observed and simulated mobility profiles. In order to compare this output with that of traditional models, we define a segmentation of the tours based solely on the main purpose of the tour. We also define a zoning system compatible for comparisons between the three surveys (55 zones). We can then define different indicators to compare the number of observed and modeled (predicted) tours by purpose for each zone.

One indicator that is regularly used in the comparison of vectors (and matrices) is the *SRMSE* [PIT 78] given in equation [5.1]:

$$SRMSE = \left(\sum i(Ni - Ti)^2 / I\right)^{1/2} / \left(\sum iNi / I\right) \qquad [5.1]$$

where i is a traffic analysis zone (TAZ) among I zones. Ni is the number of observed tours originating in i (i.e. the homes of the trip makers are located in i). Ti is the number of modeled/predicted tours. The *SRMSE* statistic is computed for 1995 and 2006, for each trip purpose and each one of the 50 simulations. As pointed out by Fotheringham and Knudsen [FOT 87], the *SRMSE* does not have a known probability distribution which would allow for hypothesis testing. It has, however, two interesting properties. The first one is that an *SRMSE* of 0 means perfect reproduction of the observed tours for every zone. *SRMSE* is usually <1 and *SRMSE*>=1 means that the mean of the errors > mean value of observations. The second one is that SRMSE is linear on the size of errors.

Another standard way of evaluating the quality of the forecasts is to regress the modeled number of tours on the observed number of tours. Then, the intercept should be as close to zero as possible and the regression coefficient should be as close to one as possible. Hypothesis testing is possible on the coefficients. We do not perform one regression per simulation but only one on data from the 50 simulations.

Finally, it is interesting to decompose the observed evolution of the number of tours between the calibration year (1985) and the year for which the forecasts are made (1995 or 2006) into two components. One component is the evolution of the number of tours as given by the evolution of the "structure" of the population (this component is totally taken into account by the model for the variables taken into account in the CHAID algorithm). The other component is the evolution of mobility behavior inside each class defined by the segmentation procedure (not taken into account by the model). In order to measure these two effects, we define the following indicators:

$$Total = (Nif - Nic)/Nic \times 100 \quad [5.2]$$

$$Structure = (Tif - Nic)/Nic \times 100 \quad [5.3]$$

$$Behavior = (Nif - Tif)/Nic \times 100 \quad [5.4]$$

where c corresponds to the calibration year and f to the year for which forecasts are made.

5.3. Results

5.3.1. *Typology from CHAID*

The decision tree resulting from the application of CHAID to the 1985 EMD in not given here because of space contraints. It is present in a previous working paper and can be requested from the authors.

The best predictor of the mobility profile is the occupation status. This was to be expected as some trip purposes are only pertinent for a specific type of occupation (e.g. workers go to work but retired people don't). After the occupation status, the mobility profile also seems to depend on the composition of the household, reflected by the presence and relative importance of variables cycleDeVie or spouse in the tree. "Young" workers without children tend to have relatively homogeneous behavior, whereas the presence of children in the household and their age seems to be quite discriminant in terms of mobility behavior. This is also reflected in the variables that explain mobility profiles of children (Statag = 2 and 3). For younger children, the most important explanatory variable is the presence of a spouse, whereas for older children, individual characteristics are more important. This reflects the fact that children tend to be more autonomous with age.

5.3.2. *Forecasting capability*

For each one of the 50 simulations (for each year and purpose) a value of the SRMSE statistic was calculated. Table 5.2 shows summary statistics of the distribution of the 50 *SRMSEs*. For 1995 and 2006, the best forecasts correspond to tours for which Work or Shop and services are the main purposes. These are also the purposes for which more tours are made. Besides College and Other, which might not be comparable across surveys, forecasts are not good at all for the Escort purpose. This might be a result of the increasing chaining of activities. Trips for the Escort purpose might thus be part of complex tours and therefore, with the typology of tours selected, might be a part of tours that have another main purpose. The overall forecasting performance is not very good, although it is worth mentioning that the *SRMSE* values found for the 2006 forecast of the Work tours are particularly interesting because they are comparable to values found by Cabrera Delgado [CAB 13] who studied traditional model forecasts on the same zoning system. For instance, forecasts using a cross-classification model with two explanatory variables (number of workers and lifecycle) for 20-year forecasts (1985–2006) gave an *SRMSE* value of 0.20. Forecasts using linear regression at the household level gave *SRMSE* values between 0.18 and 0.22. The micro-simulation approach seems comparable then in terms of forecasting performance to traditional approaches.

Purpose	Min.	1st. Qu.	Median	Mean	3rd. Qu.	Max.
1995 Forecasts						
Work	0.146	0.148	0.149	0.148	0.149	0.151
School	0.271	0.276	0.277	0.277	0.278	0.285
HighSchool	0.193	0.198	0.200	0.200	0.202	0.204
College	0.399	0.404	0.408	0.408	0.411	0.418
ShopServ	0.215	0.218	0.219	0.219	0.219	0.221
Escort	0.420	0.422	0.423	0.424	0.425	0.430
Leisure	0.164	0.166	0.168	0.168	0.169	0.172
Other	0.625	0.635	0.639	0.639	0.643	0.654
2006 Forecasts						
Work	0.206	0.209	0.210	0.210	0.211	0.213
School	0.266	0.272	0.275	0.274	0.276	0.282
HighSchool	0.203	0.206	0.208	0.208	0.210	0.213
College	0.370	0.377	0.380	0.380	0.383	0.396
ShopServ	0.174	0.177	0.178	0.178	0.179	0.181
Escort	0.452	0.457	0.460	0.459	0.462	0.464
Leisure	0.210	0.213	0.214	0.214	0.216	0.220
Other	1.266	1.290	1.299	1.300	1.312	1.330

Table 5.2. *Summary statistics for the distribution of SRMSEs across simulations for each trip purpose and forecasted year*

In order to have a better idea of the kind of error in the forecasts, we plotted the results in the usual 45° representation of observed versus predicted trips for each zone and each main purpose of the tour, for 1995 and 2006. We also represented the effect decomposition in the form of a map of the study area with the 55 zones. These are available, as before, in the working paper because of space constraints.

For work purposes, a clear overestimation is observed, both in 1995 and 2006. While the coefficients of the regression are close to 1, the presence of a positive intercept accounts for the general over-estimation. The number of Work tours increased between 1985 and 1995/2006 in the outer part of the perimeter, while it seems to have decreased in the suburbs near Lyon (especially between 1985 and 2006). The model predicts accurately the direction of variation for most of the zones but with a general overestimation. For zones where there is a decrease in the number of tours, the amount of the decrease is not as important as it should be, whereas for zones where the number of tours increased, the predicted increase was more important than the observed one. There has been a change in mobility behavior as regards to the work travel purpose. Workers are not as inclined as they were to go home to lunch [HUB 12]. This change in behavior is not taken into account by the model; hence, the general overestimation.

School and Highschool tours seem to be accurately predicted (regression coefficient near to 1) but the variability of the forecasts is large. This is partly related to changes in organizational constraints of education (for example school day schedules) and could be included in a model specifically for these travel purposes.

The ShopServ tours suffer mostly from overestimation in the zones with a relatively low number of tours and from underestimation in the zones with relatively high number of tours. The geographic distribution of the over- and underestimations is interesting because it shows a concentration of the underestimations on the west side (wealthier) side of the study area while the overestimations occur mostly on the east side (poorer municipalities). In 2006, the underestimation is concentrated in the suburbs from the west side that are near Lyon. In the rest of the study areas, the number of tours has been overestimated.

The forecasts for Escort tours are relatively poor. There has been a change in mobility behavior that isn't taken into account by the explanatory variables selected.

For Leisure tours, the forecasts are relatively good which is surprising because of the strong increase in the number of trips related to this purpose. The correlation between motorization (modeled here by accesVp) and income might hold sufficiently over time for the impact of the increase in the number of tours for Leisure purpose to be relatively well forecasted.

For purposes College and others, some well-known problems with data comparability surface as sampling methods have evolved between 1995 and 2006 for students living in college residences. The forecasts are of poor quality.

5.4. Conclusion

Although the forecasting capabilities of traditional methods of trip generation seem on par with those of micro-simulation, this study has shown that micro-simulation achieves this while also keeping a lot of information that is usually lost. Moreover, these results are encouraging because we tested the algorithm with its default settings, and some extra constraints (on the minimum number of persons per final node, for example) are likely to improve the results.

The extra information available, about trip chaining, can be useful in improving the forecasting capability of the micro-simulation procedure. [KOL 84], who analyzed the stability of trip generation rates, have shown that trip generation seems reasonable despite important behavioral changes at the trip rate level (behavior). This is due to the fact that structural and behavioral changes tend to serendipitously compensate so as to produce reasonable forecasts. With micro-simulation, travel behavior is represented in a manner that is richer than simple trip rates. Therefore, it would be possible to include some of the changes in activity organization into the modeling framework by modifying the distribution of mobility profiles from which the "donor" of travel behavior is selected, according to expected changes or specified scenarios.

The research reported in this paper served as a sort of proof of concept. Indeed, we used the CHAID algorithm with its default settings, and the results were as good in forecasting as those of traditional methods. It would be interesting to apply different segmentation procedures. For instance, we continue this research by testing separate segmentation procedures for each occupational type (thus orienting more the link between travel purpose and individual).

5.5. Acknowledgments

The authors wish to thank ADEME who funded under the PREDIT (contract 08 66 C0084) the research that was used to support this article and the Rhône-Alpes Region, which funded Jorge Cabrera Delgado's doctoral grant under its Research Cluster: Transport Territories and Societies.

5.6. Bibliography

[ADL 79] ADLER T., BEN-AKIVA M., "A theoretical and empirical model of trip chaining behavior", *Transportation Research Part B: Methodological*, vol. 13 no. 3, pp. 243–257, 1979.

[AXH 00] AXHAUSEN K.W., "Definition of movement and activity for transport modeling", in HENSHER D.A., BUTTON K.J. (eds), *Handbook of Transport Modelling*, Chapter 16, Pergamon, Amsterdam, pp. 271–284, 2000.

[BAD 98] BADOE D.A., MILLER E.J., "An automatic segmentation procedure for studying variations in mode choice behaviour", *Journal of Advanced Transportation*, vol 32, no. 2, pp. 190–215, 1998.

[BON 04] BONNEL, P., *Prévoir la Demande de Transport*, Presses de l'École Nationale des Ponts et Chaussées, Paris, 2004.

[CAB 13] CABRERA DELGADO J., Quelle prise en compte des dynamiques urbaines dans la prévision de la demande de transport?, PhD Thesis, Université Lumière Lyon 2, 2013.

[CER 08] CERTU, *L'enquête ménages déplacements Standard Certu: guide méthodologique*, Éditions du Certu, Lyon, 2008.

[DEV 11] DEVELOPMENT CORE TEAM, *R: A Language and Environment for Statistical Computing*, R Foundation for Statistical Computing, Vienna, Austria, 2011.

[FAR 13] FAROOQ B., BIERLAIRE M., HURTUBIA R. *et al.*, "Simulation based population synthesis", *Transportation Research Part B: Methodological*, 2013.

[FOT 87] FOTHERINGHAM A.S., KNUDSEN D.C., *Goodness-of-Fit Statistics*, Geo Books, Norwich, 1987.

[FOX 10] FOX J., HESS S., "Review of evidence for temporal transferability of mode-destination models", *Transportation Research Record: Journal of the Transportation Research Board*, vol. 2175, pp. 74–83, 2010.

[HUB 12] HUBERT, J.-P., MADRE J.-L., MOISSONNIER J. *et al.*, "La pause méridienne: un facteur clé de l'évolution de la mobilité en France depuis 35 ans", *Économie et Statistique*, vols. 457–458, pp. 35–55, 2012.

[KAS 80] KASS G.V., "An exploratory technique for investigating large quantities of categorical data", *Journal of Applied Statistics*, vol. 29, no. 2, pp. 119–127, 1980.

[KIT 00] KITAMURA R., CHEN C., PENDYALA R.M. *et al.*, "Micro-simulation of daily activity-travel patterns for travel demand forecasting", *Transportation*, vol. 27 no. 1, pp. 25–51, 2000.

[KOL 84] KOLLO H.P.H., PURVIS C.L., "Changes in regional travel characteristics in the San Francisco Bay Area: 1960–1981", *Transportation Research Record: Journal of the Transportation Research Board*, vol. 987, pp. 57–66, 1984.

[KOS 84] KOSTYNIUK L.P., KITAMURA R., "Trip chains and activity sequences: test of temporal stability", *Transportation Research Record: Journal of the Transportation Research Board*, vol. 987, pp. 29–39, 1984.

[LOH 08] LOH W.-Y., "Classification, regression tree methods", in RUGGERI F., KENNETT R., FALTIN F. (eds), *Encyclopedia of Statistics in Quality and Reliability*, Wiley and Sons, Chichester, 2008.

[PIT 78] PITFIELD D.E., "Sub-optimality in freight distribution", *Transportation Research*, vol. 12, no. 6, pp. 403–409, 1978.

[STO 05] STOPHER P.R., GREAVES S.P., XU M., "Using national data to simulate metropolitan area household travel data", *Journal of Transportation and Statistics*, vol. 8 no. 3, pp. 83–95, 2005.

[STO 05] STOPHER P.R., MCDONALD K.G., "Trip generation by cross-classification: An alternative methodology", *Transportation Research Record: Journal of the Transportation Research Board*, vol. 944, pp. 84–91, 1983.

[STR 98] STRAMBI O., VAN DE BILT K.-A., "Trip generation modeling using CHAID, a criterion-based segmentation modeling tool", *Transportation Research Record: Journal of the Transportation Research Board*, vol. 1645, pp. 24–31, 1998.

[SUP 83] SUPERNAK J., TALVITIE A., DEJOHN A., "Person-category trip-generation model", *Transportation Research Record: Journal of the Transportation Research Board*, vol. 944, pp. 74–83, 1983.

[TRB 07] TRANSPORTATION RESEARCH BOARD, Metropolitan Travel Forecasting: Current Practice and Future Direction, Special Report 288, National Academy of Sciences, Washington D.C., 2007.

[VAU 97] VAUGHN K.M., SPECKMAN P., PAS E.I., "Generating household activity travel patterns (HATPs) for synthetic population", *Proceedings of the 76th Annual Meeting of the Transportation Research Board*, Washington D.C., 1997.

[VAU 99] VAUGHN K.M., SPECKMAN P., SUN D., Identifying relevant socio-demographics for distinguishing household activity-travel patterns: a multivariate regression tree approach, Paper prepared for The National Institute of Statistical Sciences (NISS), Research Triangle Park, 1999.

[ZHA 02] ZHAO Y., KOCKELMAN K.M., "The propagation of uncertainty through travel demand models: an exploratory analysis", *The Annals of Regional Science*, vol. 36, no. 1, pp. 145–163, 2002.

6

Modeling and Solving International Journey Planning Problems

The work presented in this chapter aims to model and solve multicriteria international itinerary planning problems. The emerging itinerary planning problems are defined on a multimodal time-dependent transportation network that consists of hierarchically interconnected international, interurban and urban transportation networks. The itinerary planning problems under study fall into two major categories: (1) static and (2) dynamic. Static refers to journey planning decisions at the pre-trip phase while dynamic refers to the journey on-trip re-planning problem. An overall solution approach was developed for solving the proposed international itinerary planning problems. This approach involves a pre-processing stage in which a customized international sub-network is designed and stored for each pair of potential origin and destination countries of a trip request. In this context, addressing any trip request involves recalling just the corresponding customized sub-network instead of the entire international transportation network. Subsequently, the international itinerary planning problem is solved on the customized network leading to a list of alternative generic itineraries, i.e. itineraries for which only the international transportation links are fully specified.

6.1. Introduction

It is widely argued that providing travelers with door-to-door international multimodal journey planning services with real-time alerting and on-trip re-planning capabilities constitutes a major need in the international travelers' community. Based on a user requirement analysis performed recently within the research project E-WISETRIP [ZOG 10] and given journey planning limitations of the existing journey planners, the major journey planning needs for international travelers include: (1) the provision of door-to-door multimodal itineraries addressing one-way, round-trip or multiple-trip journeys taking into account multiple criteria and

---

Chapter written by Konstantinos N. ANDROUTSOPOULOS and Konstantinos G. ZOGRAFOS.

user defined constraints expressing travelers' special needs and (2) on-trip re-planning capabilities addressing potential transport disruption incidents.

Covering the above journey planning needs involves dealing with two major categories of trip planning problems: static and dynamic. The static refers to journey planning decisions at the pre-trip phase and includes: (1) the elementary itinerary planning problem defined within a given pair of origin/destination points within a given earliest departure and a latest arrival time at destination, (2) the return trip itinerary planning problem, (3) the itinerary planning problem with multiple intermediate stops and (4) the trip itinerary contingency planning problem, i.e. planning in advance alternative back-up itineraries that may resolve several anticipated trip disruptions. The dynamic refers to the journey on-trip re-planning problem. The emerging itinerary planning problems are defined on the hierarchically interconnected international, interurban and urban transportation networks (Figure 6.1) modeled as a multimodal time-dependent network.

The remainder of this chapter includes five sections. Section 6.2 presents the major features of the international itinerary planning problems addressed in this chapter. Section 6.3 provides the network model of the underlying transportation network and specifies the models that express the itinerary planning problems covered in this paper. Section 6.4 provides an overview of the previous related work on modeling and solving itinerary planning problem on multimodal networks and section 6.5 presents the proposed solution approach. Finally, section 6.6 outlines the major finding of this research.

6.2. Defining international itinerary planning problems

6.2.1. *Elementary international multimodal itinerary planning problems*

The objective of this problem is to determine the alternative itineraries between an origin and a destination for a given earliest departure time and/or a latest arrival time. In general, each itinerary should include the sequence of transport services (transport mode, point of embarking, point of disembarking, departure time, arrival time, walking transfers), connecting the origin with the destination. The core part of any itinerary consists of a sequence of international transport segments between two terminals, and the intermediate interurban and urban transport segments (including walking) for executing the transfers between international terminals or between an international terminal and the origin or destination of the trip. The criteria taken into account in solving the above itinerary planning problem are the following: (1) total travel time, (2) total number of transfers, (3) CO_2 emissions per passenger and (4) cost. Two major categories of constraints apply to this type of itinerary planning problem: (1) time based constraints, and (2) mode based constraints. The former category of constraints involves the earliest departure time from the origin and/or the latest arrival time at the destination. The second category of constraints relates to the special needs of the traveler regarding the types of modes that cannot be used for executing the journey, e.g. exclusion of ships.

6.2.2. *International round-trip itinerary planning problem*

Round trip planning involves an itinerary planning problem in a multimodal transportation network aiming to specify alternative itineraries for a return international trip given an earliest departure time from the origin, a latest arrival time at the destination, an earliest departure time from the destination back to the origin and a latest arrival time back at the origin. A major feature of this problem is that the selection of an international transport mode for the forward trip may affect the alternative travel mode options and the associated costs for the return trip and vice versa. The criteria used for planning international roundtrip itineraries are the same as in the one-way case. Concerning the time constraints, the earliest departure time from the origin and the latest arrival time at the destination and the earliest departure time from the destination back to the origin are essential scheduling parameters in order deal with the corresponding itinerary planning problem. In addition to time constraints, mode-based constraints may also be relevant expressing the special needs of the traveler regarding the types of modes that cannot be used for executing the journey.

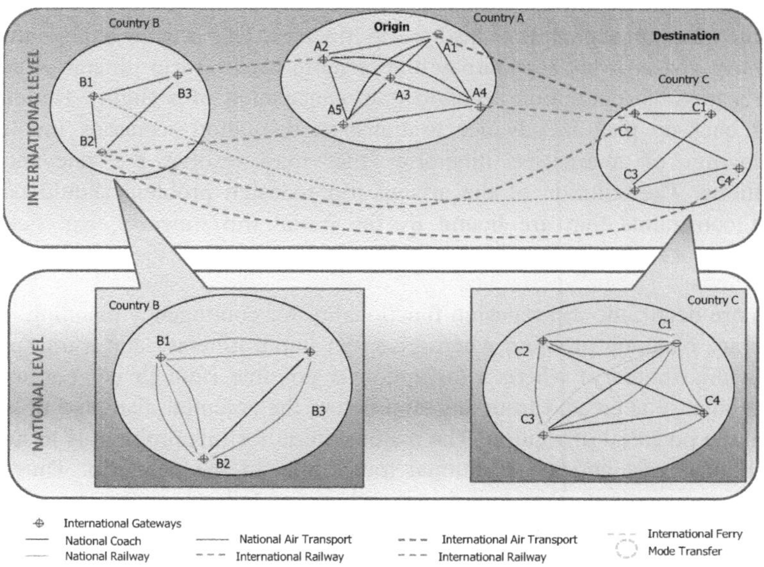

Figure 6.1. *International and national layers of the transportation network under study. For a color version of this figure, please see www.iste.co.uk/jacob/traffic.zip*

6.2.3. *International multiple-trip itinerary planning problem*

A multiple-trip includes a specified number of intermediate stops at various cities. The travelers stay at each intermediate stop and the corresponding arrival time at each stop is known in advance. These multiple trip requests may arise for both business and leisure travelers. Dealing with this type of multiple trip requests involves the solution of an itinerary planning problem aiming to determine alternative itineraries from an origin to a destination passing through a sequence of mandatory intermediate stops, each one constrained by a latest arrival time and (possibly) an earliest departure time. The departure from the origin is also constrained by an (traveler-specified) earliest departure time, while the arrival at the final destination may be also optionally constrained by a latest arrival time. Each solution to the above problem constitutes an extended itinerary which includes sub-itineraries for executing all intermediate trips defined by the mandatory intermediate stops. The criteria and constraints applicable for this problem are similar with those for the round trip itinerary planning problem.

6.2.4. *International trip itinerary contingency planning problem*

The contingency planning problem emerges from the international traveler's need to build a contingency plan for his/her trip itinerary. In this type of journey planning problem, the traveler deals in advance with the possibility of a disruption in any of the transport segments of his/her trip itinerary. This is achieved by building at the pre-trip phase of his/her journey one or more back-up trip itineraries that may resolve a possible disruption scenario (e.g. cancelation of a flight). Building this type of back-up plan in advance to transport disruption scenarios involves the determination of alternative itineraries that may resolve the relevant travel implications. The solution of the arising trip redesign problem should take into account constraints that are bound by the initial trip itinerary, e.g. tickets for transport services already booked.

In more detail, the trip redesign functionality for contingency planning involves an itinerary planning problem where given an initial itinerary and a transport link (used in this itinerary) where a disruption is possible (though not confirmed), a revised itinerary is created from any stop before the potential disrupted link, which resolves the potential disruption. The major criteria for this problem is to minimize the additional cost and the additional travel time of the new trip. Thus, if any transport segment of the initial trip itinerary following the transport link hypothetically disrupted has been booked, the alternative trip itineraries are bound to using these booked services.

6.2.5. *International dynamic itinerary re-planning problem*

Dynamic re-planning of a trip aims to provide the traveler with alternative options of fixing his/her trip in real-time after a transport segment of his/her trip has been disrupted (service canceled or departure/arrival delayed). Unlike the trip contingency planning problem, in this problem the alternative itineraries can only be built from the intermediate stop that the traveler is currently located in or heading to. The objective of the emerging itinerary planning problem is to determine an alternative sub-itinerary from its current location in the initial itinerary to the destination, taking into account that a transport service initially included in his/her trip itinerary is not accessible anymore (due to cancellation or increased delay). The alternative itinerary is calculated while the traveler is en route taking into account the updated schedules (if available) of the underlying transport services. The overall goal of this itinerary planning task is to provide an itinerary that deviates the least from the initial plan, while it minimizes any possible additional cost and travel time. It is a similar problem with the one encounter in contingency planning; however, it may address more complicated situations where the underlying network changes in two ways: (1) in terms of the schedule of the transport services, (2) in terms of the transport links (e.g. in case of cancellations).

6.3. Modeling issues

Any international trip involves a sequence of international transport services joined with transfers performed by interurban and/or urban transport services. Thus, apart from leaving the origin or accessing the destination, the interurban/urban transport services are also used for transfers between different international transportation services. It should be pointed out though that any transfer may also involve a multimodal itinerary planning problem defined on an interurban or urban public transport system. However, in practice this local itinerary planning problem is not addressed directly within the international itinerary planning problem. It is actually addressed by the traveler only after the international transport services have been determined. Thus, for planning purposes, any urban or interurban transfer is treated as a single transportation link with a proxy transfer time. Figure 6.2 presents schematically how the national/urban/local transfers are incorporated in the international network under study.

The underlying international transportation network is decomposed to a series of distinct international transport service sub-networks. An international transport service is defined as a transport service operated between two terminals located at different countries. The international transportation services are built on the basis of a set of terminals which represent the international gateways of each country defined as follows: an international gateway of a country is a station (or terminal) of a transport mode used by at least one international transport service. This definition implies that if an international transport service embarks/disembarks passengers to a series of stops within one country then in general any of these stops should be considered as international gateways.

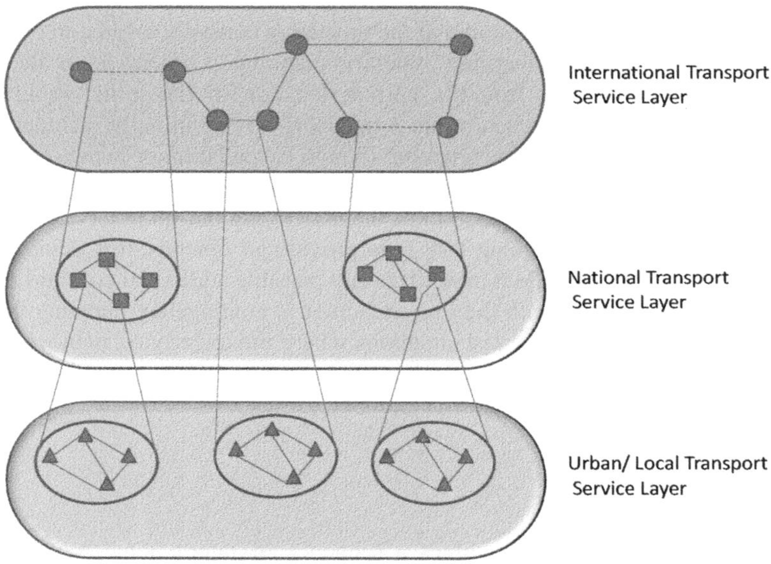

Figure 6.2. *Hierarchical structure of the international transportation network under study. For a color version of this figure, please see www.iste.co.uk/jacob/traffic.zip*

Any transportation network may be modeled by a mathematical graph $G(N, A)$ where N is the set of nodes and A the set of links, taking into account the following assumptions: (1) each link corresponds to a phase of a potential trip or an activity taking place within a trip, and (2) the nodes correspond to events delimiting the links as defined above [CAS 09]. The above analysis suggests that the underlying international transportation network may be modeled by a graph $G(N, A)$, where N is the set of international gateways (terminals of the international transportation services) and A the set of arcs defined as $A := A_i \cup A_t$ where A_i denotes the international transportation services and A_t the local or interurban transfer links. Nodes associated with the local transport services (e.g. bus stops, metro stations, etc.) are not considered in this network.

The international itinerary planning problems defined on the network described above are formulated as time-dependent shortest path problems in multimodal transportation networks with multiple criteria and time windows. In the elementary, round-trip and multiple-trip itinerary planning problems, the expected outcome relates to a set of non-dominated paths. However, in the contingency planning and dynamic itinerary re-planning problems, the expected output relates to a single or a limited number of alternative solutions. Given these solution requirements from the perspective of the user, the elementary, round-trip and multiple trip itinerary planning problems are modeled as multi-criteria time-dependent shortest path problem in multimodal transportation network while the contingency planning and dynamic itinerary re-planning problems are modeled as lexicographically time-dependent shortest path problems in a multimodal transportation network. This chapter provides an overall solution approach for addressing the former group of problems.

6.4. Previous related work

The multicriteria shortest path problem has been widely studied in the literature [HAN 80, MAR 84, WAR 87, MOT 91, PAL 97]. It has been proved in Hansen [HAN 80] that the number of non-dominated solutions may increase exponentially with the number of nodes in the network. Pseudo-polynomial forward label setting algorithms have been proposed for addressing the multicriteria shortest path problem. A key feature of this type of algorithm is that each non-dominated path identified is represented by a multidimensional label which hosts the value of the path under each of the criteria under consideration. Starting from the origin, these types of routines iteratively construct new paths by extending the already identified non-dominated paths. At each iteration, the lexicographically minimum label is selected and it is attributed as permanent. It can be proved that the lexicographically minimum label is non-dominated. Then new paths are created and by extending the corresponding recently identified permanent path with the relevant outgoing links from its tail node. It should be pointed out that the correctness of the label setting algorithms is founded on the fact that if a path is non-dominated then any of its sub-paths are non-dominated too.

The adaptation of these routines for addressing the multicriteria time-dependent shortest path problem is not valid in general. Similar forward label setting approaches have been proved to work only for non-decreasing cost attributes [GET 00, KOS 93]. If travel time is also included in the cost attributes, then it is necessary that the FIFO property holds for every arc of the network (i.e. earlier departure time from the upstream node implies earlier arrival time at the downstream node, i.e. no over-passing is allowed). However, in general Belman's principle does not hold, and therefore, a forward label setting algorithm cannot be applied [HAM 06]. Alternatively, it has been proved that a backward label setting routine is valid for determining the non-dominated solutions to a multicriteria time-dependent shortest path problem [HAM 06, AND 08].

A major issue that arises in shortest path problems defined on multimodal networks relates to the viability of a path. Each node of a multimodal network represents a transport service between two consecutive stops. Thus, each solution of a shortest path problem (with single or multiple criteria) on a multimodal network represents a sequence of transport services used in order to travel from an origin to a destination. However, in practice not all possible combinations of transport services are eligible from the perspective of performing a trip. For instance, in an urban public transport system, if a traveler uses his/her private car in order to get to a metro station, then after getting out of metro it is not possible to use the private car again. Similarly, in international trip planning, if a traveler uses his private car to get to an airport, then upon arriving at the destination airport he/she cannot make use of the private car again. Thus, in this type of problem, path viability should be considered in determining feasible itineraries.

In recent studies, the assessment of the viability of a path on a multimodal network is modeled through a deterministic finite-state automation [BAR 00, LOZ 01, GRA 10, GUE 11]. Any partial path defined on a multimodal network is associated with a state which is specified by the sequence of transport services that it includes. In expanding the partial path to a new path through concatenating an outgoing arc from its last node, then a new state is determined for the new path. If this state lies within a given set of eligible states, the corresponding new path is viable.

Gueye *et al.* [GUE 11] provided forward label setting and correcting algorithms for solving a bi-objective shortest path problem in a multimodal network, where the objective functions under consideration were travel time and number of transfers. Two alternative algorithmic approaches were presented. The first approach was based on the Chrono space algorithm [PAL 97] which aimed at determining the non-dominated paths in a multimodal network in an order of increasing number of transfers. The second approach aimed at determining the non-dominant path in an order of decreasing number of transfers and increasing travel time. Finally, Gueye *et al.* [GUE 11] propose a bidirectional label setting algorithm which improves the computational efficiency of the label setting approach. In all algorithms proposed, the authors enhance the basic dominance rule to the state-based dominance rule, where one path may dominate another even if they are not on the same state. It should be pointed out that the above algorithms are not applicable if the travel time is not FIFO or an additional criterion is included which is not a non-decreasing function of time. As a consequence, these type of routines are not valid for the itinerary planning problems under study since the cost, which is an essential criterion in international tip planning, is not in general non-decreasing.

6.5. Algorithmic approach

The itinerary planning problems envisaged to be addressed are defined on a hierarchically structured multimodal transportation network which consists of a large set of distinct transportation services falling into one of three layers. The top layer includes any international transport services connecting international transport stations and terminals. The middle layer includes the national transport services connecting the international transport stations or terminals of each country. Finally at the bottom of the hierarchical structure there are urban/local transport services which connect international stations/terminals or national station/terminals.

Based on the description of the international transportation network under consideration two major issues arise: (1) the emerging network is huge thus requiring substantial computational time for solving the relevant itinerary planning problems, and (2) a substantial part of the network would never be involved in any of the alternative trip itineraries for a given trip request.

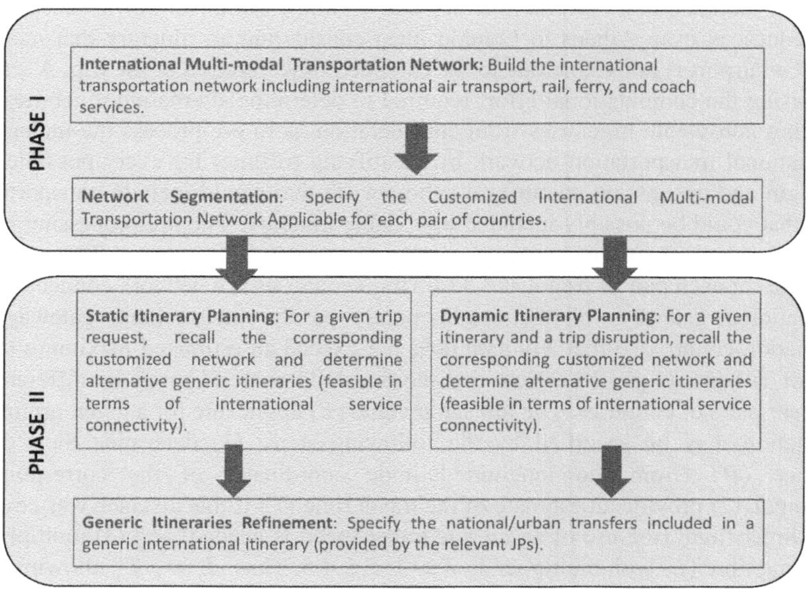

Figure 6.3. *The two-phase solution approach dealing with the international itinerary planning problems*

A two-phase solution approach was developed for solving the itinerary planning problems taking into account the above issues. In the first phase of the approach a pre-processing takes place (offline) in which a customized international sub-network is designed and stored for each pair of potential origin and destination countries. This network includes only arcs and nodes of the top layer of the transportation network (Figure 6.2) which could be potentially be included in alternative itinerary. In this context, addressing any trip request involves recalling just the corresponding customized sub-network instead of the entire international transportation network. At the second phase of the approach the international itinerary planning problem is solved on the customized network leading to a list of alternative itineraries (i.e. itineraries for which only the international transportation links are fully specified). The remainder of this section analyzes the proposed solution phases.

6.5.1. *Customized network*

It is evident that using the entire international transportation network to identify alternative itineraries leads to a computationally burdensome task. Moreover, the entire international transportation network may include transport links that would never be included in a feasible itinerary due to time constraints. For instance, if the trip request is from Athens to London, then considering an itinerary that includes Moscow airport is not viable due to the extended time duration of the trip. A way of decreasing the computational effort required to determine alternative itineraries and avoiding non-viable itineraries from consideration, is to pre-process the underlying international transportation network by identifying (offline) for every possible pair of origin and destination countries a sub-network that would include transportation links that could be possibly included in a viable itinerary. The pre-processing phase should be applied to any pair of countries covered by the system. The expected outcome for each pair of origin and destination countries is a network connecting the international gateways of the origin-country to the international gateways of destination-country, so that any path does not exceed an estimated maximum travel time of L hours (e.g. 12 h). Note that the threshold value (L) may be different for different pairs of countries. The threshold value (L) applicable for a given origin and destination may be specified by the following steps: (1) determine their direct distance (D) (from the longitude-latitude coordinates of the corresponding locations), (2) provide an estimate of the travel time (T) if this distance was covered by a direct flight (the use of an *ad hoc* travel speed is needed) and (3) multiply the emerging time (L) with multiplier λ, $\lambda > 1$, e.g. 1.5. Thus, $L := \lambda T$, allowing for a flexibility in the size of the customized network. In particular, for larger values of λ the larger the emerging customized network. The emerging customized network is static in the sense that its links correspond to transport connections with no timetable information. A major prerequisite for applying the pre-processing steps relates to the availability of a static version of the international transportation network with the following features:

– the static network involves a list of international gateways and a list of static links representing transport connections between the international gateways. The list of transfers (if any) between two gateways is also assumed available;

– each node (gateway) is identified by its unique ID, while it is also associated with the following type of information: (1) country, (2) longitude-latitude coordinates, (3) transport mode;

– each link of the static network is identified by its two end-nodes (gateways) and a transport mode.

The following approach is proposed in order to determine the static network associated to a given pair of countries:

1) Specify all international gateways of both origin and destination countries.

2) Apply a forward shortest path label setting algorithm from the origin-country gateways and a shortest path backward label setting algorithm from the destination-country gateways on the static international transportation network [PAL 98].

3) Each time a (forward or backward) label is fixed for a gateway, two issues are checked: (1) if the gateway is already (backward or forward) labeled (so a path from an origin-gateway to a destination-gateway has been found) and (2) if the labeled gateway is connected (via transfer) to another gateway which is already (backward of forward) labeled. In both cases, the travel time of the path is calculated (by adding the corresponding travel times of the constituent sub-paths). If the path travel time exceeds L then the process is terminated and the links already associated with at least one identified path constitute the customized network for any trip request associated with the specific pair of countries. In practice the above process excludes any path of the static network for which a lower bound on its travel time exceeds the *ad hoc* threshold value (L).

The travel time on any static link is calculated by the ratio of the distance (calculated from the gateways' coordinates) with the *ad hoc* set travel speed. This pre-processing phase is applied before solving any of the international itinerary planning problems.

6.5.2. *Solution algorithm for the elementary international itinerary planning problem*

The proposed solution approach for the elementary international itinerary planning problem includes two major stages. In stage I, the international part of the trip is identified, (i.e. sequence of international segments from the origin to the destination), including the walking/urban/interurban transfers where necessary (i.e. transfer from the origin to the first international gateway, the transfer from the last international gateway to the destination and the transfers between international gateways. In this stage of planning the emerging paths are not fully defined in the sense that the transfer links are not fully specified (an indicative estimate for travel time is only used in this stage). As a consequence, there are two issues pending to be specified in the second stage of the solution approach for each of the generic itineraries identified: (1) determine the optimal (possibly multimodal) local/interurban itinerary for performing each transfer while retaining the overall feasibility of the entire itinerary and (2) validate that the complete itinerary (enhanced with the detailed transfer itineraries) remains optimal/non-dominated. Thus, in stage II each of the identified generic itineraries is further refined by identifying alternative transfer itineraries for performing the corresponding urban/interurban transfers.

In general, more than one alternative international gateway may be used in order to start or finish a journey. Thus, the itinerary planning problem under study is treated as a multicriteria time-dependent shortest path problem with multiple origins and multiple destinations. Considering each destination separately, the emerging problem is decomposed to a series of multi-criteria time-dependent shortest path problems with multiple origins and a single destination. A backward label setting algorithm is proposed for solving the arising multicriteria path finding problem. The proposed algorithm incorporates the notion of buckets introduced in Pallottino and Scutella [PAL 98] for time-dependent path finding problems.

Each label $\lambda_{it}^k := (c_T(p_{it}^k), c_M(p_{it}^k), c_I(p_{it}^k), c_E(p_{it}^k))$ is associated with a node i and a point in time t and consists of the following attribute: the total travel time $c_T(p_{it}^k)$, the total monetary cost $c_M(p_{it}^k)$, the number of transfers $c_I(p_{it}^k)$ and the CO_2 emissions $c_E(p_{it}^k)$ of an itinerary p_{it}^k in the underlying customized multimodal network connecting node i with destination (d). The set of non-dominated labels λ_{it}^k for node i and time t is denoted by Λ_{it} (λ_{it}^k is the kth non-dominated label in Λ_{it}). The solution algorithm includes the following steps:

1) Initialize labels for the destination node (d): Set $\lambda_{dt}^1 := (0,0,0,0)$, for $t \in [e, l]$ (e is the earliest departure time of the trip and l is the latest arrival time) and place each label in the corresponding list (bucket) B_t.

2) For λ_{it}^k do the following:

i) For each label (λ_{it}^k) in B_t do the following:

1) Create a new label $\lambda_{i(t-1)}^{k'} := (c_T(p_{it}^k) + 1, c_M(p_{it}^k), c_I(p_{it}^k), c_E(p_{it}^k))$, which corresponds to the same itinerary with label (λ_{it}^k) including one unit time of waiting at node (i). If any of the new labels is non-dominated in $\Lambda_{i(t-1)}$ then insert it in $\Lambda_{i(t-1)}$ (disposing of any other existing label in $\Lambda_{i(t-1)}$ dominated by it) and in list B_{t-1}.

2) Create a new label $\lambda_{jt'}^{k'} := (c_T(p_{it}^k) + c_T^{t''}(j,i), c_M(p_{it}^k), c_I(p_{it}^k) + 1, c_E(p_{it}^k))$ for each transfer link (j,i) incoming to node i. Each of these labels corresponds to the itinerary from j to destination d which starts with the transfer link (j,i) at time $t' := t - \tau(j,i)$ and continues through the itinerary associated to label (λ_{it}^k). If any of the new labels is non-dominated in $\Lambda_{it'}$ then insert it in $\Lambda_{jt'}$ (disposing of any other existing label in $\Lambda_{jt'}$ dominated by it) and in list $B_{t'}$.

3) Create a new label $\lambda_{jt''}^{k''} := (c_T(p_{it}^k) + c_T^{t''}(j,i,s), c_M(p_{it}^k) + c_M^{t''}(i,j,s), c_I(p_{it}^k), c_E(p_{it}^k) + c_E^{t''}(i,j,s))$ for each link (j,i) traversed by service s at time (t'') such that $t := t'' + \tau^{t''}(j,i,s)$. If any of the new labels is non-dominated in $\Lambda_{jt''}$ then insert it in $\Lambda_{jt''}$ (disposing of any other existing label in $\Lambda_{jt''}$ dominated by it) and in list $B_{t''}$.

4) Remove (λ_{it}^k) from list B_t.

where $c_T^{t''}(j,i)$ denotes the total travel time on transfer link (j,i), while $c_T^t\ (i,j,s)$, $c_M^t\ (i,j,s)$, $c_E^t\ (i,j,s)$ denote the travel time, monetary cost and emissions for traveling from i to j through service s departing from (i) at time t.

The labels in sets Λ_{it} where i is any of the origins of the problem correspond to the non-dominated itineraries from i to d starting ant time t. Repeating the above steps for each of the destination nodes leads to the entire set of alternative generic itineraries being forwarded to stage II to specify alternative ways of realizing transfers.

6.5.3. *Solution algorithm for the international round-trip and multi-trip itinerary planning problem*

The round trip consists of two separate elementary problems. The question that arises is whether one should address this trip planning problem as a single two-stage problem or as two separate problems. Considering the total travel time of both trip phases may be misleading in assessing two alternative solutions. For instance, say that there are two solutions S1 and S2 with total travel time 14 and 12 h respectively. The second solution clearly dominates the first in terms of travel time. However, it could be the case where the two phases of S1 have a 7 h duration while S2 involves phase 1 with 9 h and phase 2 with 3 h (e.g. due to schedule issues). In such a case the traveler may have favored solution S1. Similar cases may exist for transfer time or the number of transfers. This observation implies that the two problems should be addressed as two separate problems. A similar approach is suggested for the multiple trip planning problem.

6.6. Concluding remarks

This chapter addresses modeling and algorithmic issues related to the international itinerary planning problem. The major categories of models associated with international itinerary planning have been determined. An algorithmic approach for solving the resulting large scale multicriteria optimization models has been proposed. The proposed algorithm includes a pre-processing stage that contributes to the reduction of the problem size and makes the solution of the underlying models feasible in reasonable (for the types of models under consideration) computational time. The models and algorithms presented in this paper have been integrated into the E-WISETRIP international door-to-door journey planner. The system is under laboratory testing. Results regarding the computational performance of the proposed algorithmic approach will be reported as soon as they become available.

6.7. Acknowledgments

The presented research work was partially supported by the European Commission under the project Enhanced WISETRIP (E-WISETRIP).

6.8. Bibliography

[AND 09] ANDROUTSOPOULOS K.N., ZOGRAFOS K.G., "Solving the multi-criteria time-dependent routing and scheduling problem in a multimodal fixed scheduled network", *European Journal of Operational Research*, vol. 192, no. 1, pp. 18–28, 2009.

[BAR 00] BARETT C., JACOB R., MARATHE M., " Formal-language-constrained path problems", *SIAM Journal on Computing*, vol. 30 no. 3, pp. 809–837, 2000.

[GET 00] GETACHEW T., KOSTREVA M., LANCASTER L., " A generalization of dynamic programming for Pareto optimization in dynamic networks", *RAIRO Operation Research*, vol. 34, pp. 27–47, 2000.

[GRÄ 10] GRÄBENER T., BERRO A., DUTHEN Y., "Time dependent multi-objective best path for multimodal urban routing", *Electronic Notes in Discrete Mathematics*, vol. 36, pp. 487–494, 2010.

[GUE 11] GUEYE F., ARTIGUES C., HUGUET M.J., *et al.*, Label-setting algorithms for a polynomial bi-objective multimodal shortest path problem, available at: http://hal.archives-ouvertes.fr/hal-00564447/, 2011.

[HAM 06] HAMACHER H.W., RUZIKA S., TJANDRA S.A., "Algorithms for time-dependent bicriteria shortest path problems", *Discrete Optimization*, vol. 3, pp. 238–254, 2006.

[HAN 80] HANSEN P., "Bicriterion path problems", in FANDEL G., GAL T. (eds), *Multicriteria Decision Making: Theory and Applications, Lecture Notes in Economics and Mathematical Systems*, Springer, Heidelberg, vol. 177, pp. 109–127, 1980.

[LOZ 01] LOZANO A., STORCHI G., "Shortest viable path algorithm in multimodal networks", *Transportation Research Part A: Policy and Practice*, vol. 35 no. 3, pp. 225–241, 2001.

[MAR 84] MARTINS E.Q.V., "On a multicriteria shortest path problem", *European Journal of Operational Research*, vol. 16, pp. 236–245, 1984.

[MOT 91] MOTE J., MURTHY I., OLSON D., "A parametric approach to solving bicriterion shortest path problems", *European Journal of Operational Research vol.* 53, pp. 81–92, 1991.

[PAL 98] PALLOTTINO S., SCUTELLA M.G., "Shortest path algorithms in transportation models:Classical and innovative aspects", in MARCOTTE P., NGUYEN S. (eds), *Equilibrium and Advanced Transportation Modelling*, pp. 245–281, 1998.

[WAR 87] WARBURTON A., "Approximation of Pareto optima in multiple objectives shortest path problems", *Operations Research*, vol. 35, no. 1, pp. 70–79, 1987.

[ZOG 10] ZOGRAFOS K.G., ANDROUTSOPOULOS K.N., NELSON J.D., "Identifying travelers' information needs and services for an integrated international real time journey planning system", *Proceedings of the 13th International IEEE Conference on Intelligent Transportation Systems (ITSC 2010), Madeira, Portugal,* pp. 19–22, September 2010.

7

Optimized Intermodal Roundtrips in Transport Networks

Most of the current intermodal applications for optimizing the travel costs in transport networks consider one-way journeys only, but most of the journeys do not consist of simple one-way trips, as vehicles or travelers usually return to the starting point. For the calculation of a roundtrip, two simple one-way trips are usually considered which is not an accurate approach. Also the possibility of switching between different transportation modes during a trip is rarely offered. To overcome these limitations, a mathematical model for optimized intermodal roundtrips is formulated as a linear mixed binary program. The proposed model is used to minimize the travel cost associated with a roundtrip where different transport modes between two nodes of the transport network may be available for the user. Computational experience is also included in the paper to highlight the efficacy of the proposed methodology which may have application for ensuring the optimal integration of transport modes in transport networks.

7.1. Introduction

In Europe, the trans-European transport network (TEN-T) will comprise different transport infrastructures such as road, rail, inland waterway networks, motorways of the sea, seaports, inland waterway ports and airports. It is an important transport system in the European context since it aims to guarantee its users a high, uniform and continuous level of services, comfort and safety. In addition, the multimodal possibility of the trans-European network contributes to the operation of the internal market and to strengthening economic and social cohesion. According to the European Commission [EUR 01], another key transport policy objective in the European Union is a sustainable transport which is achieved by successful intermodal transport of both goods and persons.

Chapter written by Cecília VALE and Isabel RIBEIRO.

By definition, the multimodal transport is the transportation performed with at least two different transport modes. Multimodal networks are by nature a very complex process to manage and the consideration of roundtrips increases its complexity. However, the goal of having multimodal transport is to combine the strengths of various transportation modes and to neutralize their weaknesses. Although multimodal transport is advocated by experts and environmentalists, it involves the accumulation of costs by using modal interfaces, such as transfer, handling and others [DIA 05]. The study of multimodal networks has been mainly focused on freight or urban transport as discussed by Boyce [BOY 07], Bontekoning *et al.* [BON 04] and Macharis and Bontekoning [MAC 04].

Intermodal transport may refer to freight or passenger transport, and it involves two or more transport modes in a journey. Transfers from one mode to the other are performed at intermodal terminals which may be a sea or a river port, an in land terminal or an airport. Intermodal freight transport involves the transportation of goods in an intermodal container or vehicle, using multiple transport modes without any handling of the freight itself during transfers between modes [EUR 93]. Hence, intermodal freight transport has advantages of reducing cargo handling, improving security, reducing damage and loss, allowing faster transportation of the freight and reducing greenhouse gas emissions by diminishing road trucking. A review of intermodal rail–truck freight transport is presented by Bontekoning *et al.* [BON 04]. Regarding intermodal passenger transport, although national planners already exist in many European countries, in general these tools do not allow planning optimal roundtrips in multimodal networks. Most of the current multimodal applications consider one-way journeys only, but most real journeys do not consist of simple one-way trips, as vehicles or travelers usually return to the starting point. In this case, for the calculation of a roundtrip, those applications usually consider two simple one-way trips which is not an accurate approach. Also, the possibility of switching between different transportation modes during a trip is rarely offered.

This chapter presents a mathematical model for optimized intermodal roundtrips for passenger or freight transport and some computational experiments to highlight the efficacy of the model and compare intermodal alternatives. In short, this study explores the notion of minimizing the travel costs in roundtrips with the possibility of multiple transport modes in the network.

7.2. Model description

In this section, the proposed mathematical model for optimized intermodal roundtrips is briefly described. No mathematical expressions are presented for an easier reading.

The model is formulated as a linear mixed binary program. The objective function consists of minimizing the travel cost associated with a roundtrip. Several assumptions and constraints are attained:

– a roundtrip is composed of an outward and a return trip;

– the first node is the last node of the roundtrip;

– the node may be fixed or optional visited, according to the definition of the user;

– the last node of the outward trip is first node of the return trip and it is always a fixed node;

– the model has no limitation regarding the number of the transport modes;

– the transport mode between two nodes at the return trip is the same as the one at the outward trip if both nodes are visited in the return trip.

In Figure 7.1, a graphic exemplification of the model is presented. Let us consider a direct graph $G=(V, A)$, where V is a set with four nodes, $V=\{1,2,3,4\}$, and A is the set of the allowed links between two nodes of V (see Figure 7.1(a)). In this graph, we assume that node 1 is the origin of the trip and node 4 is considered to be fixed, corresponding to the destination of the outward trip. All the other nodes are optional, except node 3. The travel cost between each node is also indicated in the Figure 7.1(a).

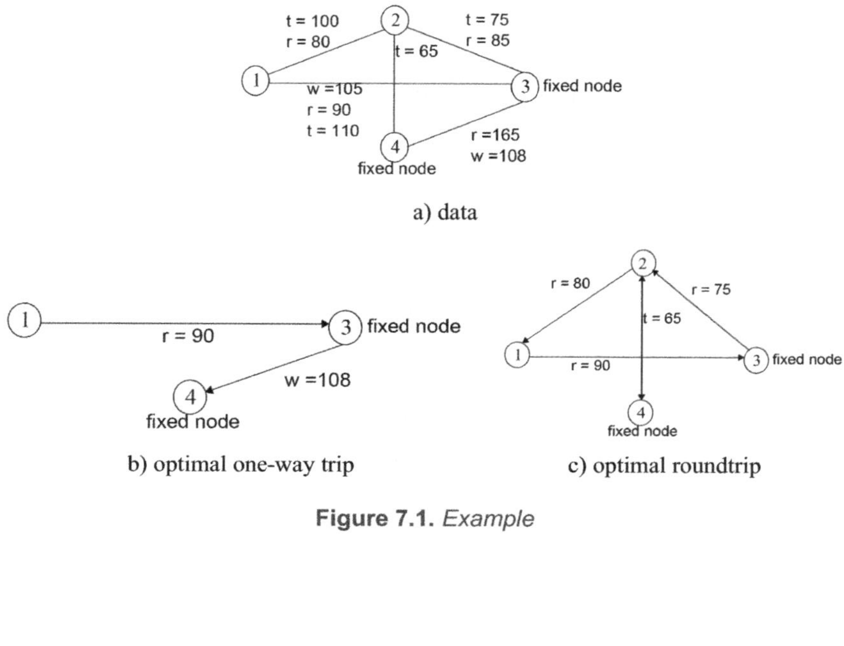

a) data

b) optimal one-way trip

c) optimal roundtrip

Figure 7.1. *Example*

The optimal one-way trip 1-3-4 indicated in Figure 7.1(b) has a total cost of 198. The optimal roundtrip, 1-3-2-4-2-1, presented in Figure 7.1(c), has a total cost of 230+145 =375 which is inferior to the cost of two one-way trips. This simple example shows the importance of planning optimal roundtrips instead of two optimal one-way journeys. From the figures, it can also be seen that the outward trip (1-3-2-4) in the roundtrip is different from the one-way trip (1-3-4) with node 4 as destination. The reason is that the model considers that if a change of transport mode occurs in node x in the outward trip the same mode has to be used when leaving node x.

Although the model described in the present paper is similar in general to the Baumann's problem [BAU 04], it is a different model because it considers the possibility of fixing some of the nodes to be visited. Baumann *et al.* [BAU 04] describe in that publication the importance of computing intermodal roundtrips and mention the constraints that may arise when considering different transport modes in a roundtrip. Unfortunately, the authors do not present any numerical application to be validated or tested with the proposed model.

7.3. Computational applications

7.3.1. *Data*

In order to show the usefulness of the proposed model, it has been applied for the calculation of optimal multimodal routes for freight transport purposes between node 1 (origin) and four destinations: nodes 15, 17, 18 and 19. In the application, there are 19 different nodes and 103 possible links between those nodes (Table 7.1). Three transport modes are also considered in the model application: road, rail and water. However, the model formulation has no limitation regarding the number of the possible transport modes between two nodes. The cost of the link between the nodes of the network is also presented for each transport mode in Table 7.1.

Link		Transport mode	Cost
Start node	End node		
1	7	Truck	286
	8	Truck	2220
	12	Truck	3792
	12	Rail	2075
2	1	Truck	895
	1	Rail	613
	11	Truck	2664
	11	Rail	806
	12	Truck	3249

	7	Truck	1217
	8	Truck	2517
3	11	Truck	1439
	11	Rail	617
	12	Truck	2438
	8	Truck	3343
4	6	Truck	1615
	6	Rail	1025
	15	Truck/water	2600
	11	Truck	885
	11	Rail	514
	12	Truck	2482
5	6	Rail	697
	15	Truck/water	2600
	11	Truck	1371
	11	Rail	716
	12	Truck	2985
6	13	Water	2100
7	8	Rail	1000
8	9	Rail	600
	16	Water	1900
	17	Truck	3363
	18	Truck	2847
	19	Truck	4165
9	17	Truck	2569
	18	Truck	1667
	19	Truck	2469
11	10	Rail	1205
	12	Rail	1106
12	17	Truck	3055
	18	Truck	2281
	19	Truck	1099
	14	Water	2050
	15	Water	2050
13	17	Truck	1119
	18	Truck	895
	19	Truck	829

Table 7.1. *Travel cost between the nodes of the network*

Link		Transport mode	Cost
Start node	End node		
14	17	Truck	1119
	18	Truck	895
	19	Truck	829
15	17	Truck	1421
	18	Truck	1529
	19	Truck	782
16	17	Truck	437
	18	Truck	1161
	19	Truck	1381
11	17	Truck	2381
	18	Truck	475
	19	Truck	1657

Table 7.1. *Travel cost between the nodes of the network (continued)*

7.3.2. *Results*

In Table 7.2, the total cost and the number of transport modes changes are presented for a one-way journey and roundtrip considering the four different destinations (nodes 15, 17, 18 and 19). Each transport mode is represented by a letter: *r* for rail; *t* for road by truck and *w* for water.

Origin/ Destination	One-way journey			Roundtrip		
	Total cost	Number of changes	Optimal route	Total cost	Number of changes	Optimal route
1-15	3956	1	1r12t19t15	7912	2	1r12t19t15t 19r12r1
1-17	3623	3	1t7r8w16t17	7246	6	1t7r8w16t17t 16w8r7t1
1-18	3099	2	1r2r11t10r18	6198	2	1r2r11r10t 18t10r11r2r1
1-19	3174	1	1r12t19	6348	2	1r12t19t12r1

Table 7.2. *Total cost and number of transport modes changes for a one-way journey and roundtrip*

From the results and as expected if there are no fixed nodes, the cost of the roundtrip is twice the cost of a one-way journey and the outward trip is equal to the return one, which proves that the one-way journey provided by the model is the optimal one. Hence, the application of the model for roundtrips is interesting only when there are fixed passage nodes that are not comprised in the optimal route or differences (in costs, etc.) in the outward or in the return trip.

To illustrate this characteristic, some variants to the origin/destination 1–17 are presented in Table 7.3. In all cases, three fixed nodes are considered. Furthermore, in case 2, the return cost between nodes 10 and 1 (1,205) is lower than the sum of the costs of the outward journey (2,500). In case 5, only road links between the nodes are allowed for the trip to be planned.

Case	Fixed nodes	One-way journey			Roundtrip		
		Total cost	Number of changes	Optimal route	Total cost	Number of changes	Optimal route
1	16,18 and 19	6496	1	1r12t19t14 t18t16t17	10926	2	1r2r11r10t18t 16t17t14t19t10r 11r2r1
2	16,18 and 19 with different cost in the return journey	6496	1	1r12t19t14 t18t16t17	10802	2	1r2r11r10t18t 16t17t14t19t 10r1
3	13,15 and 18	7026	1	1r2r11r10t 18t13t19t15t17	11162	2	1r2r11r10t18t 13t17t15t18t 10r11r2r1
4	4, 5 and 11	7676	2	1r2r-11r4r 6r5w15t17	13832	4	1r2r11r4r6r5w 15t17t15w5r 11r2r1
5	4,5 and 11 and all links by road	–	–	–	19360	0	1r2r8r17r16r19r 12r4r11r5r112r1

Table 7.3. *Variants to the origin/destination 1–17*

In cases 1–3, the optimal routes consider two types of transport modes: road and rail. The three modes (road, rail and water) appear in the optimal routes of case 4: the link between nodes 5 and 15 is done by water.

From the table, the total cost of the roundtrip in case 1 is 84% of the cost of two one-way journeys between the origin and the destination because the return trip (from node 17 to node 1) is different than the outward one as shown in Figure 7.2. The similar situation can be observed in case 3. It should be mentioned that with the proposed model, the fixed nodes are visited obligatory in the outward journey and they might be visited a second time in the return one (cases 3, 4 and 5).

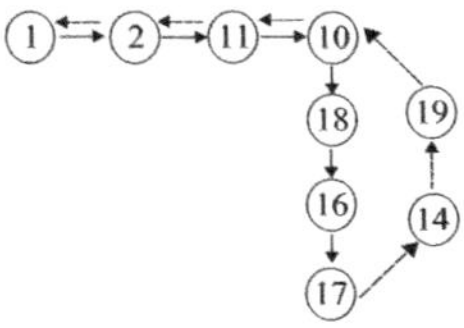

Figure 7.2. *Case 1: roundtrip*

Also when comparing the first two cases, the return journey is different from the outward one in case 2. As a consequence and due to the difference in the return journey, the number of visited nodes is not the same in the two situations being respectively 13 and 10.

In case 4, the cost of the roundtrip is 90% of the cost of two one-way journeys. Also, in this case, the total number of changes in the roundtrip is the double of the one in the one-way journey. When comparing the roundtrip of cases 4 and 5, the return trip in this one in longer than the one in case 4 due to the imposition of considering exclusively road links (case 5); also, the total cost is substantially higher in case 5.

7.4. Conclusions

For planning optimized and efficient journeys, the possibility of considering roundtrips and multimodal voyages should be regarded.

This chapter presents a mathematical model for optimized intermodal roundtrips and computational experiments. The model is formulated as a linear mixed binary program and allows minimizing the travel cost associated with a roundtrip where different transport modes between two nodes of the transport network may be available for the user. The model can be used for freight or passenger transport. The model application shows the importance of planning optimal roundtrips instead of two optimal one-way journeys when fixed nodes should be visited and the consequences of having restrictions when planning a trip. In the future, the authors aim to develop the model to consider other restrictions, such as time windows, route duration and capacity constraints, in order to widen its applications.

7.5. Bibliography

[BAU 04] BAUMANN D., TORDAY A., DUMONT A.-G., "The importance of computing intermodal roundtrips in multimodal guidance systems", *4th Swiss Transport Research Conference*, Monte Verità, Acona, March 25–26, 2004.

[BOY 07] BOYCE D., "Forecasting travel on congested urban transportation networks: review and prospects for network equilibrium models", *Networks and Spatial Economics*, vol. 7, no. 2, pp. 99–128, 2007.

[BON 04] BONTEKONING Y.M., MACHARIS C., TRIP J.J., "Is a new applied transportation research field emerging? – A review of intermodal rail–truck freight transport literature", *Transportation Research Part A*, vol. 38, pp. 1–34, 2004.

[DIA 05] DIAS J.C.Q., *Logística Global e Macrologística*, Edições Sílabo, 2005.

[EUR 01] EUROPEAN COMMISSION, European transport policy for 2010: time to decide, White Paper, 2001.

[EUR 93] EUROPEAN CONFERENCE OF MINISTERS OF TRANSPORT, "Terminology on combined transport", Internet site http:/www.cordis. lu/transport, 1993.

[MAC 04] MACHARIS C., BONTEKONING Y.M., "Opportunities for or in intermodal freight transport research: a review", *European Journal of Operational Research*, Management of the Future MCDA: Dynamic and Ethical Contributions, vol. 153, no. 2, pp. 400–416, 2004.

8

Modeling Traffic Hindrance Caused by Road Construction as Part of a Multicriteria Assessment Framework

One of the consequences of increased utilization of road infrastructure is more frequent maintenance work. Since road works generally result in less available road capacity, we can witness an increase in traffic hindrance, which involves delays and externalities such as noise, air quality, safety and the emission of GHG. Hindrance is generally not restricted to the location where maintenance work takes place, but can occur in a wider area as a result of altered flow patterns in the whole network.

The type of reconstruction and the way the work is executed determines not only the cost and service life but also the amount of hindrance. We present a multi-objective framework where for a longer period of time, cost and hindrance of specific road maintenance works can be determined as part of a decision support tool for the optimal planning of maintenance work. For this, we developed an alternative traffic assignment method that is able to predict traffic flow in a network in the presence of road works.

8.1. Introduction

The use of the road infrastructure has substantially increased during the last decades. As a result road, maintenance work must be executed more frequently. Since road work generally results in less available road capacity, the impact of road work on delays and externalities as safety, noise, air quality and the emission of GHG is also increasing. This impact is generally called hindrance, which is defined as the objectively measurable externalities induced by road works, while nuisance is defined as the way in which hindrance is perceived.

Chapter written by Eric VAN BERKUM and Henny TER HUERNE.

To create fair competition between contractors, national Dutch authorities implemented the EMAT (Economically Most Advantageous Tender) principle [DRE 09]. This implies that the success of a bid for construction work is determined not only by its cost but also by its quality, which involves the actual improvement of the road pavement, and the amount of hindrance and nuisance. The latter is largely determined by the way in which the work is executed and how it is communicated with all parties involved.

Ideally a successful road maintenance project keeps the balance between all those aspects. It improves the quality of the pavement significantly at a reasonable price and hindrance remains below an acceptable level.

In this chapter we present a framework that can help us find the right balance between all these parameters. For a road maintenance project, decisions must be made regarding the type of work, e.g. top layer only or replacing the complete pavement (top layer, binder and base layer), type of execution, such as working hours, e.g. 24/7 or night time only, or blockage (complete or partial) and additional traffic management measures such as a rerouting strategy. All decisions determine the cost of the project, while decisions on the type of work determine the service life, and hindrance is mainly determined by type of execution and additional traffic management measures. Obviously external circumstances may implicate that not all combinations are feasible, e.g. when the width of a road is not large enough, the combination of partial blockage and complete pavement reconstruction (i.e. top layer, middle layer and base) is not feasible.

8.2. Framework

The transport network is a public private commodity; it is of high value for society. It is important to properly maintain this commodity, in order to avoid capital loss by poor functionality of the system. The quality of the road network should be kept above a predefined threshold level for supporting safe traffic and transport. This can only be achieved by adequate and timely scheduled road maintenance activities. The starting point for road maintenance work is the quality of the pavement. Most administrations base their strategy for road maintenance on the present state of the pavement (condition based maintenance) and the importance of the pavement within the road network [BUR 05]. The network structure and its robustness are important features when determining a maintenance strategy [SCO 06]. Based on maintenance schemes we can try to restrict traffic obstacles. "Planning" consists of a series of road works, including the time of execution and the pavement materials (e.g. concrete, hot mix asphalt, a combination of those two). These are decision variables in relation to road maintenance work.

Maintaining the infrastructure of pavements can be cyclic or condition based. Cyclic maintenance makes it more possible to plan road works within a network and simultaneously reduce (manage) the hindrance. In 2008 BB&C studied three cases of maintenance on motorways [BB 08]; they found that differences in costs between cyclic maintained pavements and condition based maintained pavements are relatively small. Cuelho *et al.* [CUE 06] concluded that in the US almost no pavements are maintained "cyclic". Reducing traffic hindrance during road construction work can be done by choosing the most appropriate way of (re-)construction and considering the desired level of throughput of the road link [BRO 07]. By choosing a specific method of constructing we can tune on the amount of remaining traffic throughput, which also determines the level of service of the remaining traffic flows. Choosing the method of constructing means deciding on the amount of equipment needed, planning of phases and choosing construction methods, procedures and work hours.

Roadblocks and hindrance are almost unavoidable during road construction. This statement proves that both variables "workspace", "mobility management" and "regulations regarding safety" are strongly interdependent [CRO 03].

The framework is a tool that may help in deciding what the best construction measure would be in relation to the traffic situation. It determines what the effects of a specific road maintenance works are and supports the trade-off between the different effects. Figure 8.1 shows effects on two criteria (cost and hindrance) that were calculated using the tool, for different solutions of a road maintenance project.

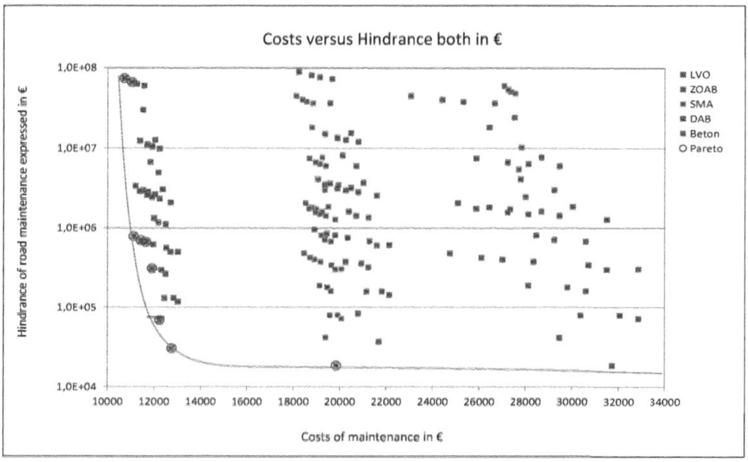

Figure 8.1. *Different solutions of road maintenance work with values for cost and hindrance, both expressed in €. For a color version of the figure, see www.iste.co.uk/jacob/traffic.zip*

As mentioned before, capacity reduction occurs just along or on the road link under construction. A complete blockage will obviously reduce the capacity to zero, whereas road work on the shoulder reduces the road capacity only partially [RIJ 02]. The use of traffic management measures may affect the capacity of the network both at the location of the road works as well as at other places within the network [BEN 03].

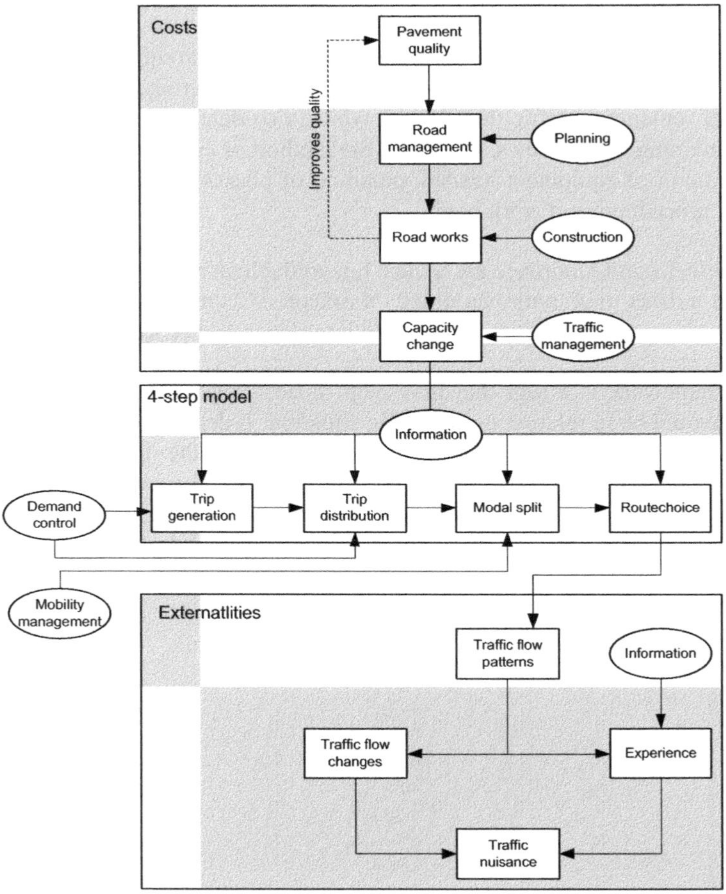

Figure 8.2. *Framework*

In our framework we assume that a maintenance project involves a number of decisions, which in turn result in an altered road network in terms of capacity reductions and traffic management measures. The decisions are listed in Table 8.1.

	Decision	Possible values
d_1	Construction type	Top layer only, complete reconstruction, i.e. top layer, binder and base layer
d_2	Pavement type	Concrete, hot mix asphalt, …
d_3	Blockage hours	24/7, daytime, night time, …
d_4	Blockage type	Partial, complete
d_5	Working zone	Location and size
d_6	Alternative routes	Set of consecutive links for each detour-route

Table 8.1. *Decisions in a maintenance project*

Not all combinations of decisions may be feasible, e.g. in some cases a complete reconstruction may not be combined with a small working zone or a partial blockage. Further, decisions for construction type, pavement type, type and hours of blockage determine D, the total duration of the project, so $D = D(d_1,d_2,d_3,d_4)$, or more general $D = D(\boldsymbol{d})$. Then, the construction type and the pavement type determine the service life S of the new road segment, so $S = S(\boldsymbol{d})$. All decisions imply an altered road network where capacities are time-dependent. For example, a project with a complete blockage during night time and no blockage during the rest of the day will result in a network where in the night period the blocked road segments have capacity of 0, where for the other time periods the network remains unchanged. For this, a road transport network G is defined as a time-dependent directed weighted graph so $G = G(t) = G(N, L(t))$, where N is the set of nodes, and $L(t)$ is the time-dependent set of links. To each link a in the network, a travel time function $\tau_a = \tau_a(x_a)$ is assigned, where travel time is assumed to be dependent of the flow x_a and a predefined value of capacity c_a. Often the so-called BPR function is applied $\tau_a(x_a) = \tau_a^0(1 + \alpha(x_a/c_a)^4)$, where τ_a^0 is the free-flow travel time and α is a parameter. Furthermore, in our case we assume capacity to be time dependent so $c_a = c_a(t)$. Obviously, since the network is determined by all decisions it holds that $G = G(\boldsymbol{d})$.

Now we want to use the altered network, i.e. supply of infrastructure, to determine an altered demand, which in the end results in altered flows in the network. We use these flows to determine not only new travel times (using the travel time functions τ_a), but also the so-called externalities as safety, emission of GHG, air quality and noise [WIS 11]. When we compare the current situation in terms of travel times and externalities with the situation that would occur when decisions $\boldsymbol{d}$ would be implemented, the amount of hindrance can be determined, obviously taking into account duration D and service life S.

To determine the altered demand we use the traditional four-step model, which consists of: trip generation, trip distribution, mode choice and route assignment [ORT 01]. First the so-called modeling area must be defined. The modeling area usually is a large geographical area; in case of the Netherlands, it may be the whole country. The modeling area is subdivided in zones, with each zone having a central point, a so-called centroid. In our case, we want to model the effects of road works. As a result, the size of the zones is chosen in such a way that zones near the work area are relatively small, and zones that lie far from the work area are relatively large. The available capacity in the network determines whether an individual makes a trip or not (trip generation), which destination is chosen (trip distribution), which transport mode is chosen (mode choice) and what route is chosen (route choice) [ORT 01]. The first three steps determine the so-called origin-destination matrix, or OD-matrix, with entries q_{ijt}, indicating the number of trips from origin zone i to destination zone j, departing in time slot t. In the network, these trips depart from the centroid of zone i and arrive in the centroid of zone j. The fourth step is also referred to as traffic assignment since not only routes are chosen but also the network is loaded with traffic, where cars are assigned to the chosen routes. When the transport system is at equilibrium (no road works/blocks) it is assumed that no traveler has an advantage by unilaterally changing his or her route. However, when construction work is taking place, the way travelers respond, and, as a result, the way the situation on the network will change depends largely on the impact the work has on capacities in the network [BRÖ 99] and on additional traffic measures, such as a rerouting strategy.

To conclude, we have developed a framework where decisions on route maintenance are captured in a decision vector $\boldsymbol{d}$, which includes decisions on the type of construction and pavement, the type and duration of the blockage (e.g. percentage of remaining road capacity) and additional traffic measures. Then for each $\boldsymbol{d}$, total cost and total hindrance are determined, considering not only the area where maintenance work takes place, but the whole network. Now this framework can be used to determine the impact on cost and hindrance of a maintenance strategy that consists of a series of decisions for all road segments in a network over a longer period of time, e.g. 20–30 years.

We argue that the best option in terms of cost and quality for maintenance work is not always chosen mainly because it is not always clear what the criteria and values for quality are. Our framework can support making those decisions since it is capable of providing these values for quality in terms of delays and externalities. Traditionally in pavement maintenance planning, road authorities often only take direct cost into account. Effects of fuel consumption, travel times and hindrance are often not accounted for.

Yu and Lu [YU 13] were, among others, early in including these indirect costs in the road maintenance planning process. They also included delays and pollution into their assessment framework. Their approach is based on integration of the lifecycle assessment (LCA) model and lifecycle cost analysis (LCCA) model in order to optimize the pavement maintenance schedule in terms of energy consumption, the emission of GHG or cost. However, their approach is mainly restricted to the area where the reconstruction work takes place. Our approach is more focused on a wider area since network effects as a result of changes in route choice may not be restricted to the local area. Then, since in the end we perform a pareto analysis we are able to determine the trade-off between the several objectives. An example is the potential outcome that a reduction of x% of the emission of GHG implies a cost increase of y%. It is then the task of the road authority to select the best decision based on a.o. these type of observations.

8.3. Route choice during road works

The equilibrium situation as defined in the previous section may be valid in regular conditions, but when network maintenance work takes place, which is a clear disturbance of the regular situation this may no longer hold. Generally road authorities install temporal road signs to indicate an officially preferred detour route. Hermelink [HER 11] has shown that a traditional traffic assignment approach is not always able to correctly describe traffic patterns during road works. He concluded that traffic flows over local detours routes are generally overestimated, while traffic flows on the officially (signed) detours are generally underestimated. To overcome these problems we developed an alternative traffic assignment approach.

8.3.1. *Traditional traffic assignment*

Route choice in a traditional traffic assignment is based on the notion that every road user will try to minimize his (or hers) travel cost. This travel cost is composed of travel time, but also distance and costs (for instance toll fees) can be part of this generalized cost equation. In order to minimize its generalized travel cost (GTC) a road user is constantly looking for alternative, more economic routes. Although in practice this behavior is not so pliable, road users are not constantly checking the alternative routes. In the long run, road users will distribute over the various alternative routes. After all, when road users are queuing at the same intersection every day, they either will look for faster alternatives, or accept the queue because they cannot find suitable alternatives.

A very important assumption in this "user equilibrium" behavior is that every road user has full knowledge of the road network. This assumption does not only entail that road users know each and every road, it also implies that road users know exactly how busy these roads are, and what corresponding travel times are. The assumption of knowing all roads is already questionable, but could, given the abundant availability of maps and route planners, be given the benefit of the doubt. It is, however, highly unlikely that road users are indeed familiar with the traffic flows and travel times on all roads. In normal traffic simulation, the effects of these assumptions are only marginal. Some road users are more familiar in region *a*, whereas others know all the roads in region *b*. And, over time, road users will experience how busy traffic is on particular routes, thereby gathering knowledge on the, for them, relevant part of the road network. Using this information, each road user will eventually find the best alternative and equilibrium is found.

However, during road maintenance, road users do not have the time to gather all this knowledge and then move into equilibrium. Road users are suddenly confronted with a closed (or partially closed) road and may have to find immediate alternatives. To aid road users in their decision, road managers often provide signed detours. These detours are usually not the shortest routes (in terms of GTC) and traditional traffic assignment algorithms are therefore unable to estimate the traffic patterns correctly.

8.3.2. *Alternative traffic assignment*

Drawbacks of traditional traffic assignment force us to develop an alternative approach that abandons the assumption of road users with full knowledge. Consequently, it is no longer possible to assign all traffic simultaneously and distinctions between groups of road users have to be made.

The most obvious distinction that can be made is based on the familiarity of road users with a specific area. Road users that are familiar in an area will tend to use all types of roads, including the local detours, whereas road users that are unfamiliar are likely to avoid residential and rural roads. Therefore, a project area is defined. It comprises the road work sections, the signed detours and the surrounding roads that are expected to be directly affected by the road works in terms of rerouting. Obviously the project area is part of the total modeling area. Within this project area, each type of road user is presented with their own unique road network, whereas outside the project area the assumption of full knowledge is maintained. This ensures that outside the project area traffic patterns are more or less similar to those of traditional traffic assignment, whilst inside the project area the effects of differences in information availability can be modeled.

Within the project area, "local traffic" is assumed to have full knowledge of the road network in the project area and behaves accordingly, by always searching for the best routes available, even if that means using residential or rural roads. "Non-local traffic", on the other hand is assumed to not be familiar with the road network in the project area. They therefore favor the use of through roads and will use residential or rural roads only if necessary (when there is no reasonable alternative). In order to achieve this behavior, all roads inside the project area that are classified as residential or rural, are penalized by adding a fixed GTC. Tests have shown that adding approximately 50% of the original GTC value ensures that "non-local traffic" behaves according to expectations. In practice, this means that "non-local traffic" prefers to take a 2.5-kilometer detour over through roads in order to avoid 1 kilometer of residential/rural road.

The altered road network that is presented to "non-local traffic" ensures that most road users will avoid the local detours and follow the main through roads instead. If however, these through roads are not the officially signed detours as prescribed by the road authorities, it is necessary to perform some additional measures. Then a reduction of GTC on all links of the signed detours for all non-local users will ensure a sufficient use of the signed detour.

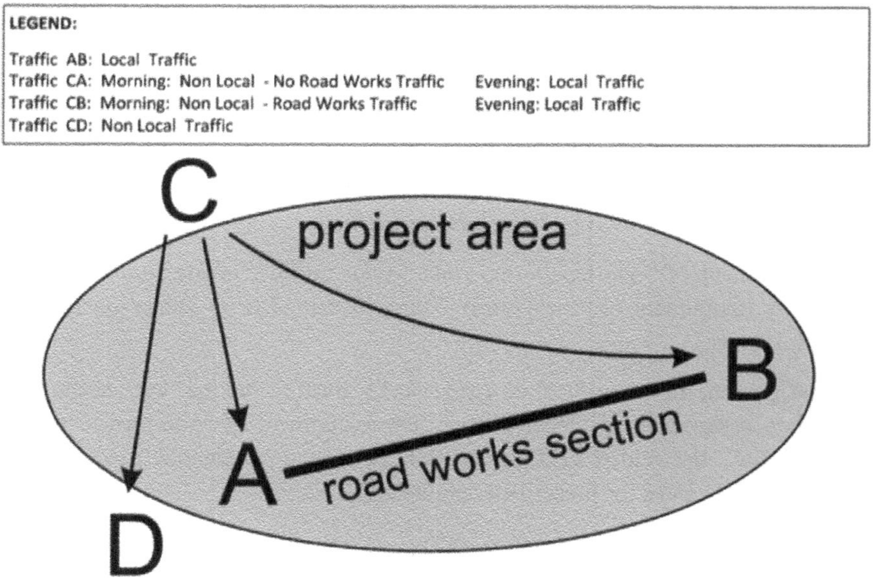

Figure 8.3. *Definitions of local and non-local, road works and no road-works traffic*

In Figure 8.3 an example is provided to explain the definitions of "project area", "local traffic", "non-local – road work traffic" and "non-local – no road work traffic". Because A and B are both inside the project area they are thus considered as "local", the traffic from A to B, which clearly has the opportunity to use the road work section, is classified as "local traffic". C and D are both outside the project area and there is no a good reason why traffic between the two would use the road work section, the traffic from C to D is therefore classified as "non-local – no road work traffic" even though they cross the project area. In the morning rush hour most people leave from home (where they feel familiar) and drive to work (where they feel not so familiar). In the morning rush hour, traffic from C to A is therefore classified as "non-local – no road work traffic", for they live outside the project area. In the evening rush hour, however, people from C to A are familiar with the project area (they live at A and drive home from work) and are therefore classified as "local traffic". The traffic from C to B might normally use a part of the road works section. In morning rush hour, traffic from C to B is therefore classified as "non-local – road works traffic". In the evening rush hour, people drive back home, are therefore familiar around B, and are therefore classified as "local traffic".

So we state that the alternative traffic assignment provides each road user the network he/she deserves. Users that are not familiar with the roads are penalized if they are using rural or residential roads and, if faced with road works, will tend to follow the directions of the road managers. The alternative traffic assignment model contains in fact a four-step approach itself:

1) preliminary assignment of "local traffic", to ensure that most ordinary roads in the project area are not used by "non-local traffic";

2) assignment of "non-local – no road works traffic", taking into account the preliminary assignment of "local traffic" and the added GTC values on rural and residential roads;

3) assignment of "non-local – road works traffic", taking into account the preliminary assignment of "local traffic", the assignment of "non-local – no road works traffic", the added GTC values on rural and residential roads and the subtracted GTC values for prescribed detours and;

4) definitive assignment of "local traffic", taking into accounts the assignments for "non-local – no road works traffic" and "non-local – road works traffic".

8.4. Example

We now want to demonstrate the working principles of the framework, and specifically how hindrance can be determined based on the outcomes of the alternative traffic assignment.

8.4.1. *Demand*

We use so-called OD-matrices from an existing transport model that was developed for this region. Had this not been the case, such an OD-matrix would have had to have been developed from scratch, using socio-economic data of all zones (i.e. number and type of inhabitants, number and type of workplaces, specific activities as schools, hospitals, etc.) and a so-called distribution function that denotes for each GTC the willingness to make a trip of a certain impedance, i.e. GTC. Obviously such a function is strictly decreasing. Using local and non-local centroids, local and non-local OD-flows can be defined, as was specified in the previous section. Demand is further defined for generally four time periods, i.e. morning peak, off-peak day, evening peak and off-peak night.

8.4.2. *Network*

All road segments in the direct neighborhood of the area where maintenance work takes place are defined as "local roads", for which we assigned altered GTC values. Then, altered GTC values for the signed detours are added as was explained in the previous section. Centroids directly around the road works area were defined as "local centroids". Obviously the blocked segments, i.e. the locations where the actual maintenance work takes place, are deleted from the network model.

For each road segment, a travel time function is defined, where travel time is assumed to be a function of flow and the capacity of that road segment. This information is sufficient to run the alternative traffic assignment, which yields for all roads in the network a value for traffic flow. With this flow it is possible to determine externalities of traffic, e.g. noise levels, CO_2 emission, travel times and air quality.

For the regular situation without any maintenance work a traditional equilibrium assignment yields traffic flows for all roads in the network. If we now compare the new situation with maintenance with the current, equilibrium situation, hindrance as a result of the maintenance works can be determined.

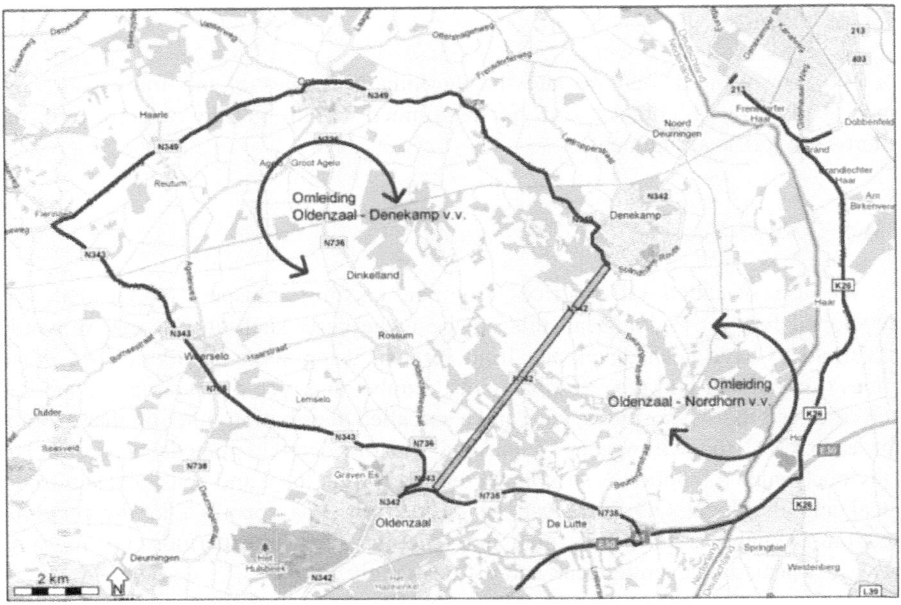

Figure 8.4. *Project area with rerouting options. For a color version of this figure, please see www.iste.co.uk/jacob/traffic.zip*

8.4.3. *Case N342*

As a case study, we choose a maintenance project that was executed on the N342, a regional road in the eastern part of the Netherlands between the cities of Oldenzaal and Denekamp. During a period of 4 weeks, the N342 was completely blocked. Road authorities installed a number of rerouting options that were indicated using temporary road signs, as depicted in Figure 8.4. Before and during the period of blockage, traffic counts were executed to be able to monitor traffic flows in the network. Locations are depicted in Figure 8.5. For this area, a regional transport planning model is operational. Part of this planning model is a series of OD-matrices, for the different time periods. These OD-matrices are calibrated using traffic counts in the before-situation, i.e. the situation without road work.

During the blockage traffic counts took place to be able to monitor the changes in traffic flows in the network. It appeared that a new equilibrium situation occured after approximately one week. So, it takes the car driver approximately one week to get adjusted to the new situation, and find an alternative route that they are comfortable with. If we now perform a regular equilibrium assignment for the altered road network, i.e. the situation with N342 blocked, we find that the traffic flows differ significantly when compared to the traffic counts. Then, we introduce the alternative traffic assignment method, and we calibrate the adjustments of GTC as was indicated earlier by penalizing rural roads and decreasing GTC for signed alternative detour routes. So for this situation our framework could have been applied to determine altered traffic flows as a result of complete blockage of N342. However, for other blockage-regimes and other detour-strategies also, the alternative assignment method will yield altered flows and as a result hindrance can then be determined. So using the framework for the whole area for any long-term maintenance strategy can determined the amount of hindrance and total amount of delays.

8.5. Conclusion

Traditionally, road maintenance strategies have been developed using some form of lifecycle cost analysis and/or lifecycle analysis. The cost function generally contains costs that are directly linked to the actual construction of the road. We have argued that in an increased utilized environment, where spare road infrastructure capacity generally decreases, hindrance in terms of delays and externalities such as noise, safety and air quality, become more and more important assessment criteria. Since hindrance is generally not restricted to the area where the maintenance work takes place, we have further argued that delays and externalities should be determined on a network level, rather than on a local level. Externalities and delays are determined as a result of a different level of service on the network and altered flows.

For this we have developed a decision framework where, given a set of decisions regarding the execution and construction type and some additional traffic management measures, delays and hindrance on a network level can be determined. It appeared that traditional route choice models are not applicable in a situation where roads are fully or partially blocked as a result construction work. For this, we have designed an alternative assignment method where local and non-local drivers and network elements are distinguished, and signed detour routes can be input for such a model. The alternative assignment model was calibrated for a real-life situation and in an example it is demonstrated how the framework should be applied. As result, using the framework, any strategy on road maintenance can be assessed for multiple objectives, i.e. cost, delay and hindrance. Applying a Pareto analysis may then support the decision of selecting the best maintenance strategy.

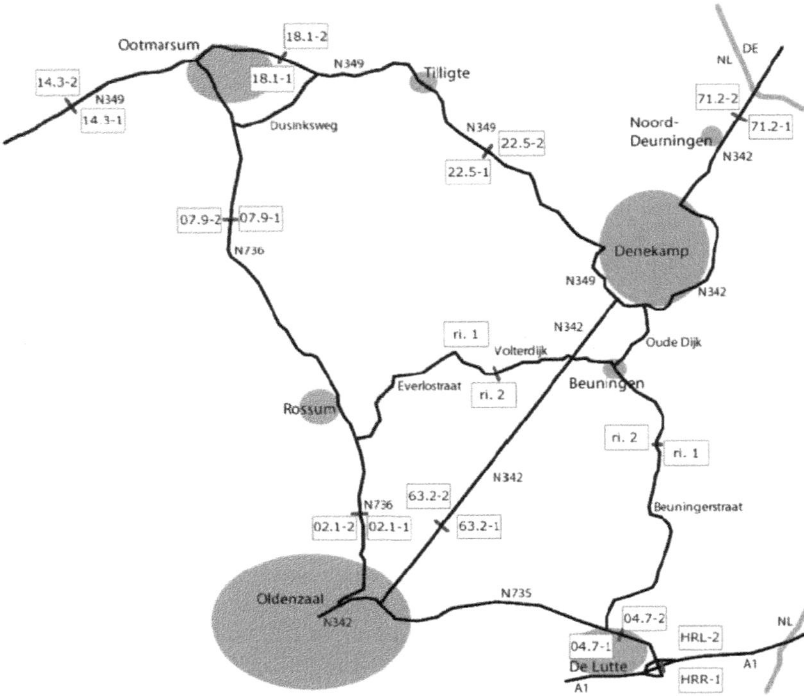

Figure 8.5. *Location of the traffic counts. For a color version of this figure, please see www.iste.co.uk/jacob/traffic.zip*

8.6. Acknowledgments

We would like to thank the Province of Overijssel, Region of Twente and Rijkswaterstaat Oost-Nederland for funding this research.

8.7. Bibliography

[BB 08] BB&C & Vital, Link Beleidsanalyse, Vergelijkingsonderzoek Cyclisch Onderhoud (VOCO) (in Dutch), BB&C & Vital Link, Den Hoorn, 2008.

[BRÖ 99] BRÖG W., SCHÄDLER M., "More passengers, higher profits for public transport – (im)possible expectation!?", *Proceedings of the 53nd UITP-Congress*, Toronto/Canada, May 1999.

[BRO 07] BROWN R., "Importance of Good Construction on Reducing Maintenance Costs", *Proceedings of the Fifth International Conference on Maintenance and Rehabilitation of Pavements and Technological Control, Mairepav'05*, Park City, Utah, USA, 8–10 August 2007.

[BUR 05] BURNINGHAM S., STANKEVICH N., Why road maintenance is important and how to get it done, Transport Note No TRN-4 World Bank, Washington DC, 2005.

[CRO 03] CROW, Onderzoeksresultaten CROW-werkgroep Evaluatie Onderhoudservaringen Betonwegen (in Dutch), Ede, 2003.

[CUE 06] CUELHO E., MOKWA R., AKIN M., "Preventive Maintenance Treatments of Flexible Pavements: a Synthesis of Highway Practicee", Final Report, US-FHWA, Springfield, Virginia, 2006.

[DRE 09] DRESCHLER M., Fair competition, How to apply the "Economically Most Advantageous Tender", (EMAT) award mechanism in the Dutch construction industry, PhD Thesis, Delft University of Technology, Delft, 2009.

[HER 11] HERMELINK W.A.H., Evaluatie verkeerspatronen groot onderhoud N342 Oldenzaal – Denekamp (in Dutch), University of Twente, Enschede, 2011.

[ORT 01] ORTUZAR J., DE D., WILLUMSEN L.G., *Modelling Transport*, Wiley & Sons, Chichester, 2001.

[RIJ 02] RIJKSWATERSTAAT, Capaciteitswaarden Infrastructuur Autosnelwegen, (in Dutch), Adviesdienst Verkeer en Vervoer, Rotterdam, 2002.

[SCO 06] SCOTT D.M., NOVAK D.C., AULTMAN-HALL L. *et al.*, "Network Robustness Index: a new method for identifying critical links and evaluating the performance of transportation networks", *Journal of Transport Geography*, vol. 14, no. 3, pp. 215–277, 2006.

[WIS 11] WISMANS L.J.J., van BERKUM E.C., BLIEMER M.C.J., "Modeling externalities using dynamic traffic assignment models: a review". *Transport Reviews*, vol. 31, no. 4, pp. 521–545, 2011.

[YU 13] YU B., LU Q., "An improved maintenance optimization methodology: integrated LCA and LCCA". *Proceedings of the TRB 2013 Annual Meeting*, Washington DC, 2013.

PART 3

Traffic Management, Monitoring and Routing

9

Behavioral Responses to Traffic Congestion – Findings from Paris, São Paulo and Mumbai

This paper focuses on the impacts of traffic congestion in the daily life of metropolitan motorized households. It is based on a doctoral dissertation defended in 2012, which studied three megacities: Paris in France, São Paulo in Brazil and Mumbai (ex-Bombay) in India. A stated-adaptation survey (simulation game) was conducted in 2009–2010 to identify behavioral responses to traffic congestion and to estimate the threshold of household tolerance to congestion in the three regions. A universal set of 15 responses was available to households in the three regions. Respondents first selected marginal adjustments then behavioral alterations. They managed to preserve their activity pattern, thanks to a few marginal adjustments, until they reached a threshold and started to alter their behavior. The threshold was almost the same from one region to another. It fluctuated around 30 additional minutes of travel time. This result was consistent with Zahavi's conjecture of travel time budget stability.

9.1. Introduction

9.1.1. *Traffic congestion*

In megacities, dealing with traffic congestion is a daily effort for many inhabitants [DOW 04]. Traffic congestion happens because road supply cannot satisfy travel demand. Congestion mainly extends travel times and/or increases their variability. We wonder how households maintain their travel time budget and preserve their lifestyle in case of intense congestion. We suppose that households have to alter their travel patterns and even their activity patterns to maintain their travel time budget and preserve their lifestyle. The first aim of this paper is to draw

Chapter written by Gaele LESTEVEN.

up an exhaustive set of responses to traffic congestion. Do the responses differ from one megacity to another? Which responses are the selected most often? And for which reasons? The second aim is to estimate the threshold of tolerance to traffic congestion. We test the hypothesis that the higher the stress due to congestion, the more households have to modify their activity patterns to maintain their lifestyle.

The remainder of this chapter is organized as follows. Section 9.1.2 describes the context of the three studied megacities. Section 9.2 explains the method and details the sample. Section 9.3 shows the results. Section 9.4 presents a final discussion.

9.1.2. *The context of the three studied megacities*

We selected three concentrated megacities presenting three stages of motorization and belonging to three different cultural areas: the Paris metropolitan region (PMR) in France, the São Paulo metropolitan region (SPMR) in Brazil and the Mumbai metropolitan region (MMR) in India. First, we present the three metropolitan regions, and then we focus on motorization, daily mobility and traffic congestion in these regions. The data presented in this section, otherwise noted, were obtained from three household travel surveys: the *2010 Global Transport Survey* [DRI 12] for the PMR, the *2007 Origin-Destination Survey* [MET 07] for the SPMR and the *2008 Comprehensive Transport Study for Mumbai Metropolitan Region* [MMR 08].

Figure 9.1 presents the three regions. Paris is the capital of France. The PMR (or *Ile-de-France*) covers an area of 12,000 km^2 where 11.8 million people live. Its population (57%) and employment (67%) are mainly concentrated in the city of Paris and its suburbs. São Paulo is the capital of the State of São Paulo and the economic center of Brazil. It is located in the south-east part of Brazil. The SPMR has a population of 20.1 million inhabitants within an area of 8,000 km^2. 60% of the metropolitan region population lives and/or works in the city of São Paulo. Located on the west coast of India, Mumbai is the capital of the State of Maharashtra and the economic capital of India. The area of the MMR is divided into a peninsula (which forms the municipality of Greater Mumbai) and the continent. It is smaller (4,350 km^2) than the PMR and the SPMR. 60% of 20.7 million residents of the MMR live in the peninsula. In the center of the MMR, located at the end of the peninsula, the population density is very high (43,000 inhabitants/km^2). It is more than twice the density of the city of Paris. Densities progressively decrease in the north of the peninsula and in the continent. At the same time, jobs have been dispersed in the metropolitan region.

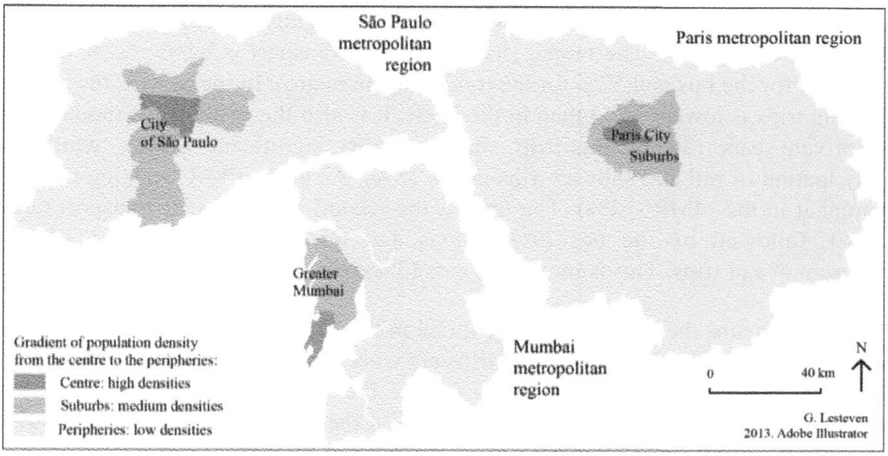

Figure 9.1. *Maps of the three metropolitan regions*

There are 430 cars for 1,000 inhabitants in the PMR. In the city of Paris, only 45% of the households own a car compared to 71% for the whole PMR. In the SPMR, motorization is lower: there are 180 cars per 1,000 inhabitants. Only 49% of the households of the SPMR own a car. Households who live in the center of the SPMR are wealthier and more motorized (59%). Motorization in the MMR remains modest, with 33 cars per 1,000 inhabitants. Only 5.4% of the households in the MMR are motorized. The motorization rate is still higher in the center where many wealthy households are found.

	Paris (PMR)	São Paulo (SPMR)	Mumbai (MMR)
Population (millions of inhabitants)	11.8	20.1	20.7
Area (km^2)	12,000	8,000	4,353
Motorization (private automobiles/1,000 inhabitants)	430	180	33
Trips/day (million)	41	38.1	34.3
Trips/day/car (million)	15.5	10.4	2.9
General mobility (trips/inhabitant)	3.87	2.0	1.65
Motorized mobility (motorized trips/inhabitant)	2.4	1.3	0.65
Average travel time by transit (min)	41	67	58 (train)/38 (bus)
Average travel time by car (min)	22	31	29

Table 9.1. *General data*

In the PMR, the modal share of walking (39%) slightly exceeds the modal share of private cars (38%). The modal share of public transport is below (20% divided into 43% for the bus and 57% for the train plus the metro). In the SPMR, the ratio of walking trips is lower (34%) than in the PMR. It is also the case for the modal share of private cars (30%). Bus trip share remains significant (29%), while the participation of rail and subway trips is low (8%). As in the PMR, walking trips are dominant in the MMR (61%). The train is the second more popular transport mode (20%), followed by the bus (9%). Taxis and three-wheelers (auto rickshaws) represent 4% of trips. This is more than cars (3%) or motorized two-wheelers (3%).

Work remains the main purpose of travel in the PMR (29%). But its share is less important than in the SPMR (44% of trips are work trips) and in the MMR (46%). In the PMR, work-purpose travelers mainly use the car (44%), followed by public transport (40%). In the SPMR, they equally use the bus (31%) and the car (30%), while in the MMR, they mainly go by foot (40%) or by train (37%). Only a few of them go to work by car (2%).

Mobility in the PMR (3.87 trips per inhabitant per day) is greater than in the SPMR (1.97) and in the MMR (1.65). A PMR inhabitant spends 92 min travelling per day, instead of 83 min in 2001. In the SPMR, the travel time budget is 77 min per day. It is 15 min more than a decade before. A wealthy inhabitant spends 95 min travelling per day, while a poor inhabitant spends 48 min. 24% of wealthy people do not travel against 46% of poor people and 55% of very poor people. The situation is similar in the MMR and the reasons are the same: lack of money and lack of jobs. Still, 40% of wealthy MMR people do not travel, which is a lot more than in the PMR and the SPMR.

Traffic congestion is characterized by low average travel speeds during peak periods. The daily average speed in Paris city has dropped from 20.9 to 15.7 km/h in 20 years [MAI 11]. At the same time, traffic congestion on the PMR expressways remains moderately constant, even with the motorization growth in the suburbs, thanks mainly to traffic management. Nevertheless, the average traffic speed during the day (7 am – 9 pm) has seen a little decrease. In the SPMR, the average traffic speed has decreased, especially during the evening peak hours: from 24.9 km/h in 1980 to 14.8 km/h in 2008 [CET 10]. In the MMR, the minimum average speed had also decreased: from 18 km/h in 1962 to 8 km/h in 2008 in the center and from 30 to 5 km/h in the suburbs (MMRDA & LEA, 2008).

In the PMR, traffic congestion mostly occurs in the suburbs and in the peripheries where most of motorized households are found. Real estate prices in the city of Paris force both people and jobs to move from the city to the suburbs. The public transport system is efficient within the city of Paris and for linking the suburbs to the city. But it is less efficient when travelling within the suburbs, for which the private car is preferred. The situation is very different in the SPMR. There, the growth of private cars is concentrated in the center, where traffic congestion is worse. The higher income groups that intensively use private cars invest much more time, space and money to travel around and so contribute to traffic congestion. In the peripheries, where poor and middle-income people live, mobility is seriously affected. In the MMR, the situation is similar: traffic congestion is the highest in the peninsula, while the road network has not changed much in four decades. Encroachments, mixed traffic and lack of traffic discipline also play a role.

9.2. Methodology

An earlier paper [LES 11] presented the methodology more in depth and started to analyze the results of the survey conducted in Paris and in São Paulo.

9.2.1. *Presentation of the method*

In order to understand how metropolitan motorized households cope with daily congestion, we referred to simulation methods. Using a simulation game offers an opportunity to put the respondents in situations and to observe all the possible behavioral responses without any *a priori* knowledge [FAI 08]. [JON 80] was the first to explore the possibilities of gaming-simulation with Household Activity and Travel Simulator (HATS). Called the stated-adaptation survey (or interactive stated response survey), this type of survey focuses the observation not only on the stated choices of the respondents, but also on their choice process [LEV 14, HAB 13].

The stated-adaptation survey we implemented was based on increasing the travel time. It allowed us to comprehend behavioral responses to traffic congestion. The survey had two stages: (1) the respondent filled up a trip diary listing all trips, no matter what purpose or mode of transport was used, for several consecutive days. (2) During an in-depth face-to-face interview, the respondent reacted to hypothetical situations based on selected trips from the trip diary. The aim of the method was not to reach behavioral representativeness, but to explore and understand the choice process. For this reason, the sample (15–20 households) for each case study was small [FAI 08], but it had to be as diversified as possible. All the interviews were tape recorded.

9.2.2. *The sample*

The survey took place in the three case studies over a year: the SPMR in September–October 2009, the MMR in March–April 2010 and the PMR in January–

February and May–June 2010 [LES 12]. We tried to avoid adverse weather conditions (snow or monsoons) and school holidays as much as possible. The selected households had to own at least one car and use it several times a week.

A sample of 64 motorized households (91 respondents) participed in the survey. Among them, 23 households (16 men and 15 women) lived in the PMR, 17 households (12 men and 11 women) in the SPMR, and 22 households (22 men and 15 women) in the MMR. Respondents were recruited by indirect relationships. At the end of the interview, they received a gift. In order to have a diverse sample, respondents first filled in a recruitment form. Too similar households were rejected. The sample was well diversified in terms of family status and places of residence. It came with no surprise that it was hard to interview low-income motorized households. The socio-economic characteristics of the sample are detailed in Table 9.2.

		PMR		SPMR		MMR		Total	
		No.	%	No.	%	No.	%	No.	%
Households		23	*100*	17	*100*	24	*100*	64	*100*
Family status	Single	4	*17*	4	*24*	3	*13*	11	*17*
	Family without children	5	*22*	4	*24*	1	*4*	10	*16*
	Family with children <15	9	*39*	4	*24*	6	*25*	19	*30*
	Family with 15<children<30	4	*17*	5	*29*	14	*58*	23	*36*
Occupation	Working (or student)	20	*87*	16	*94*	23	*96*	59	*92*
	Retired	3	*13*	1	*6*	1	*4*	5	*8*
Education level	University	19	*83*	15	*88*	23	*96*	58	*91*
	High school	4	*17*	2	*12*	1	*4*	6	*7*
Place of residence	Centre	5	*22*	10	*59*	9	*38*	24	*38*
	Suburbs	8	*35*	2	*12*	12	*50*	22	*34*
	Periphery	10	*43*	5	*29*	3	*13*	18	*28*
Car ownership	One car	15	*65*	9	*53*	14	*58*	38	*59*
	Two cars	7	*30*	4	*24*	9	*38*	20	*31*
	Three cars	1	*4*	3	*18*	1	*4*	5	*8*
	Two Wheels	3	*13*	0	*0*	6	*25*	9	*14*
	Driver	0	*0*	1	*6*	9	*38*	10	*16*
Yearly income in the PMR (in euros)	< 20,000	0	*0*						
	20,000–40,000	8	*35*						
	> 40,000	15	*65*						
Yearly income in the SPMR (in euros)*	< 11,583			1	*6*				
	11,583–23,166			3	*18*				
	23,166 – 46,332			3	*18*				
	>46,332			10	*59*				

Yearly income in the MMR (in euros)**	4,250–8,500					3	*13*		
	8,500–17,000					0	*0*		
	17,000–42,500					14	*58*		
	> 42,500					7	*29*		
Average household size		2,9		2,6		3,1		2,9	
Average respondent age		43		39		44		42	

*Converted from Brazilian real to euros in 2009: 1€ = 2.59 Br Rs

**Converted from Indian rupees to euros in 2010: 1€ = 58.8 Indian Rs

Table 9.2. *Sample characteristics*

9.2.3. *Implementation of the survey*

All the adult members of the household had to fill a trip diary during a whole week in the PMR and three consecutive days in the SPMR and the MMR. The aim was to identify travel and activity decisions which had been taken within the household during the trip diary filling process. A few days after completing the trip diary, the regular car drivers of the household were interviewed.

First, the interviewer reminded the respondents about the contents of the trip diary and asked the respondents to select three representative types of trips. Most of the time, the selected *compulsory trip* was a commute or drop children off at school trip. The selected *non-compulsory trip* occurred mostly for shopping or leisure. The selected *travel time budget* (TTB) was the sum of the trips made during the most representative day of the trip diary. The *revealed travel time* (RTT) referred to the amount of time the respondents quoted for these trips in their trip diary.

Then, two scenarios were tested. Scenario 1 considered non-recurrent congestion, presenting variable and unpredictable travel times. Scenario 2 concerned recurrent congestion, presenting longer travel times. For both scenarios, the compulsory and the non-compulsory trips were taken into account. The TTB was considered only for scenario 2. Each scenario was built in three levels corresponding to three levels of change in travel conditions (+50%, +100% and +200%). Each level was presented as a pre-filled playing card. The playing cards were progressively shown to the respondent (levels 1, 2 and 3 of scenario 1; then levels 1, 2 and 3 of scenario 2). Respondents were expected to state how the change in travel conditions described in each card would affect their travel and activity patterns. Most of the time, other members of the household were present during the interview to check the validity of the respondent's stated responses. Table 9.3 presents the two scenarios.

	Scenario 1	Scenario 2
	Non-recurrent congestion	Recurrent congestion
Under test	Variable and unpredictable travel times	Longer travel times
Data selection	A regular and compulsory trip (e.g. a commute trip)	A regular and compulsory trip
	A regular and non-compulsory trip	A regular and non-compulsory trip
		Whole day trips (travel time budget)
Three levels of changes in travel conditions	1/between the revealed travel time (RTT) and +50% RTT	1/+ 50% of revealed travel time
	2/between the RTT and 2× RTT	2/ RTT × 2
	3/between the RTT and 3× RTT	3/ RTT × 3

Table 9.3. *The two scenarios*

9.3. Results

9.3.1. *A universal list of responses to traffic congestion*

The first aim of the survey was to establish an exhaustive list of behavioral responses to traffic congestion. According to [MOK 97], motorized households deal with a universal set of behavioral responses to traffic congestion, regardless of their family status, their level of income or their place of residence. This does not mean that a household takes all strategies listed but it selects some of them, according to its constraints and resources. [MOK 97] proposed a universal set of 16 possible responses based on a survey conducted in San Diego, California, in 1992. [BOV 02] summed up the universal set into 11 strategies: (1) accept travel costs (do-nothing situation); (2) reduce travel costs (a more comfortable vehicle); (3) adapt departure time; (4) change route; (5) buy time (e.g. use toll roads); (6) temporal changes (flexitime, compressed work week); (7) change mode; (8) telecommute; (9) relocation of place of workplace or home; (10) start a home-based business; (11) quit work. We wonder if this universal set is still accurate today in these different metropolitan regions.

In the survey that we conducted in the PMR, the SPMR and the MMR, the 91 respondents answered up to 15 playing cards (six playing cards for the compulsory trip related to scenarios 1 and 2; six cards for the non-compulsory trip; three cards for the TTB related to the scenario 2). Sometimes, it was not relevant for the respondent to answer to all the cards (such as the respondent who did not use his/her car to go to work). Respondents were free to answer what they wanted. Most of the time, they stated one response per card, i.e. per level of scenario. Some of the respondents gave a second response (14% of responses), even a third (2%). 1.142 responses were counted and listed. At the end, 15 different behavioral responses were found. Regarding their impacts on household travel and activity patterns, the 15 behavioral responses can be organized in five groups. Group 1 consists of *marginal adjustments* ("maintain behavior", "leave earlier" and"change route"). They could slightly affect household travel patterns. The four remaining groups are composed of behavioral alterations. Group 2 is *modal alterations* ("other mode", "carpool" and"have a driver"), group 3 *temporal alterations* ("reschedule", "decrease frequency" and "change shift") and group 4 *spatial alterations* ("change destination", "chain trips" and "work from home"). *Modal, spatial* and *temporal alterations* affect household travel patterns and even activity patterns. *Radical alterations*, named as group 5 ("quit", "quit job" and "move"), are ruptures in the activity pattern.

All the 15 responses occurred in the three regions, except for the response "driver" which was specific to the MMR. In Mumbai, many people have a driver. They drive about 70% of the private cars on weekdays [DAT 10]. In the sample, which was not representative, 9 of the 24 MMR households employed a driver. They considered the private car as a "private taxi" [MMR 08]. Having a driver is probably the most efficient response to traffic congestion. The activity pattern remains well and comfortably preserved. Nevertheless, having a driver in Mumbai is a privilege of high-income households, even more than owning a private car. We assume that the number of drivers will progressively reduce with the democratization of private cars, as is already the case in Delhi.

Table 9.4 presents the number of times each response was stated per type of trip (compulsory, non-compulsory, TTB), without specifying the scenario. The assumption of a universal set is confirmed.

		Paris (PMR)			São Paulo (SPMR)			Mumbai (MMR)		
Groups	Responses	C	NC	TTB	C	NC	TTB	C	NC	TTB
Marginal	1) Maintain behavior	60	55	23	23	39	5	42	52	16
Adjustments	2) Change departure time	44	0	2	49	17	8	55	8	6
	3) Change route	14	7	0	18	8	3	7	5	1
Modal	4) Other mode	23	16	8	18	12	8	24	13	24
Alterations	5) Carpool	0	0	2	2	0	1	4	1	1
	6) *Have a driver*	*0*	*0*	*0*	*0*	*0*	*2*	*0*	*1*	*8*

Temporal	7) Decrease frequency	2	9	1	1	17	0	1	11	4
Alterations	8) Reschedule	3	14	1	7	15	4	14	11	4
	9) Change shift	5	1	1	6	0	1	1	4	2
Spatial	10) Chain trips	1	2	3	0	1	5	1	1	1
Alterations	11) Change destination	0	15	2	5	18	4	5	17	3
	12) Work from home	3	0	0	5	0	3	9	0	1
Radical	13) Quit	6	8	2	6	13	15	9	19	6
Alterations	14) Quit job	8	0	5	1	0	2	10	0	7
	15) Move	5	0	7	15	0	7	4	0	7
	Number of times	174	127	57	156	140	68	186	143	91

C: Compulsory trip (scenarios 1 and 2); NC: Non-compulsory trip (scenarios 1 and 2); TTB: Travel Time Budget (scenario 2).

Table 9.4. *The 15 behavioral responses and the number of times they occurred*

Interestingly, the universal list [MOK 97] created and [BOV 02] completed has not changed a lot in the last 20 years. The differences come from the technological progress and the comfort it offers: it is today common to own a car radio and a cell phone.

9.3.2. *Different frequencies*

We would like to know which responses among the 15 listed were stated the most often and for which reasons. We suppose that households try to preserve their activity pattern as much as they can, because it is the result of negotiations within the household [VOV 04]. This is why we assume that the responses that are the most common are marginal adjustments, followed by alterations. Table 9.5 details the frequency of the responses.

	Paris (PMR)	São Paulo (SPMR)	Mumbai (MMR)
Compulsory trip (scenarios 1 and 2)	*n=174*	*n=156*	*n=186*
Marginal adjustment (%)	68	57	56
Alteration (%)	21	28	32
Radical alteration (%)	11%	15	12
Non-compulsory trips (scenarios 1 and 2)	*n=127*	*n=140*	*n=143*
Marginal adjustment (%)	49	46	46

Alteration (%)	45	45	41
Radical alteration	6	9	13
Travel time budget (scenario 2)	*n=57*	*n=68*	*n=91*
Marginal adjustment (%)	44	24	25
Alteration (%)	31	41	53
Radical alteration (%)	25	35	22

Table 9.5. *The frequency of the responses per trip for each region*

The hierarchy in which the most common responses are those that affect the least the activity pattern was respected for compulsory trips. The hierarchy began to fail for non-compulsory trips. Respondents tended to alter their behavior a little more. They mostly changed their destination (spatial alternation) and reschedule their trip (temporal alteration). Contrary to compulsory trips, destination and time were not forced. They preferred to keep their car and cancelled the trip if travel times were too long (levels 2 and 3). In the case of travel time budget, the hierarchy was confirmed in the PMR, but not in the SPMR and in the MMR. The respondents from the São Paulo and Mumbai regions altered their behavior more than the respondents from the Paris region.

A second analysis focused on the length of the trips and if it affected the hierarchy. First, we looked at short trips, for which the revealed travel time was equal to or shorter than 30 min, then at long trips (a revealed travel time of more than 30 min). The hierarchy was confirmed, except for long compulsory trips. Marginal adjustments were more common for short trips and alterations for long trips. In fact, the length and the compulsory nature of the trips were two factors that determine the behavior of respondents.

9.3.3. *The threshold of household tolerance to congestion*

The second aim of the survey was to estimate the threshold of household tolerance to traffic congestion. We wanted to test the hypothesis that the higher is the stress due to traffic congestion, the more households have to modify their travel and activity patterns to maintain their lifestyle.

We created a new variable called variable of delay. The delay was the time added to the revealed travel time at each level of scenario 2. For example, the delay would be 15 min if a revealed travel time of 30 min were increased by 50% (level 1). It would be 30 min if the revealed travel time were doubled (level 2) and 60 min if it were trebled (level 3). We wondered what the delay was when the respondents started to alter their behavior. Figure 9.2 shows the results for compulsory trips in the second scenario. The threshold matched with the intersection of the curve of marginal adjustments and the curve of alterations. Delays were grouped in three periods of time: less than 30, 30–60 and 60–90 min. The threshold of household tolerance to congestion was almost the same from one region to another. It fluctuated around 30 min in addition to the revealed travel time. When travel times exceeded these 30 min, respondents started to alter their behavior. Up to an additional half-hour travel time, people managed to preserve their activity pattern, thanks to a few marginal adjustments (e.g. changing the departure time, changing the itinerary route, using the phone).

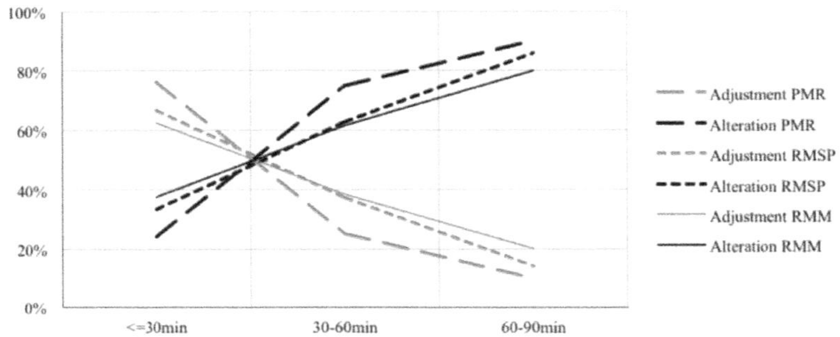

Figure 9.2. *Behavioral thresholds according to the delay in scenario 2 for compulsory trips*

This result cannot be applied generally because of the non-representative sample. But it appears to be consistent with the surveys of [STE 02, WEI 10]. Stern conducted a revealed preference survey of 160 car drivers in the Tel Aviv region [STE 02]. He estimated a threshold for compulsory trips that varied between 30 min and 1 h. Weis *et al.* conducted a stated-adaptation survey of 84 households living in the Zurich region about their behavioral responses to longer travel times, not only regarding car trips but all the trips they did [WEI 10]. Their scenario was close to our scenario 2. There were four levels (+50%, +100%, +200% and −50%). Their respondents preserved their activity pattern as much as they could. First, they reduced time at home before cancelling non-compulsory trips, then compulsory trips. From 90 min of additional travel time, they adopted radical alterations. In our survey, radical alterations started to be more common in the period of "60–90 min".

9.4. Conclusions

According to the results of the survey, most of the metropolitan motorized households manage to preserve their lifestyle when facing growing congestion.They continue to use their car, even when the degrees of congestion increase. It confirms the fact that the effects of congestion on the automobile-based system are minor compared to the effects of car dependence [DUP 11]. What is even more interesting is that the households are aware of this fact. In order to preserve their lifestyle, households first select marginal adjustments that affect their activity pattern the least. Then they use alterations and, only at the end, radical alterations. These results are based on a simulation game. We suppose that in real life, respondents are even more resistant to traffic congestion. In the game, respondents knew that the scenarios were composed of successive increases of travel time. This increasing bid may have prompted them to diversify their responses.

Respondents from the Paris region appeared to be less flexible than respondents from the SPMR and the MMR. They seemed to keep marginal adjustments as long as they could. Then they directly switched to radical alterations. There are several possible explanations for their lower degree of flexibility. For the last 30 years, median and average travel times to work in the PMR have remained stable at around 30 min. At the same time, the PMR has sprawled and commute distances have doubled. The existence of a threshold of tolerance in terms of travel times for commute trips seems to be constituent of a metropolitan way of life. It could explain this stability of travel times to work [MAS 10]. Moreover, some commuters may be unwilling to sacrifice the comfort and flexibility of the private car just to save a minor amount of time [KEN 14].

Most of the respondents in the survey stated that even in bad traffic conditions, they would manage to preserve their activity terms thanks to short-term strategies (marginal adjustments) or medium-term strategies (modal, spatial and temporal strategies) until they reach the threshold of tolerance of about 30 additional minutes of travel time. It is interesting to notice that the threshold of tolerance to traffic congestion was more or less the same in the three regions. This is consistent with the Zahavi's conjecture of travel time budget stability [ZAH 74]. People maximize their travel distances according their travel time budget (TTB) and their travel cost budget. Motorization increases the speed and reduces travel times (it takes less time to drive the same distance with higher speed). With the access to motorization, the TTB decreases. Then it stops to a level where the average TTB does not reduce itself anymore and becomes stable [CRO 04]. The daily TTB is about 1 hour. It tends to increase with the size of the city (about one hour 30 min in megacities). The fact that respondents eventually reached a threshold supports the Zahavi's conjecture.

To conclude, we noticed that from one region to another, the set of behavioral responses was similar. But household behaviors were not identical. The differences in behavior could be explained by the place of residence of the motorized households in the metropolitan regions and the location of traffic congestion. As the origin-destination analysis pointed out in section 9.1, motorized households in the

SPMR and the MMR are most of the time wealthy households which are mainly found in the center of the metropolitan regions. In the PMR, motorized households mostly live on the periphery of the metropolitan region and tend to belong to middle or low-income classes. Traffic congestion mainly occurs in the center of the SPRM and the MMR. The picture of the Paris region is very different. There, the traffic congestion has sprawled out of the city to the main roads of the suburbs and the peripheries. Because of the lack of efficient public transport from periphery to periphery, the PMR households continue to use their cars. Even when travel times increases a lot, the private car remains competitive compared to public transport. This assumption should be tested in further research. Moreover, it would be interesting to estimate the probability of switching to public transport according to higher private car travel times in these megacities. For example, research about Berlin showed that the probability of using public transport was more than 50% when the travel time by public transport was inferior to 1.5 the travel time by car [REI 08].

9.5. Acknowledgments

This chapter is based on a recent doctoral research funded by Renault and defended at the University of Paris 1 Pantheon-Sorbonne in 2012, under the supervision of Prof. Gabriel Dupuy. The survey in the SPMR was conducted with the help of Prof. O. Strambi and B. Alves from the Polytechnic School of the University of São Paulo. The survey in the MMR was conducted with the help of A. Datar and S. Kelkar the from Mumbai Environmental Social Network.

9.6. Bibliography

[BOV 02] BOVY P., SALOMON I., "Congestion in Europe: Measurements, patterns and policies", in STERN E., SALOMON I., BOVY P. (eds), *Travel Behaviour: Spatial Patterns, Congestion and Modelling* , Edward Elgar Publishing, Cheltenham, pp. 143–179, 2002.

[CET 10] CET, *Relatório de Atividades Operacionais*, Companhia de Engenharia de Tráfego, São Paulo, 2010.

[DOW 04] DOWNS A., *Still Struck in Traffic. Coping with Peak-hour Traffic Congestion*, The Brookings Institution, Washington, D.C., 2004.

[CRO 04] CROZET Y., JOLY I., "Budgets temps de transport: les sociétés tertiaires confrontées à la gestion paradoxale du 'bien le plus rare'", *Les Cahiers Scientifiques du Transport*, vol. 45, pp. 27–48, 2004.

[DAT 10] DATAR A., *Why* mumbai faces traffic jams more frequently in more areas? And car centric policies are making it worse, Working Paper, Mumbai Environmental Social Network, Mumbai, 2010.

[DRI 01] DRIEA&STIF, Enquête Globale Transport, Direction régionale et interdépartementale de l'équipement et de l'aménagement d'Ile-de-France, Paris, 2001.

[DUP 11] DUPUY G., *Towards Sustainable Transport: The Challenge of Car Dependence*, John Libbey, Montrouge, 2011.

[FAI 08] FAIVRE D'ARCIER B., "Explorer les changements de comportements et d'attitudes face à l'automobilité", in CLOCHARD F., ROCCI A.S. (eds.), *Automobilités et altermobilités. Quels changements?*, L'Harmattan, Paris, pp. 133–145, 2008.

[HAB 13] HABIB K.N., SASIC A., WEIS C. *et al.*, "Investigating the nonlinear relationship between transportation system performance and daily activity–travel scheduling behaviour". *Transportation Research Part A,* vol. 49, pp. 342–357, 2013.

[JON 80] JONES P., "Experience with household activity-travel simulator (HATS), consumer perspectives", *Transportation Research Board*, vol. 765, pp. 6–12, 1980.

[KEN 14] KENT J., "Driving to save time or saving time to drive? The enduring appeal of the private car", *Transportation Research Part A*, vol. 65, pp. 103–115, 2014.

[LES 12] LESTEVEN G., Les stratégies d'adaptation à la congestion automobile dans les grandes métropoles: analyse à partir des cas de Paris, São Paulo et Mumbai, Thesis, Université Paris 1 Panthéon-Sorbonne, Paris, 2012.

[LES 11] LESTEVEN G., ALVES B., "Behavioral responses to traffic congestion. Findings from São Paulo and Paris", *Transportes*, vol. 19–2, pp. 42–48, 2011.

[LEV 14] LE VINE S., ADAMOU O., POLAK J., "Predicting new forms of activity/mobility patterns enabled by shared-mobility services through a needs-based stated-response method: Case study of grocery shopping", *Transport Policy*, vol. 32, pp. 60–68, 2014.

[MAS 10] MASSOT M.-H. (eds), *Mobilités et modes de vie métropolitains. Les intelligences du quotidien*, L'Œil d'Or, Paris, 2010.

[MAI 11] MAIRIE DE PARIS, Paris transport and travel 2011 report, Mairie de Paris, Paris, 2011.

[MET 07] METRÔ, Pesquisa Origem e Destino 2007 – Região Metropolitana de São Paulo, Metrô, São Paulo, 2007.

[MMR 08] MMRDA, LEA, Comprehensive transport study for Mumbai metropolitan region, Mumbai Metropolitan Region Development Authority, Mumbai, 2008.

[MOK 97] MOKHTARIAN P.L., RANEY E., SALOMON I., "Behavioral response to congestion: Identifying patterns and socio-economic differences in adoption", *Transport Policy*, vol. 4 no. 3, pp. 147–160, 1997.

[REI 08] REINHOLD T., KEARNEY A.T., "More passengers and reduced costs – the optimization of the Berlin public transport network", *Journal of Public Transportation,* vol. 11, no. 3, pp. 57–76, 2008.

[STE 02] STERN E., "Congestion in Europe: measurements, patterns and policies", in STERN E., SALOMON I., BOVY P. (eds), *Travel behaviour: Spatial Patterns, Congestion and Modelling*, Edward Elgar Publishing, Cheltenham, pp. 218–232, 2002.

[VOV 04] VOVSHA P., BRADLEY M., BOWMAN J., "Activity-based travel forecasting models in the United States: progress since 1995 and prospects for the future", *EIRASS Conference on Progress in Activity-Based Analysis*, Maastricht, 2004.

[WEI 10] WEIS C., DOBLER C., AXHAUSEN K., "Stated adaptation survey of activity scheduling reactions to changing travel conditions: field work and preliminary results", *12th WCTR Conference*, Lisbon, 2010.

[ZAH 74] ZAHAVI Y., Travel time budgets and mobility in urban areas, report, United States Department of Transportation, Washington D.C., 1974.

10

Empirical Analysis of Lane Changing Behavior at a Freeway Weaving Section

This chapter presents an empirical analysis on lane changing behavior using a trajectory data set collected at a weaving section in Grenoble (France). A detailed literature review shows that a strong empirical understanding of weaving mechanisms is still lacking. This chapter is devoted to filling this gap by investigating lane changing behavior at a microscopic level. We distinguished lane changes from the main road and those towards the main road. We conducted a descriptive empirical analysis of the weaving lane changes examining the positions of the lane changes and the accepted gaps. Our results show that under heavy congested traffic conditions weaving vehicles tend to change lane as soon as possible after the start of the weaving section. When the traffic conditions are fluid, the lane changing positions spreads more along the first part of the weaving section. Weaving vehicles coming from the main road tend to change lane earlier than weaving vehicles coming from the auxiliary lane. As weaving vehicles change lane at the beginning of the studied weaving section, our findings ask whether the length of the weaving section is a key variable to estimate its capacity. Our findings also raise some questions about the more relevant micro-simulation models to reproduce the operation of weaving sections.

10.1. Introduction

Congestion on freeways mainly occurs at discontinuities of the road network such as merges and weaving sections. Merges have been intensively investigated from both empirical and simulation points of view. The most recent version of the Highway Capacity Manual [HCM 10] defines weaving sections as discontinuities of the road network formed when merge segments are closely followed by diverge segments. Because of their geometrical configuration, a lot of lane changes occur at weaving sections. Those lane changes lead to a reduction of the discharge flow, even

---

Chapter written by Florian MARCZAK, Winnie DAAMEN and Christine BUISSON.

if the total demand is lower than the capacity of the weaving section. As for many traffic phenomena, traffic in weaving sections has been examined either through data analysis, through deriving analytical expressions that permit to estimate level of capacities or through simulation.

Empirical analyses at a macroscopic level have been conducted in Lee and Cassidy and Skabardonis and Kim [LEE 09, SKA 10] to identify bottleneck activation for two weaving sections. The authors used oblique cumulative vehicle counts from loop detectors placed inside the weaving zone at 400 m away (first site) or 500 m away (second site). From the oblique curves analysis, Lee and Cassidy [LEE 09] conclude that "bottleneck activations at both weaving sections were triggered by disruptive freeway to ramp lane changes". They also explain that the location of the lane changes from the freeway to the ramp along the weaving section play a role in the discharge flow. To complete this overview of empirical data analysis, we need to mention that [SAR 11] presents a very small sample of microscopic data (130 weaving maneuvres from ramp to freeway in total). Nevertheless, the authors conclude that "the surrounding freeway vehicles significantly affect the weaving vehicle acceleration behaviour".

The operation at weaving sections is also estimated through analytical procedures. Lertworawanich and Elefteriadou [LER 03] used a combination of analytical formulas and a model of gap acceptance and linear optimization to forecast the capacity of weaving sections. The proposed methodology expresses the capacity as a function of the demands, the ratios of weaving vehicles and the speeds of weaving and non-weaving vehicles. A simple analytical model for estimating the capacity of weaving sections is proposed in Rakha and Zhang [RAK 06]. The model includes three variables: the weaving section length, the weaving section volume ratio and the weaving ratio. This paper also demonstrated that the proposed model is consistent with field data. Roess and Ulerio [ROE 09a, ROE 09b] analyzed very short periods, but for a set of locations representing a rather large variety of infrastructure characteristics. Those two papers present the updated methodology, integrated in the current version of the HCM [HCM 10], to design weaving sections. The first paper describes the first step of the methodology dealing with the estimation of the number of lane changes from and towards the freeway and the average speed of weaving and non-weaving vehicles [ROE 09a]. The second paper aims at determining an analytical expression of a weaving segment capacity [ROE 09b]. It is worth mentioning here that the guidelines on weaving section differ from one country to another. For example, the Dutch guidelines for the design of weaving segments are compared with those of the U.S. [MIN 03]. The main difference is that the methodologies do not use the same variables to estimate the capacity. To summarize the guidelines for weaving section design and capacity estimation, one can say that they rely on the assumption that lane changing in weaving sections do have an impact on capacity but without a clear proof from empirical observations.

Simulation is widely used to estimate the capacity of weaving sections. The simulations can be conducted either at a microscopic level or at a macroscopic level. A review of previous studies and existing models about weaving sections is proposed in Shoraka and Che Puan [SHO 10]. Simulation studies of weaving sections use existing models, such as in Simone [CAL 12] or Vissim [ROE 08]. In this context, the capacity is expressed as the maximum rate of vehicles that are expected to traverse the weaving section. The capacity is a function of numerous factors such as the number of lanes, the length of the weaving section, the proportion of heavy vehicles, the length of deceleration and acceleration lane, the speed, etc. Existing models appear to not or hardly capture the specific traffic phenomena occurring at weaving sections. Therefore, some authors attempt to adapt existing models for the specific case of a weaving section. A generic continuous gas-kinetic traffic flow model is for example proposed in Ngoduy [NGO 06]. The lane changing probability is expressed as a function of the density, the speed, the weaving flow fraction and the vehicle compositions on the target lane. The proposed model is useful in supporting the geometry design for weaving sections by estimating their necessary length.

All previously mentioned papers agree on the conclusion that the fraction of lane changing vehicles, in relation to the total flow, strongly impacts the capacity of any weaving section. From that, design guidelines recommend long weaving sections. The underlying assumption is that, for a given proportion of lane changing vehicles, the longer the weaving section the higher the capacity, meaning that when the lane changes may be performed along a longer portion of road their impact is spread, lane changes may have separate and not a cumulative impact and the total capacity of the weaving section is higher.

However, the literature review presented above has illustrated that the longitudinal position of weaving lane changes is not sufficiently examined to confirm this assumption. The analysis of lane changing positions is also still lacking. In addition, the link between the road configuration and traffic flow characteristics and the longitudinal position, has not been investigated extensively. Therefore, this paper will examine the longitudinal position of lane changes using empirical observations of somewhat less than 1,000 weaving maneuvers. The data are trajectory data, collected through a helicopter.

This chapter is organized as follows. A description of the data collection site and technique is given in the first section. Then, the methodology and the definitions to analyze the data are presented. The next section contains the results of the analysis, while the paper ends with conclusions and recommendations for future research.

10.2. Data collection site and technique

The weaving study site is shown in Figure 10.1 [MOC 11]. It is situated at the junction between the urban freeways RN87 and A480 in the southeast part of Grenoble. We chose this weaving site because this junction is located in a dense urban network. The two traffic flows on this weaving section have different natures, competing for the same space:

– The medium distance traffic (few kilometers) corresponds to drivers using the RN87 and the A480 (either to the north – Exit 1– or the south direction – Exit 2) entering the weaving section in one of the two leftmost lanes and exiting via one of the two rightmost exists;

– The local traffic originating from the rightmost entry lane and/or exiting on the leftmost lanes.

The mix of traffic (local and medium distance), the total observable number of lane changes and the variety in lane changes at this weaving section allow us to have a combination of traffic situations reflecting various mechanisms of occurrence of congestion.

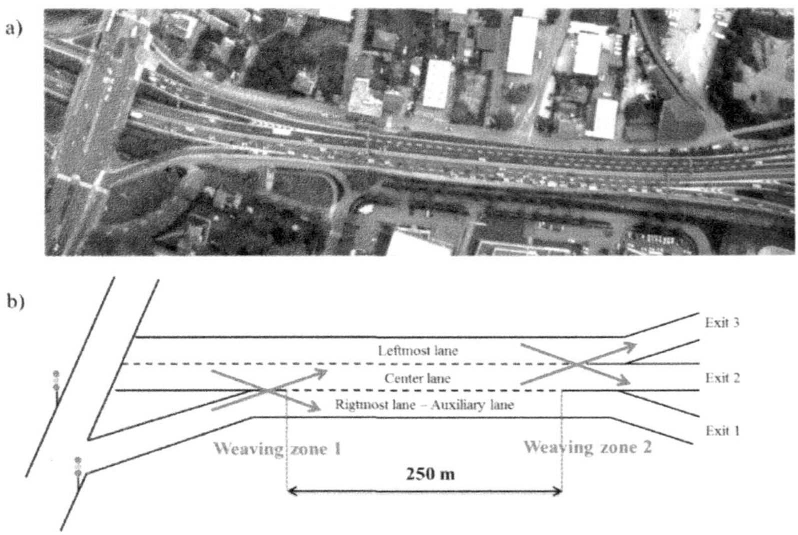

Figure 10.1. *a) Aerial view of the site; b) sketch of the site. For a color version of this figure, please see www.iste.co.uk/jacob/traffic.zip*

The weaving section consists of two lanes on the main road and one auxiliary lane. The maximum allowed speed on the main road is 70 km/h (20 m/s). The block marking between the main road and the auxiliary lane has a length of 250 m. The arrival pattern at the auxiliary lane is specific to the studied weaving section. A traffic signal, just upstream of the auxiliary lane, creates platoons of vehicles. Note that the infrastructure configuration is not simply a 2+1 $\rightarrow$ 2+1 weaving classical type A configuration, but more a double type A weaving section with two entries and three exits.

The video recordings have been collected using a high-resolution video camera mounted underneath a helicopter. For a further explanation of the procedure to stabilize the images and extract the trajectories, the authors redirect their readers to Knoppers *et al.* [KNO 12]. The weather conditions were dry and sunny.

In this chapter we consider two 1-hour samples respectively recorded on Thursday, September 15, 2011 and Friday, September 16, 2011. The traffic during these 2 hours is either in free-flow or in congestion with prevalence of congestion. The congestion forms downstream of the weaving section at Exit 1 or Exit 2 (see Figure 10.1). Then, the congestion moves upstream on both the main road and the auxiliary lane. There are also some periods for which the studied weaving section is an active bottleneck. The congestion forms at the start of the weaving section and moves upstream on the center lane and the auxiliary lane.

The traffic mainly consists of passenger cars. The quality of the data is, at the time of writing, not good enough to have a complete trajectory for every vehicle passing the weaving section. Therefore, we selected from the datasets, the weaving vehicles for which we have a complete trajectory, as well as the complete trajectories of their putative leader and follower on the target lane. This results in the following samples:

– 544 lane changing trajectories from the auxiliary lane to the main road and 995 lane changing trajectories from the main road to the auxiliary lane for Thursday, September 15, 2011;

– 450 lane changing trajectories from the auxiliary lane to the main road and 705 lane changing trajectories from the main road to the auxiliary lane for Friday, September 16, 2011.

10.3. Methodology and definitions

Figure 10.2 presents the conceptual framework describing the weaving behavior based on a previous work on merging behavior [MAR 13]. In this model, the input of the decision to change lane consists of the offered gaps, the weaving section configuration and the characteristics of the drivers. The output of the decision to change lane are the accepted and the rejected gaps. The offered gaps are a result of the traffic conditions on the weaving section, the characteristics of the vehicles composing the gap on the target lane (acceleration, speed, length of the vehicles) and the previous lane changing maneuvers. The interactions at weaving sections are very complex due to the involvement of weaving and non-weaving flows in the mandatory lane changing process. One of the objectives of this paper is to obtain insights of these interactions between weaving vehicles.

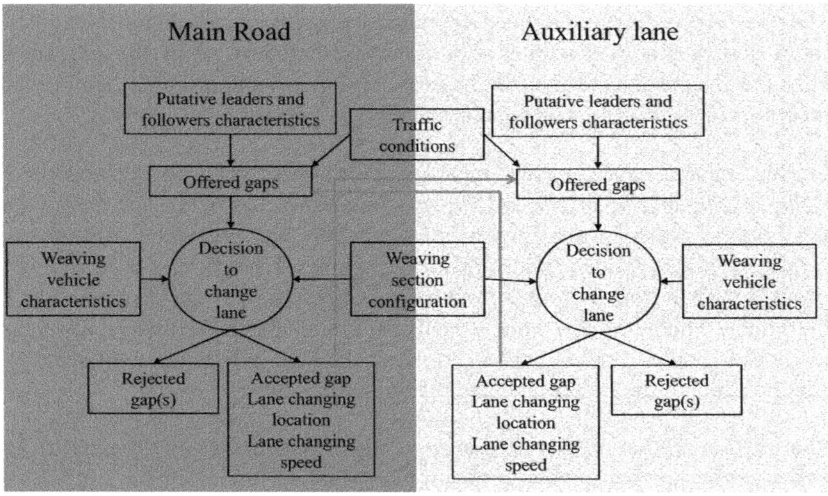

Figure 10.2. *Conceptual framework to study lane changing behavior at weaving section*

In the remaining part of this chapter, we focus the analyses on the results of the decision to change lane. We consider the lane changing positions, the lane changing speeds and the accepted gaps as key variables to explain the lane changing behavior.

In the two 1-hour video datasets considered in this work, the congestion mainly occurs because of the interaction between the lane changing vehicles coming from the center lane and those coming from the auxiliary lane (weaving zone 1 in Figure 10.1). The lane changing vehicles reduce their speed (1) to find a gap and change lane or (2) to let a vehicle coming from the adjacent lane perform its lane changing maneuver. As they are a part of the mechanism that causes the congestion, we distinguish two types of mandatory lane changes: (1) the lane changes from the rightmost lane on the main road to the auxiliary lane and (2) the lane changes from the auxiliary lane to the main road. We call "main road weaving vehicle" a vehicle coming from the main road and changing lane to reach the auxiliary lane. We define inversely a "ramp weaving vehicle" as a vehicle coming from the auxiliary lane and changing lane to reach the main road.

We define the lane changing moment as the moment that the centerline of the weaving vehicle crosses the block marking between the main road and the auxiliary lane. The accepted gap is the net distance between the putative leader and the putative follower at the lane changing moment. The speed of the weaving vehicle at this moment is defined as the lane changing speed. A weaving vehicle could not change lane at a high speed if the traffic conditions on the target lane are congested. The longitudinal position of the weaving vehicle at the lane changing moment is defined as the lane changing position. It is measured from the start of the block markings. One of the objectives of the paper is to assess whether there is an interaction between lane changes of the different weaving flows. To this aim we consider the lane changing positions. We assume that if there are interactions between weaving vehicles, a correlation should exist between the lane changing positions.

10.4. Results

We aim to determine whether a relation exists between the lane changing position of main road weaving vehicles and the lane changing position of ramp weaving vehicles. We therefore start the empirical analyses by considering the lane changing position. Figure 10.3 shows the cumulative distributions of the lane changing positions of both vehicle groups on both days.

As traffic conditions during observations were essentially congested, not many observations can be found when the speed is bounded by 15 and 20 m/s. Approximately 25% of the lane changes occur before the start of the block markings when the traffic conditions are very congested (speed below 5 m/s). The analysis of the video recordings shows that this situation occurs when the congestion comes from downstream (from Exit 1) and moves upstream on the center lane and the auxiliary lane (see Figure 10.1). When the traffic conditions are very saturated, the drivers, independently of their direction, tend to change lane as soon as possible to reach their target lane. When the speed is below 10 m/s, the lane change locations are spread out along the block marking. In this case, more than 95% of the lane changes occur in the first 150 m of the weaving section. Even then, the total length of the weaving section is not used.

The cumulative distributions for the lane changes of the main road weaving vehicles and the ramp weaving vehicles are very similar. It appears however that the lane changes towards the auxiliary lane occur, in general, earlier than the lane changes in the other direction when the speed becomes higher than 5 m/s. Because of the specific arrival pattern at the auxiliary lane, there are occasionally fewer vehicles on this lane. The vehicles coming from the main road towards the auxiliary lane can find larger gaps to change lane. Those observations are similar for both days.

To test whether or not two datasets are from the same distribution, one classical way is to use a Kolmogorov-Smirnov test to assess if the null hypothesis (the two datasets are from the same distribution) is accepted or not. We chose this method to evaluate first, if for each day the distributions of the lane changing positions for the two directions are similar, and second if the distributions for one direction inside a speed class are similar. At a 5% significance level, the null hypothesis is always rejected. One can thus conclude that there is, for a given day, a significant difference between the lane change locations for the two different directions. A second conclusion is that the distributions for one direction are also different from one day to another.

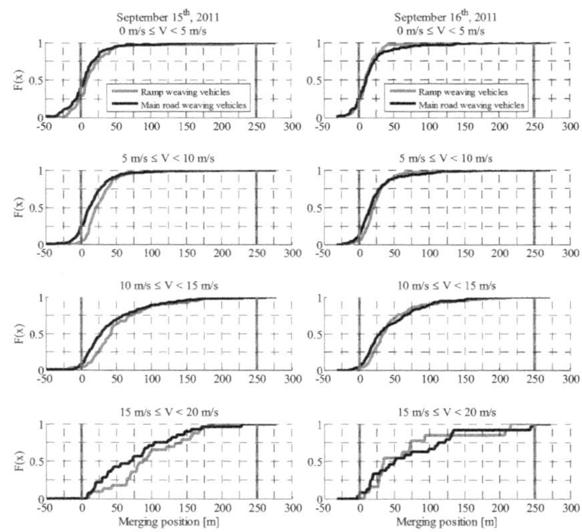

Figure 10.3. *Cumulative distributions for lane changes positions for different classes of speed. We consider the lane changes of main road weaving vehicles (black line) and the lane changes of ramp weaving vehicles (gray line)*

The next analyses deal with a cross-comparison of the lane changing positions and the speed to determine whether a relation between the lane changing positions, in relation to the prevailing traffic conditions, exists or not. Figure 10.4 shows the mean value of the lane changing positions from the main road as a function of the lane changing positions towards the main road. The data have been aggregated using the lane changing speed. The confidence intervals at the 5% significance level are also represented. The color of the dots gives the speed: the darker the dot, the lower the speed.

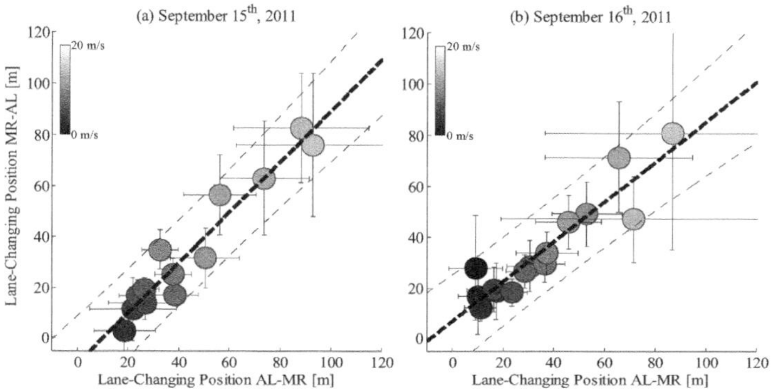

Figure 10.4. *Mean lane changing positions for main road weaving vehicles as a function of the mean lane changing positions for ramp weaving vehicles. The color of the dot gives an indication of the speed of the weaving vehicle at the time of the lane change: the darker the dot, the lower the speed*

Figure 10.4 shows that there seems to be a relation between the lane changing positions in both directions. The linear regression lines that fit the data for September 15 and September 16 are respectively:

$$x_{MR-AL} = 0.99x_{AL\text{-}MR} - 10.36 \quad (R^2 = 0.92; PCC = 0.96) \qquad [10.1]$$

$$x_{MR-AL} = 0.78x_{AL-MR} + 6.46 \; (R^2 = 0.86; \quad PCC = 0.93) \qquad [10.2]$$

$x_{\text{MR-AL}}$ (respectively $x_{\text{AL-MR}}$) is the lane changing position for the main road weaving vehicles (respectively the ramp weaving vehicles). The coefficient R^2 equals the square of the Pearson correlation coefficient (PCC) that is a measure of the linear correlation between two variables. The PCC is higher than 0.51, the threshold at a 5% significance level for 13 degrees of freedom (15 observations – 2). Therefore, we can conclude that there is indeed a linear correlation between the lane changes from and towards the main road for the prevailing traffic conditions.

Previously, the data have been aggregated using the speed. This gives an insight into the relation between the lane changes from and towards the main road for the prevailing traffic conditions. However, it does not illustrate a temporal relation between the lane changes. We assume that the lane changes at a weaving section are the results of a collaboration/competition between the weaving vehicles. If this assumption is true, the lane changes for a given time period should occur at the same location.

The confidence intervals for each speed class overlap. Therefore, Figure 10.5 shows that the mean lane changing position for the lane changes from the main road is not significantly different from the mean lane changing position towards the main road. Figure 10.5 confirms that, for a given time period, the lane changes occur at the same location. Figure 10.5 also shows that the temporal mean lane changing position is constant with the time.

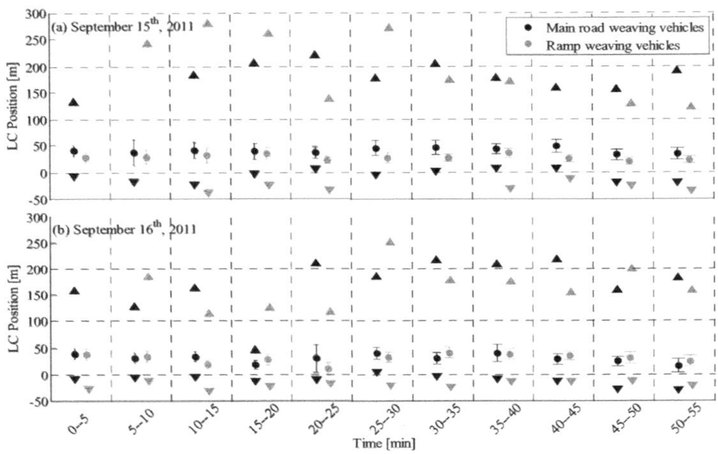

Figure 10.5. *Temporal aggregation of the lane changing positions: the mean temporal lane changing positions and their corresponding confidence intervals at à 5% significance level are represented. The upward-pointing triangles represent the maximum observed lane changing position. The downward-pointing triangles represent the minimum observed lane changing position*

The empirical analyses continue with an analysis of the accepted gaps, see Figure 10.6. Hereafter, we focus on the accepted gaps. Our findings show that the weaving vehicles change lane at the start of the weaving section. The choice process seems very simple: the drivers take the immediate neighboring gap. Figure 10.6 also shows that the ramp weaving vehicles accept smaller gaps than the main road weaving vehicles. Only the largest accepted gaps are similar for both types of lane changes. The pattern is similar for both days. The specific arrival pattern on the auxiliary lane could explain those observations. The main road weaving vehicles accept larger gaps because temporarily fewer vehicles are driving on the auxiliary lane when the traffic light is red. So basically, the difference is not in their choice process, but in the offered gaps, which are simply larger (on average).

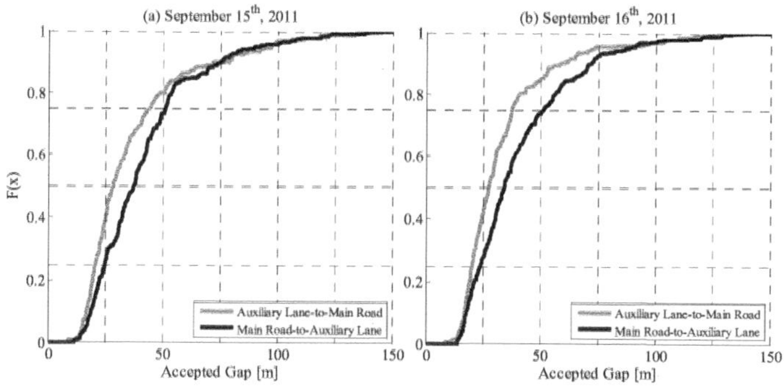

Figure 10.6. *Cumulative distribution for accepted gaps*

10.5. Discussion and conclusion

A descriptive analysis of the lane changing behavior using trajectory data measured at a weaving section is presented in this paper. This work is the first part of a larger research project about weaving sections. The aim of this analysis is twofold. We analyze the trajectories at a microscopic level to better understand the mechanisms that occur at a weaving section (especially the lane changes). The empirical analysis will be also useful to develop/improve and validate accurate models to estimate the capacity of weaving section. We analyzed the longitudinal lane changing positions and the accepted gaps to determine whether a relation exists between the lane changes from and towards the main road.

The trajectories have been collected using a high-resolution camera mounted underneath a helicopter. The studied weaving section is located in Grenoble (France). We do not have sufficiently accurate data to be able to analyze the lane changing activity in the whole weaving section. In particular, we are not able to study the acceleration of the weaving vehicles. However, we can give first insights into the lane changing behavior at a weaving section.

The main findings of this chapter are the following:

– the analysis of the longitudinal position shows that the lane changes occur in the first part of the weaving section, even before the start of the block line, when the congestion is heavy;

– when the traffic conditions are very congested the lane changes from and towards the main road occur at the same location near the beginning of the weaving section;

– with increasing speed, the lane changing positions are more spread out along the block line between the main road and the auxiliary lane. Under free-flow conditions, more than 95% of the lane changes occur in the first 150 m of the weaving section. The first 60% of the total length of the weaving section are used to change lane;

– with increasing speed, the weaving vehicles coming from the main road change lane earlier than the weaving vehicles coming from the auxiliary lane;

– there is a linear correlation between the lane changes from and towards the main road. One can therefore reasonably assume that the weaving vehicles interact in a competition/collaboration behavior especially when the traffic conditions deteriorate;

– the vehicles changing lane from the auxiliary lane to the main road accept smaller gaps and headways than the vehicles changing lane from the main road to the auxiliary lane.

Previous studies using micro-simulation to estimate the capacity of a weaving section pinpointed that the length of the weaving section is a key factor influencing its capacity. However, our empirical study shows that only 60% of the total length of the weaving section is used to change lane. This length reduces in case congestion occurs.

Recent studies considered the gap acceptance theory to model the lane changes at a weaving section. Our empirical observations raise the question whether the gap acceptance theory is the most appropriate theory to reproduce the lane changing behavior at a weaving section. In our view the lane changes at a weaving section are more the results of a competition/collaboration between the weaving vehicles. This will be subject for future research.

10.6. Bibliography

[CAL 12] CALVERT S.C., MINDERHOUD M.M., "Generic freeway weaving section capacity estimation trough microsimulation", *TRB 91th Annual Meeting Compendium of Papers*, Washington D.C., 2012.

[HIG 10] HIGHWAY CAPACITY MANUAL, Transportation Research Board, National Research Council, Washington D.C., 2010.

[KNO 12] KNOPPERS P., VAN LINT H., HOOGENDOORN S.P., "Automatic stabilization of aerial traffic images", *TRB 91th Annual Meeting Compendium of Papers*, Washington D.C., p. 13, 2012.

[LEE 09] LEE J.H., CASSIDY M.J., An empirical and theoretical study of freeway weave bottlenecks, California PATH Research Report, University of California, Berkeley, p. 70, 2009.

[LER 03] LERTWORAWANICH P., ELEFTERIADOU L., "A methodology for estimating capacity at ramp weaves based on gap acceptance and linear optimization", *Transportation Research Part B: Methodological*, vol. 37, no. 5, pp. 459–483, 2003.

[MAR 13] MARCZAK F., DAAMEN W., BUISSON C., "Merging behaviour: empirical comparison between two sites and new theory development", *Transportation Research Part C: Emerging Technologies*, 2013.

[MIN 03] MINDERHOUD M.M., ELEFTERIADOU L., "Freeway weaving: comparison of Highway Capacity Manual 2000 and Dutch guidelines", *Transportation Research Record: Journal of the Transportation Research Board*, Washington D.C., pp. 10–18, 2003.

[MOC 11] MOCOPO., Measuring and modelling traffic congestion and pollution", available at: http://mocopo.ifsttar.fr/, 2011.

[NGO 06] NGODUY N., "Derivation of continuum traffic model for weaving sections on freeways", *Transportmetrica*, vol. 2, no. 3, pp. 199–222, 2006.

[RAK 06] RAKHA H., ZHANG Y., "Analytical procedures for estimating capacity of freeway weaving, merge and diverge section", *Journal of Transportation Engineering*, pp. 618–628, 2006.

[ROE 08] ROESS R.P., Analysis of freeway weaving sections, NCHRP Final Report, Transportation Research Institute, Polytechnic University, Brooklyn, N.Y., 2008.

[ROE 09a] ROESS R.P., ULERIO J.M., "Level of service analysis of freeway weaving segments", *Transportation Research Record: Journal of the Transportation Research Board*, Washington D.C., pp. 25–33, 2009.

[ROE 09b] ROESS R.P., ULERIO J.M., "Capacity of freeway weaving segments", *Transportation Research Record: Journal of the Transportation Research Board*, Washington D.C., pp. 34–41, 2009.

[SAR 11] SARVI M., ZAVABETI A., EJTEMAI O., "Modelling freeway weaving manoeuvre", *Proceedings of 34th Australasian Transport Research Forum (ATRF)*, Adelaide, South Australia, Australia, p. 16, 2011.

[SHO 10] SHORAKA M., CHE PUAN O., "Review of evaluating existing capacity of weaving segments", *International Journal of Civil and Structural Engineering*, vol. 1, no. 3, pp. 683–694, 2010.

[SKA 10] SKABARDONIS A., KIM A.M., Weaving analysis, evaluation and refinement, California PATH Research Report, University of California, Berkeley, p. 70, 2010.

11

Applying and Testing a New Generation Traffic Management with Multi-objectives

Policy makers face difficult challenges within urban areas with respect to increasing traffic and increasing congestion, related air quality problems and noise pollution on street level. This chapter provides an innovative solution to tackle the challenge by multi-objective traffic management and describes the differences in decision-making on a traffic management level when the different policy goals are taken into account. The considerations between the different policy goals can be put into a multi-criteria instrument and algorithm to allow for a flexible while consistent means of managing traffic within urban areas.

This chapter shows two case applications in the cities of The Hague and Helmond which provide an example of the extent of the potential that multi-criteria traffic management can offer. The key performance indices are also computed to showcase the benefit and potential of the new generation of traffic management. The combination of the case applications together with different possibilities to realize multi-criteria traffic management will show the flexibility of this new generation traffic management. It also indicates that the steps are location specific to be able to solve the local problems.

11.1. Introduction

There exist many challenges for policymakers with respect to increasing traffic and increasing congestion, related air quality problems and noise pollution on street level. A wide range of measures is available to tackle these problems: for example setting up environmental zones, electrification of the vehicle fleet or investing in roadside equipment to inform road users. But the question remains if this is sufficient to reach the multiple targets from policy goals and more importanly whether these measures are robust for the future. One key area of measures is

---

Chapter written by Martijn DE KIEVIT and Yusen CHEN.

traffic management for which we try to provide operational tools for traffic operators to realize these goals. The focus is on air quality from pollutants like NO_x and PM_{10} for which a fining system will come into place in the coming years putting extra pressure on the effectiveness of policy measures that will in turn also become robust for the future.

This chapter provides an innovative solution to tackle the above challenges on multi-objective traffic management and describes the differences in decision-making on a traffic management level when different policy goals are taken into account. The considerations between the different policy goals can be put into a multi-criteria instrument and algorithm to allow for a flexible while consistent means of managing traffic within urban areas.

Furthermore there are five major transitions that are taking place in the field of traffic management that influence the position of the road operator in the field of traffic management and for which multi-criteria traffic management provides new opportunities. These transitions are:

– the shift in focus from a single stretch of road to network-wide management;

– next steps focus on automation of local traffic control operators;

– a connection with in-car developments is essential for more effective traffic management;

– the step from pure public traffic management to a combined public and private partnership; and

– the paradigm shift for which other objectives than pure throughput or efficiency are becoming more and more important.

The combination of these five major transitions are currently taking place in the Netherlands where in a group of both road operators and private companies the next steps for traffic management are formulated. Part of the work being done on the policy level needs to be translated into realization projects for which proof is still necessary, but also the concepts that are developed need to be robust enough to be able to deal with the abovementioned transitions.

This chapter starts by describing the concept of multi-criteria traffic management, followed by a literature review of relevant research focusing on both the multi-criteria aspect and the simulation aspects that are necessary for implementation. This is followed by a description of two case studies performed in which the concept of multi-criteria traffic management is applied. These concern the cities of The Hague and Helmond in which the focus has been slightly altered to understand different aspects of multi-objective traffic management. The impacts are shown of the solution that is provided in these cases. This chapter concludes by providing a set of conclusions but more importantly next steps that need to be taken in order to create the potential that is shown in simulation in real life.

11.2. Definitions

First, we start with the definition of Dynamic Traffic Management (DTM), which focuses on ways of combating rising congestion and the environmental problems associated with it. DTM is a way of making better use of the existing road network capacity and managing traffic flows. With DTM the traffic flows are informed, regulated and facilitated on the basis of up-to-date traffic information. This makes use of resources that are dependent on the situation, like route information panels, traffic merging filtering, traffic lights and rush-hour lanes. The core in DTM is the control and routing strategy.

Multi-criteria Dynamic Traffic Management (mcDTM) is a solution to pro-active traffic management, involving multiple departments (traffic and environment) at a public agency. Its technological core consists of traffic control and consistent traffic information by estimation and prediction, providing an expandable and module framework to future-proven technology and to open data technology platforms.

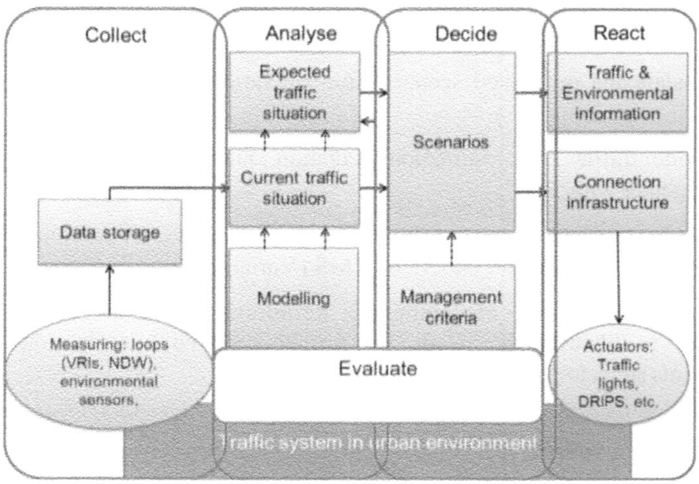

Figure 11.1. *The concept of multi-criteria management explained based on the case study in The Hague*

mcDTM is the next step in traffic management, making it:

– multi-criteria (for decision support of green and sustainable traffic management, this includes prediction of traffic flows and environmental issues), enabling multiple organizations or departments to work efficiently and together;

– real-time and predictive (based on the fusion of different data sources to allow for better decision-making) with historical profile and real-time data, enabling pro-active intervention;

– public private partnership (setting targets for service providers in cooperation with the Traffic Management Center (TMC));

– user-centric: providing consistent traffic information based on multi-criteria, to both roadside and in- car devices, as well as control equipment;

– in-car: providing information in vehicles (or better to travellers) to increase the impact and understanding of traffic management measures.

In other words, this means that multi-criteria traffic management (or green traffic management) can:

1) Formulate a framework for consistent consideration of alternative criteria such as congestion/throughput, environment and safety.

2) Provide an optimization for a given control strategy, regarding considered criteria.

3) Translate the optimized scenario into operational traffic control setting and guidance.

4) Provide traffic state estimation/prediction, allowing pro-active intervention, working together with TMC.

5) Provide consistent travel information, to travellers, in-car and by DRIP's (Dynamic Route Information Panel), or VMS/(Variable Message Signs).

Having set this target, explaining the definition a literature review has been carried out to identify the needs for further research but also the necessary developments steps to realize implementation.

11.3. Literature review

Concerns over any environmental impacts and efforts to include it in traffic modeling led to early research in both traffic assignment methods and in multi-criteria. In the early 1990s, static models were modified and applied to accommodate such efforts. In recent years, dynamic models as well as traveler's information are taken into account. This section will review a few most relevant papers related to the topic.

Tzeng and Chen [TZE 93] developed a framework for multi-objective traffic assignment, including travel time, air pollution and travel distance. A Pareto set of solutions is formulated and the Lagrangian is constructed of equivalent minimization problems with these weighted objectives. The highlighted objective with large weight will direct the assignment towards the improvement of the objective. Then a Frank-Wolfe algorithm is used to obtain the set of weight that is closest to the optimal solution of an objective.

Wismans *et al.* [WIS 10] used a Pareto set to derive solution. The objective function consists of (1) congestion: minimization of total travel time; (2) traffic safety: minimization of total number of injuries, by exposure and risk per road type; (3) climate: CO_2 calculated with average speed; (4) air quality: using CAR model; (5) noise: using the Dutch National regulations for calculations model. The optimization is carried out with a bi-level approach. At an upper level: Network Design Problem (NDP), with the multiple objectives (measures) as described above. At a lower level: Dynamic Traffic Assignment (DTA), reaction to these measures. Solution methods include Generic algorithm (GA), simulated annealing and grid search. Solution dominance is found if one alternative scores better than the other. It is difficult to select a better solution with multiple targets. The solution method does not seem to be suitable for a large network, given its implementation with GA.

Nagurney [NAG 00] showed the emission paradoxes by illustrating examples on a Braess network: (1) addition of a road may result in an increase in total emission with no change in travel demand; (2) total emission may increase with a decrease in total travel demand. She concludes that the network topology, cost structure and the travel demand structure are key aspects to consider in making policy towards emission improvement. In her examples, emission is output and is not a component in cost function.

In a recent paper, Nagurney and Dong [NAG 02] developed an equilibrium model for multi-class and multi-criteria including travel time, travel cost and emission. They proved that a solution exists and is unique, provided the cost function is monotone. They apply this framework to bi-criteria (cost and emission) and lead to the same conclusion that a unique solution exists as long as the generalized cost satisfies the monotonicity condition.

All these three models are analytical and try to provide a proof of solution existence, given the cost monotonicity.

Gaker *et al.* [GAK 10] investigated how to predict and influence travelers' behavior regarding personalized information and social influence. Different surveys and experiments have been carried out. It is found that there is much from behavioral economics to be able to transfer and to have great potential for influencing transport behavior. Also person and trip-specific information regarding greenhouse gas emission has significant potential for increasing sustainable behavior. Based on a number of surveys and experiments, it is discovered that social influence positively impacts a decision in, for example, purchasing a hybrid car as well as in route choice. This research and publication is among the first of the kind to indicate that people (travelers) show sustainable behavior when provided with context- and person-specific information on the environmental impact of their actions. What is even interesting is that a Value of Green is of $0.50 per pound (of green gas).

Gao [GAO 12] established a dynamic traffic assignment model where users make strategic route choices in response to real-time information in a stochastic time-dependent network with correlated link travel times. A routing policy is defined as a decision rule which specifies what node to take next out of the current node based on the current time and real-time information, essentially a mapping from network states to decisions on next nodes. A routing policy can manifest itself as different paths depending on the underlying stochastic process that drives a traffic network. A path is purely topological and is a special case of a routing policy where any decision on the next node is not dependent on the current time or real-time information. This analytical and mathematic DTA method needs large number of computing time for large scale network. Simulation method is adopted for this reason.

Chen *et al.* [CHE 13] carried out revealed and stated preference surveys to find out what relation exists between a traveler's profile and ecological routing. User interviews and online surveys with 650 interviewees show that results are comparable to Gaker's. Travelers would be able to choose for a more environmental friendly route when available alternatives are comparable in travel time and out-of-pocket costs.

The introduction of traveler's perception of environmental concern into route choice implicates also a new dimension in modeling traffic performance. As vehicular emission can only be computed after a journey is finished, the introduction of emission into the cost function needs a traffic model or simulation to take into account this delayed feedback and thus necessary convergence. This feedback is not taken into account in any of the above-reviewed models, meaning that each reviewed model computes all criteria at the same time. Further questions arise as to whether conventional traffic laws would still hold when introducing a delayed feedback? We will investigate these properties in the following sections.

11.4. Methodology

To be able to calculate the key performance indices (to showcase the benefit and potential of the new generation traffic management) two approaches could be used, as described in Chen *et al.* [CHE 13a]. As described the analytical approach by both WIS [WIS 10] and Gao [GAO 12] have been unsuccessful since a simultaneous optimization view of a real network is still impossible. Therefore, a different approach was used which will be further elaborated in the case studies. The combination of the case applications together with different possibilities to realize multi-criteria traffic management shows the flexibility of the new generation traffic management. It also indicates that the steps are location specific to be able to solve the local problems.

The steps that the concept of mcDTM takes are explained by means of the example of CO_2 emissions (any other factor could be added in this concept):

– construct a composite cost function, including multi-criteria such as travel time, cost, and emission.

The weight to each criterion is comparable to [GAK 10] and [CHE 13]. The cost is Value of Time and Value of Green to travel time and CO_2 emission:

$$COST = VoT\ TravelTime + VOG\ CO_2\ Emission$$

– perform a dynamic simulation with this composite cost (equilibrium assignment) and obtain relevant output for control objectives, such as throughput, cost and emission;

– check the convergence of each objective as well as a combined and weighted value of all objectives. If convergence continues, iterate from the previous step;

– update the network criteria based on newly performed simulation and compute an averaging composite cost of initial one and newly-updated one;

– compare this new cost with input one and exit the iteration if the difference falls within the convergence;

– iterate from the first step.

These steps have been elaborately described in [CHE 13b].

11.5. Application cases and results

The two case applications can be best described as follows:

– Den Haag DVM: a dynamic traffic simulation model is built and contains traffic control and traffic information panels (DRIPS). The model is calibrated using traffic controllers data (flow and timing plans) and is applied to evaluate impact of traffic management measures (control and DRIPS);

– EcoMove: a real-time dynamic simulation that is designed to evaluate impact of vehicle guidance on energy saving. A modeling suite containing several modules is developed, allowing to fetch both offline and online traffic data, and providing traffic prediction for a given horizon (30 min to hours ahead). An ideal traffic situation could be generated to provide to TMC an optimal scenario for multi-criteria routine and traffic management.

11.5.1. *Simulation*

The tests consist of implementing the methodology as presented above and checking the algorithm. To enable this, the following elements have been prepared:

1) a calibrated simulation model that outputs traffic indicators as well as emission;

2) the possibility of updating routing cost by feeding the output objective values back to the cost component;

3) evaluation at whole network, polygon/sub-area and individual route levels.

11.5.1.1. *Simulation setup*

Two calibrated mesoscopic simulation models in DYNASMART have been used. One is a town-size (Helmond, 78 zones/31000 vehicles) and the other is a large city-size (The Hague, 340 zones/70000 vehicles). These different cases were selected to see if the size of a network would influence results and thus the findings. Each model has 2 hours of OD matrix, at the time-slice of 10 min. Each model contains junction control as well as VMS as in operation now. Emission calculation has also been implemented with Versit+ [SMI 06] that is based on individual vehicle trajectory. Both models are calibrated based on loop detector data at junction using an OD estimator [CHE 13].

The design of simulation runs consists then mainly of sensitivity of relative weight of each components of the composite cost, i.e. travel time and emission, and feedback of control objective values to the composite cost. This leads to the following setting:

1) VoT (value of time, Euro/hour): 0.2. 2.0, 5.0, 20 and 50;

2) cost update with either fully new cost or moving average;

3) evaluation based on whole network, polygon/sub-area and individual routes, with objective values such as traffic throughput, travel time and emission.

11.5.1.2. *Routing strategies*

To test the routing strategies two alternative routes were designed, with possibility to change both junction control settings and VMS routing message, to cope with the intended strategy. The base situation is compared with the multi-criteria strategy to see whether and how travelers would react to traffic control and routing information.

For both use cases a calibrated version of the mesoscopic model DYNASMART was used to identify the potential. For each of the cities also the junction control and current availability of VMS was implemented. For each case study the above mentioned VoT has been iterated to investigate for the defined key performance indicators (sensitivity VoT, network performance, network bearing capacity and optimal routing results) the impacts.

11.5.2. *Results and analysis*

Following the above settings, experiments are carried out with a series of VoT, with the dynamic simulation models and with equilibrium assignment.

11.5.2.1. *Sensitivity of VoT*

Average travel (per vehicle) time, emission and trip distance are taken as KPI (key performance index) and they are plotted per VoT in successive iterations, depicted in Figure 11.2. Iteration and cost updating are explained in the following paragraph.

Lower (<2.0) and higher (>50.0) VoT indicates that the KPI show much variation and do not lead to a sign of convergence. A VoT around 10 to 20 shows more stable trends and leads to convergence.

The convergence in this case means:

1) the total output indicators (flow, emission and travel time) are stabilizing during consecutive iterations;

2) input cost components are matching output cost components within the same iteration.

For the latter, this means that the original amount of emission input would need to be the same as the output emission.

Travel cost updates during iterations have been designed as follows:

1) initial cost contains only travel time as emission is not available at this moment yet;

2) the output to this iteration contains now emission too;

3) this composite cost is taken and generates new travel and emission;

4) the average previous input cost and output cost is averaged and taken as next input;

5) repeat until both convergences in output objectives and input costs has been reached.

Key results from one of the scenarios are depicted below.

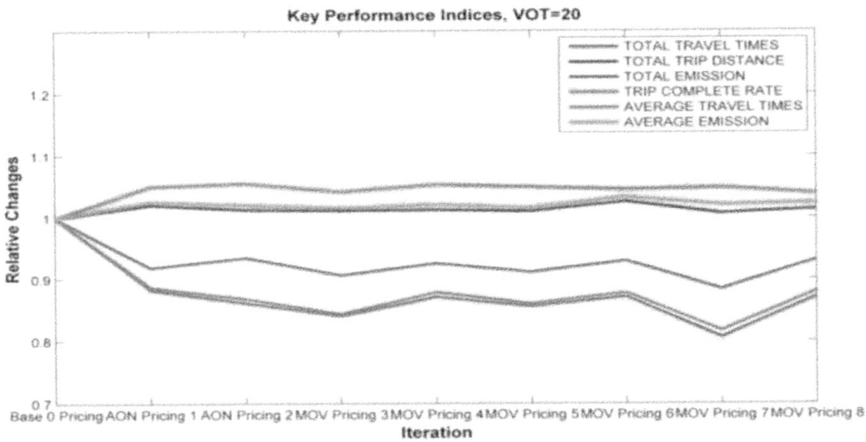

Figure 11.2. *Key performance indices with value of time for a color version of the figure, see www.iste.co.uk/jacob/traffic.zip*

As can be seen in Figure 11.2, the key performance indicators decrease compared to the base scenario in which no multi-criteria traffic management is performed.

11.5.2.2. *Optimal routing results*

Another aspect to be examined is the impact of route choice for two alternative routes. For this, the simulation with a VoT equal 10 is taken and the green time for one alternative is increased (route 2 in yellow, keeping the competing one unchanged), depicted in Figure 11.3. The KPI (travel time and emission) are reduced, given an increased green time for this route. In the Helmond case, similar results can be found:

– increase the green time for one alternative (keeping the competing one unchanged);

– reduce the KPI such as travel time and emission, given an increased green time for this route.

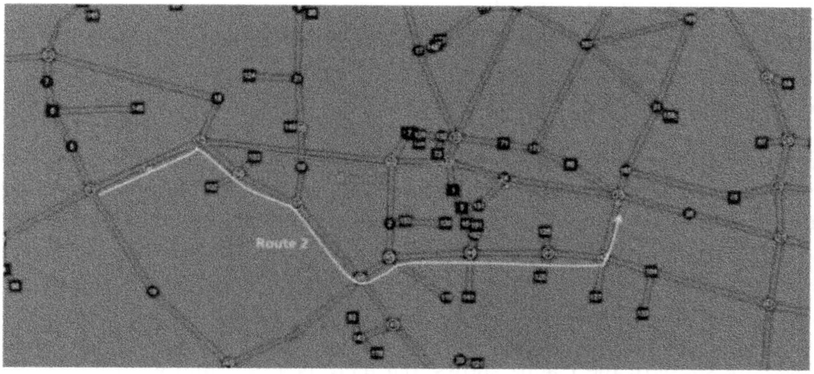

Figure 11.3. *A picture of the Helmond alternative routes*

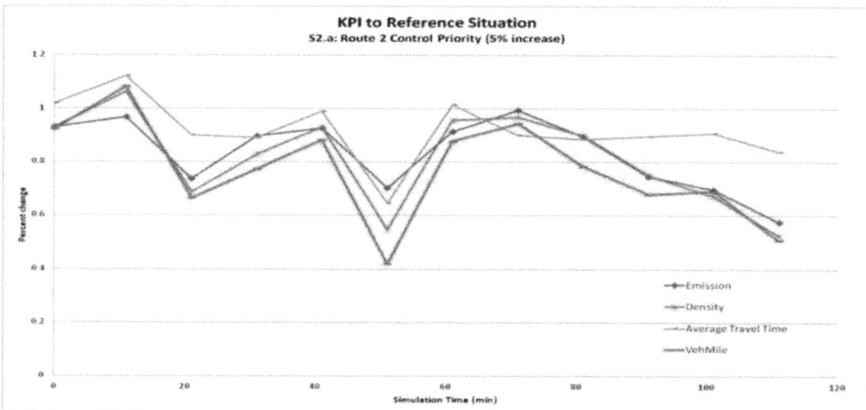

Figure 11.4. *Route choice for a given increase of green time*

Figure 11.4 shows the longer the simulation runs, the more convergence is realized for the different objectives without impacting travel time or vehicle mileage in a negative way. The question still remains if there are other downturn effects in the network that need to be taken into account.

11.6. Concluding remarks

We can see the positive impact of multi-criteria on whole network performance as well as route choices on alternative routes. The introduction of emission into a multi-criteria composite cost function and its application on a Braess network break the paradox and improve the network performance with reduced travel time and emission provided the experienced emission is fed into the cost function as feedback. A careful design of multi-criteria cost can help improve network routing efficiency towards either UE or SO.

Extending these experiments into a real urban network confirms the above findings and the total network could perform towards SO with a well-tuned multi-criteria including both travel time and emission. Further research is ongoing. We want to understand further and explain theoretically how multi-criteria routing could effectively change the network traffic performance and how we could operate this, in a real traffic control center.

The case study of the Hague shows that there is a potential when implementing multi-criteria traffic management, this however has only been based on simulation. The next step is to check and verify what the possibilities are to demonstrate the feasibility in order to be able verify the simulated potential. One step could be to test and verify current scenarios that are implemented in the TMC to tackle incidents, and based on the used data to identify potential improvements. Another potential step is the collection of real-time data in order to be able to predict the status of the network in advance to be able to anticipate the expected traffic situation. Another interesting option towards implementation is to combine and extent (like in The Hague) the sensor network to allow for better monitoring of different objectives in real-time.

Another aspect is to focus on potential elements that could be added in order to take expand the potential of multi-criteria dynamic traffic management. Objectives such as robustness of the network or the reliability of the travel time through the network are first aspects that come to mind. In turn a focus could also be put on specific groups of users, like trucks, public transport or emergency vehicles and how these can be served best using the possibilities mcDTM has to offer. All in all many possibilities arise when looking forward, but it is important to take a first step towards implementation to be able to realize the next ones.

11.7. Acknowledgments

The authors would like to thank the EU FP7 Framework Research Programme (E-Wisetrip and EcoMOVE), and the city of The Hague, for financing our research projects. The authors acknowledge the contributions of the team members.

11.8. Bibliography

[BAR 08] BARTH M., BORIBOONSOMSIN K., "Real-world carbon dioxide impacts of traffic congestion. Transportation research record", *Journal of the Transportation Research Board*, vol. 2058, pp. 163–171, 2008.

[BOY 04] BOYCE D., XIONG Q., "User-optimal and system – optimal route choices for a large road network", *Review of Network Economics*, vol. 3, no. 4, 2004.

[CHE 13a] CHEN Y.S., JONKERS J., VONK NOORDEGRAAF D., "Understanding and estimating travelers' choices towards international multimodal journey planning", *92nd Annual Meeting of the TRB*, Washington D.C., 2013.

[CHE 13b] CHEN Y.S., REMODE Dynamic OD Estimation Manual, Software Manual, 1993–2013.

[CHE 13c] CHEN Y.S., ZHANG F., "Empirical findings with multi-criteria routing for dynamic traffic management", *16th International IEEE Conference or Intelligent Transport Systems*, The Hague, 2013.

[FRE 08] FREY H., ROUPHAIL C., NAGUI M. *et al.*, "Link-based emission factors for heavy-duty diesel trucks based on real-world data", *Transportation Research Board*, vol. 2058, pp. 23–32, 2008.

[GAK 10] GAKER D., ZHENG Y., WALKER J., "Experimental economic in transportation: a focus on social influence and the provision of information", *Transportation Research Board*, vol. 2156 2010.

[GAO 12] GAO S., "Modeling strategic route choice and real-time information impacts in stochastic and time-dependent networks", *Intelligent Transportation System IEEE Transactions*, vol. 13, no. 3, pp. 1298–1311, 2012.

[MAH 06] MAHMASSANI HANI S., HAYSSAM SBAYTI, Dynasmart-P User Manual, 2006.

[NAG 00] NAGURNEY A., "Congested urban transportation networks and emission paradoxes", *Transportation Research Part D: Transport and Environment*, vol. 5, no. 2, pp. 145–151, 2000.

[NAG 02] NAGURNEY A., DONG J., "Traffic network equilibrium and the environment: a multicriteria decision-making perspective", *Computational Methods in Decision-Making Economics and Finance*, vol. 5, pp. 501–523, 2002.

[SMI 06] SMIT R., SMOKERS R., SCHOEN E. *et al.*, A new modeling approach for road traffic emissions: Versit+ LD – Background and Methodology, TNO Report 06.OR.PT.016.1/RS, 2006.

[TZE 93] TZENG G.H., CHEN C.H., "Multiobjective decision making for traffic assignment", *IEEE Transactions on Engineering Management*, vol. 40, no. 2, 1993.

[VEN 09] VENDRIK E., A dynamic traffic assignment model with social cost, MSc Thesis, Delft University of Technology, 2009.

[WIS 10] WISMANS LUC J.J., ERIC C. VAN BERKUM, MICHIEL C.J. BLIEMER, "Optimization of externalities using DTM measures: a Pareto optimal multi objective optimization using the evolutionary algorithm SPEA2+", *11th TRAIL Congress*, Netherland, 2010.

12

ON-TIME: A Framework for Integrated Railway Network Operation Management

Railway traffic management today is mainly performed by human controllers according to national rules. Traffic Control Systems which are used to manage data coming from the field are expensive and have proprietary interfaces. Under these circumstances, managing cross-border traffic and dealing with future increases in capacity demand represent difficult challenges for railway operators.

In the ON-TIME (Optimal Networks for Train Integration Management across Europe) project funded under the 7th Framework Programme of the European Union, railway infrastructure managers, railway technology developing industry and academic institutions strive to regard the process of railway traffic management in a holistic way in order to meet these challenges.

This chapter presents the overall approach of the project and presents the approaches on real-time perturbation management and real-time train speed regulation in detail.

12.1. Introduction

12.1.1. *Motivation and fundamental principles*

The European railway network will face growing traffic demand in the next years, encouraged by the European Commission's will to shift important parts of freight and passenger traffic to this eco-friendly transport mode [EUR 11]. Existing bottlenecks in the network will impact more negatively on traffic quality and new capacity bottlenecks are expected to occur and persist. At the same time, expectations of railway customers (passengers and freight) are growing as well, as they expect shorter

Chapter written by Thomas ALBRECHT and Meena DASIGI.

travel times and greater system reliability. These challenges can only be met using innovative traffic management solutions at the short and medium term.

For many years, algorithms for real-time conflict detection and resolution have been described in the scientific literature, see Corman and Meng [COR 13] for a review. But only recently have these algorithms been able to solve problem instances of practical relevance in real-time. They are however still not applied in daily operation in large railway networks for two main reasons:

1) The benefit of the algorithms is difficult to predict;

2) Operational traffic control systems (TCSs) as currently in operation at the nationally acting railway infrastructure managers are not easy to extend.

The main goal of the ON-TIME (Optimal Networks for Train Integration Management across Europe) project partially funded by the European Commission is an increase of the capacity in the railway networks through an integrated approach to railway operation optimization [ROB 13]. The project is driven by the railway infrastructure managers (IM) from the United Kingdom (NetworkRail), Italy (RFI), Germany (DB Netz), France (RFF) and Sweden (Trafikverket), supported by 14 partners from academia and industry. A major project goal was to define and test a flexible system architecture for railway traffic management based on extensible interfaces. That shall lead to a system design, where hardware equipment installed on the track-side remains usable for long periods of time, but software optimization components and hardware used for non-safe calculations as traffic management can easily be exchanged and extended depending on the current and future state of the art. This shall reduce the risk of investments into monolithic products with long product lifetimes.

A set of principles was defined at the start of the project (see project website [ONT 13]) to guide the development process, among those were:

– by managing railway traffic more efficiently, it is possible to increase the capacity of the railway;

– trains can operate with reduced headways if the effect of a single perturbed train can be mitigated so as to not cause a knock-on or ripple effect on other trains;

– instructing drivers to vary train speeds will allow traffic managers to manage the separation between trains and so increase the used capacity of the infrastructure;

– ON-TIME will study the management of railway disruption in two major subsets:

- those that can be resolved through correction by an ICT system with no human intervention. IM/RU communication will be between systems,

- those that can only be resolved with human intervention to make decisions about the redisposition of assets (paths, crew, rolling stock). These decisions will need RU/IM communication both human and system which will be only supported by an ICT system;

– train re-planning will be both:

- event driven (e.g. track blockage, power supply not working),

- clock driven (e.g. small perturbations: cyclic delay update).

12.1.2. *Railway control loop*

The project guidelines can best be understood if railway operation is described as nested (cascading) control loop, as can be seen from Figure 12.1. In this figure, different control levels of railway operation are given as white rectangular blocks, the background color of the figure indicates which of the major players – infrastructure manager (IM) or railway undertaking (RU) – are involved in the respective control task.

Driven by the transport demand for both passengers and freight, railway undertakings determine the demand of train paths and plan the usage of rolling stock. The infrastructure manager allocates paths in such a way that demands are fulfilled in the best possible way for the railway undertakings and no conflicting paths are provided, i.e. the timetable should be conflict-free and feasible in practice [GOV 13]. The timetabling demand is seen as a starting point of all operation measures. In particular the project tackled the problem of different planning horizons: today, timetabling starts more than a year before the actual operation. Changes in demand – in particular for freight traffic – require significant re-planning effort. Within the project it was assumed that passenger traffic would need to be planned a long time in advance (at least three months to allow seat reservation). Regular train headways are preferred for passenger trains in order to facilitate access for the passengers. On the other hand, freight traffic has very short planning horizons (less than a week) and train departure times do not need to be cyclic. The project aimed at defining timetabling tools and processes, which can fulfill both requirements and create a feasible and robust timetable which is resilient to delays when executed in real operation with active traffic management.

For managing traffic in real-time, the technical state of the infrastructure and vehicles must be monitored to detect faults or failures which might impact the control options. Furthermore, the traffic state (train positions and speeds) as captured by trackside and onboard sensors must be detected and continuously compared with the timetable in order to identify deviations.

Once deviations are detected, different control measures can be applied to get back to the normal state. They can be categorized according to the degree of involvement of the railway undertakings (RU).

RU must be involved, e.g. if planned commercial stops cannot be reached due to track blockage or if trains shall be deviated over diversionary routes (where it is not sure that drivers have route knowledge). These measures are usually only taken in case of large disruptions and must thus be considered in close cooperation between RU and IM, which is illustrated in Figure 12.1. These problems and solution processes also frequently require the active involvement of human experts. This process results in a set of (modified) rolling stock or crew constraints as well as the (modified) commercial timetable, where the commercial stops for each train and the lines the train shall take are listed.

There are other measures which do not require any intervention of the RU, these are e.g. re-routing of trains in stations or on parallel tracks, re-timing of departures and arrivals at stations, re-ordering of trains at junctions or in stations, cancelling or adding non-commercial stops, e.g. for overtaking purpose. These measures are performed frequently, even in case of slight delays and shall be summarized under the term "perturbation management". Actions taken here are manifold and might be needed with high frequency. Therefore, methods and algorithms must be developed which are able to work without human involvement. An important prerequisite for a successful introduction of this kind of system is a high level of automation and predictability of the route setting process as part of the underlying Traffic Control System.

Furthermore, traffic fluidity can be improved by controlling the speed of trains in such a way that their distance corresponds to the optimal value through system bottlenecks, i.e. that stops due to red signals protecting other trains shall be prevented. This control measure requires input from the IM side – which must be provided automatically depending on the result of the perturbation management process. However, this must be implemented by the train driver belonging to the RU, e.g. with the help of driver advisory systems, or by Automatic Train Operation (ATO) systems.

In passenger railways in particular, delays require active connection management in order to determine whether connections must be held or can be broken [DOL 13]. This might be handled by the RU, in collaboration between RU and IM or by the IM, depending on the traffic structure in a network.

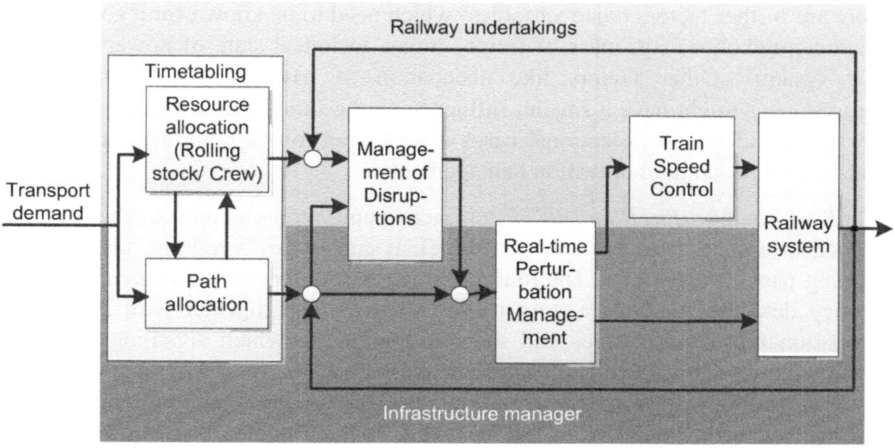

Figure 12.1. *Project approach as control loop*

12.1.3. *Structure of the chapter*

This chapter presents in detail the part of the framework dealing with the management of small perturbations in section 12.2 and the approach chosen for speed control by Driver Advisory Systems (DAS) in section 12.3. The focus of the description will be on the interfaces between different functions and systems which are proposed in the project ON-TIME in order to enable seamless integration of different algorithms into traffic control systems of different suppliers. Section 12.4 explains the approach taken for the validation of the project results.

12.2. Real-time perturbation management

12.2.1. *Important data structures*

In order to have a clear understanding of the interaction of functional modules, some terms are defined first:

– *traffic state*: in control theory the state of a system is defined as the set of variables, which is sufficient to predict future system behavior, if all external input variables are known. To predict the future behavior of a complex railway system, first of all current train positions must be known. Train speed as an additional state variable can be measured directly or derived from the progress of position over time.

There are further factors (state variables) which need to be known for a good system behavior prediction, e.g. mass of freight trains, technical state of power supply or ATP systems. Other factors like occupation of passenger trains or weather circumstances might have a smaller influence on the future traffic behavior and thus only be considered in exceptional cases in order not to overcomplicate the system models used in standard operation management;

– *Real-Time Traffic Plan* (RTTP): a microscopic representation of the timetable planned in a short term horizon (real-time). It consists of a routing, timing and a stopping part for each train. The routing/timing part describes the sequence of track vacancy detection sections a train passes along its run together with its planned occupation and release times and the information, to which signaling route this section belongs. This description has been chosen because most available algorithms for track conflict detection and resolution work with such a sectional representation of the infrastructure [CHE 10, COR 13a, PEL 13], and they are also important, for example, when blocking time diagrams or time-distance-graphs shall be displayed for human dispatchers. The stopping part of the RTTP describes, on which of these sections, where exactly (along the section) and when the train will stop (be it a commercial stop or an operational stop, e.g. to permit overtaking).

From the above described train view of the RTTP, the infrastructure view can be generated, i.e. the sequence of trains which will pass a certain track section and when this is expected to happen. From this so-called infrastructure view of the RTTP, the sequence in which the routes are to be set can be determined by the Traffic Control System.

– *Train path envelope*: used to describe the time windows [ALB 13] along a train run, within which a train can move without getting additionally hampered by the signaling system, e.g. due to train following or crossing conflicts. Time windows can be defined along the train run for any signal passing point (usually conditions of the kind "earliest suggested passing time") or track vacancy detection section release point (usually conditions of the kind "latest permitted passing time"). Further time windows can be defined around station arrival and departure times.

12.2.2. *Important functions and their interaction*

The Perturbation Management Module (PMM) can be divided into four main sub-modules that allow to effectively manage traffic in real-time when disturbances are observed in the network.

The first sub-module is the so-called Traffic State Monitoring (TSM) which is responsible for monitoring current traffic conditions by collecting, via track-side and train-side sensors, all the information relative to both the traffic and the infrastructure. This sub-module runs continuously in order to provide the current traffic state and/ or the traffic state history each time that this is requested by the other modules, e.g. the conflict detection and resolution module or the HMI.

The second sub-module is called Conflict Detection and Resolution (CDR) and represents the most important part of the PMM. Its components and their interaction are illustrated as a simplified SysML activity diagram in Figure 12.2. In the project, an integrated approach of solving track occupation and connection conflicts is proposed. However, in this chapter, the focus shall be exclusively set to track conflicts. The execution of this sub-module first involves a call to the Traffic State Prediction function which forecasts the state evolution of both traffic (e.g. train positions and speeds) and infrastructure (e.g. route settings, signal aspects, switch positions) within a certain time period ahead named "prediction horizon". For this prediction, the last computed order/routing part of the RTTP and the current traffic state as well as further constraints (crew, rolling stock, connection constraints) are used as input. Based on this prediction, the Track Conflict Detection function is activated to identify the presence of potential track occupation conflicts, i.e. multiple trains trying to use one of the infrastructure resources (track detection sections) at the same time. If conflicts exist, the Track Conflict Resolution function is executed, which computes a new RTTP.

There might be extraordinary events (like line blockages), which might require that any ongoing optimization process be terminated immediately and a new optimization cycle start afterwards. This is illustrated with the flash symbol inside the dashed area in the SysML diagram.

The module of CDR is triggered cyclically in normal operating conditions. If infrastructure becomes unavailable or returns from unavailable to available, an event is sent from the Traffic Control System (TCS). In such a case, the routes of the trains which were originally planned over the element and those which are currently routed over diversionary paths must be checked and the routing part of RTTP be adapted accordingly. There might be trains for which no feasible routing can be found within the constraints of the commercial timetable/rolling stock/crew schedule. In such a case, the module of disruption management must be invoked which involves interaction between RU and IM.

The third sub-module is the Train Path Envelope Computation (TPEC) which aims to identify the time allowances available in real operation that can be exploited by a train to adopt an energy-saving driving strategy without running late with respect to the timetable [JAE 13]. This module is executed whenever the RTTP is updated (timing, routing or stopping part). A train path envelope must be computed for each moving train in the area. It should be noted that the computation of train path envelopes does not require conflict resolution of any kind, i.e. it runs with any RTTP available in the PMM. The train path envelope is input to all computations related to train speed control.

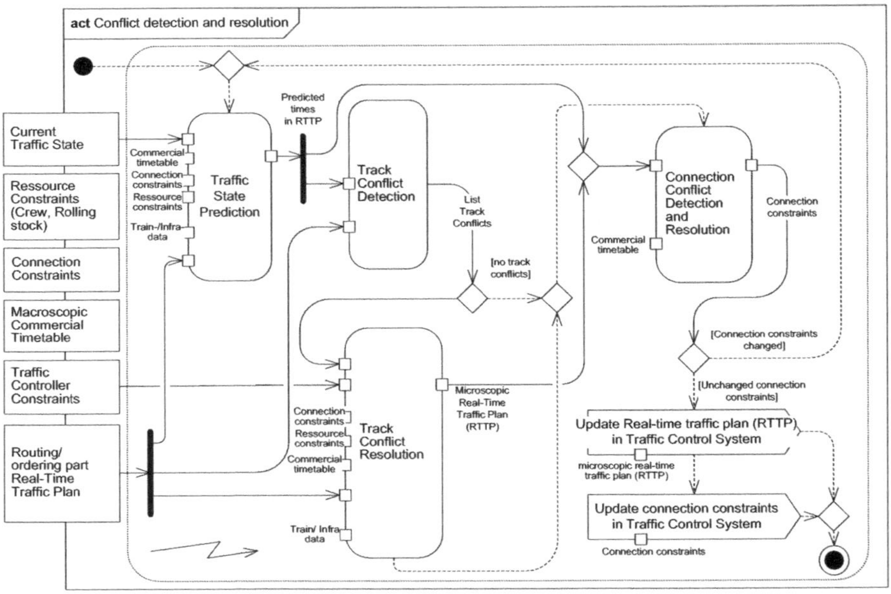

Figure 12.2. *Conflict detection and resolution module as simplified SysML diagram*

A fourth sub-module is the Human-Machine Interface (HMI) which is focused on giving real-time information to the operators (dispatchers/traffic controllers) by a screen visualization of the current traffic state, e.g. through schematic infrastructure view, as well as past and the predicted traffic state, e.g. through the so-called train graph (time-distance diagram). Whatever visualization is used, it will be generated from the RTTP. Giving information continuously to the human operators is seen as one of the key success factors for the project in order to guarantee acceptance of the tools by the operators and avoid the "turn it off" syndrome. The approach of involving humans in the process is called “control by awareness” [GOL 13], an antonym to the commonly used concept of “control by exception”.

The HMI also allows for a direct interaction between the operators and the RTTP, which is particularly useful in events where the operator might have more information from the field than the traffic management system (TMS). In such a case, operators can modify the RTTP which will then be implemented as such by the TCS. The differences to the last automatically computed RTTP are recorded as so-called operator constraints, which will be considered in the next automatic execution runs of the CDR sub-module as described above.

In order to execute this approach in the context of a real railway system it is most important that the route setting process of the TCS follow exactly the intentions of the PMM, i.e. the system for automatic route setting must strictly execute what is described in the RTTP.

12.3. Train speed control

12.3.1. *Functional description*

In order to enable maximal throughput through bottlenecks, active control of train speed is necessary. Therefore, information from the perturbation management must be considered. In European railway networks, trains are predominantly controlled by human drivers. They can be technically supported in this task by so-called Driver Advisory Systems (DAS). There exist different options for distributing and sharing the intelligence between central units (CUs) and on-board units (OBU). The most reasonable architecture alternatives can be derived from the functional analysis of the speed control problem as given in Figure 12.3.

The starting point for all computations is the train path envelope as provided by the perturbation management module. Within this envelope, a trajectory (speed over distance or time) must be computed, which is drivable (feasible) and fulfills a given objective function to the best possible extent, e.g. minimal energy consumption or number of required control actions.

Driving advice or control commands for an Automatic Train Operation System (ATO) can then be derived from this trajectory. The execution of this advice or command determines speed and position of a train (its state). They can be measured by on-board or trackside sensors and fed back into each of the controllers previously described.

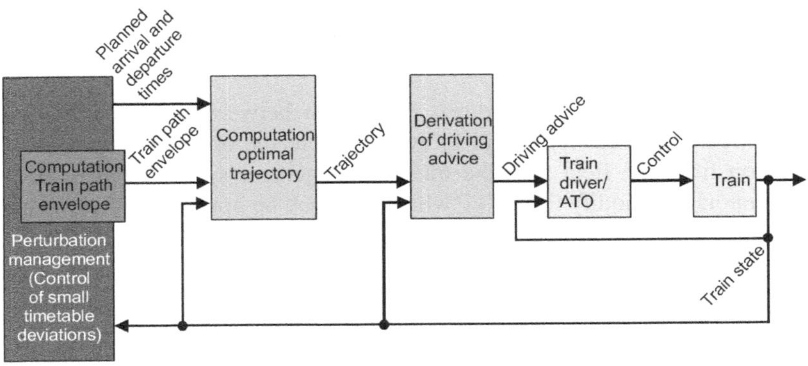

Figure 12.3. *Nested structure of speed control loop*

12.3.2. *Distribution of intelligence and possible interface definitions*

All controllers in the cascading control loop in Figure 12.3 can be realized either on-board (in the OBU) or off-board (in the CU). There therefore exist three basic architecture alternatives given in Table 12.1.

Computation of trajectory	Computation of advice	Name architecture
Central unit	Central unit	DAS-C
Central unit	On-board unit	DAS-I
On-board unit	On-board unit	DAS-O

Table 12.1. *Comparison of the proposed architecture alternatives for Driver Advisory Systems (DAS)*

In the approach DAS-C (C = Central) all computations are done centrally. The driving advice (e.g. "coasting" or "drive at 120 km/h") is transmitted from central to OBU every time a new control action is required. On-board, this advice must only be presented to the driver (visually or acoustically). The requirements on the on-board system are therefore quite simple, e.g. the cab-radio display could be used. Therefore this system has a high potential for interoperable application. However, the communication channel must be highly available, as no advice can be given without active connection between CU and OBU. Furthermore, train state information should be available in the CU with high accuracy. Both requirements are met, for example, on lines equipped with the European Train Control System (ETCS) level 2. One component of the system of fundamental importance is a highly available (and safe) communication channel between train and radio block center, which is among others used to transmit train position and speed several times per minute from train to RBC. This channel could also be used for transmitting driving advice in the other communication direction. An example for an existing system of this architecture is "Admirail", which is installed and in full commercial operation since 2007 in the Lötschberg base tunnel in Switzerland [MON 09].

In the architecture called DAS-I (I = Intermediate) the optimal trajectory of the train within the envelope is computed centrally and transmitted to the OBU. Based on the trajectory and state information available on-onboard the advice is computed in the OBU. With such a system design, the communication channel is only needed when the trajectory changes, i.e. the requirements on availability of the communication channel are smaller, but the individual message size becomes bigger (trajectory compared to a single advice). If certain track and train data (track resistances, acceleration and braking behavior, running resistances of train) are available both centrally and on-board in a consistent way, the communication message content can be reduced significantly, e.g. to the points where the driving regime changes. The OBU should react on deviations of real train behavior from the optimal trajectory, e.g. by displaying an advice different from the one foreseen in the optimal trajectory or by displaying the deviation from the trajectory to the driver. If large deviations are detected track-side in the CU, a new trajectory is computed and then sent to the OBU.

In the architecture DAS-O (O = On-board), both trajectory and driving advice are computed in the OBU. The train path envelope must be transmitted to from central to OBU. The OBU software must not only have access to current train data as suggested for DAS-I, but also have access to a complete set of valid track data including e.g. line speed, temporary speed restrictions, signal positions and track resistances. This can be provided either in an on-board track atlas or as part of the communication from CU to OBU. Further requirements for a system of such architecture are on-board sensors for position and speed, synchronized timing information in OBU and CU and available computation power in the OBU. These requirements are often fulfilled by existing DAS for energy-efficient driving as operational in several European railway undertakings, e.g. InLineFAS [ALB 12] or CATO [LAG 11]. These systems need to be extended with an interface for the train path envelope, while the algorithms might also need to be changed in order to guarantee that the trajectory and advice fulfill the constraints of the train path envelope.

It still needs to be decided how far regulatory and commercial aspects of the relationship IM-RU need to be adapted if this kind of DAS is widely implemented. The reason is that energy consumption of trains will be significantly determined by the calculations of the IM, although in most European countries, energy cost has to be paid by the RU depending on actual energy consumption.

12.4. Demonstration and validation approach

The proposed systems and algorithms have been tested in simulation scenarios for different locations throughout Europe:

– East Coast Main Line in the UK;

– Iron Ore Line in Sweden;

– a part of the Dutch network.

Therefore, the microscopic railway operation simulator HERMES brought in and significantly extended by project partner Graffica had been used (see Figure 12.4).

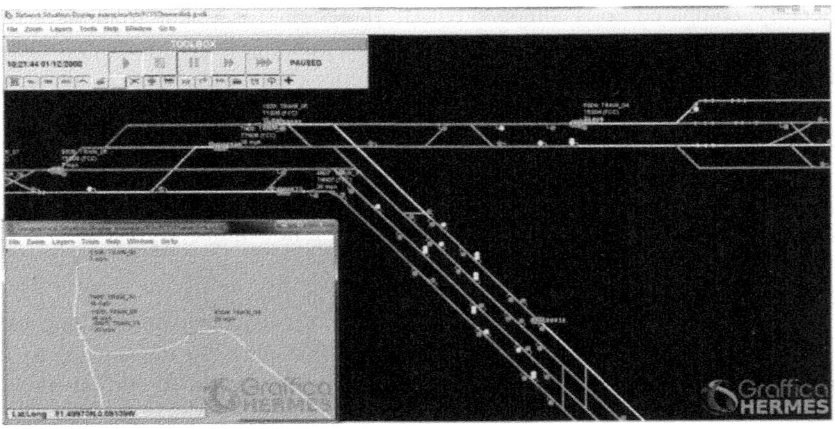

Figure 12.4. *Screenshot of the microscopic train simulator HERMES for a part of the East Coast Mainline. For a color version of this figure, please see www.iste.co.uk/jacob/traffic.zip*

HERMES was extended to import data on timetable, infrastructure and rolling stock given by the different infrastructure managers in different formats. They were transformed into the current version of the railML format [RAI 13], which is used throughout Europe for railway specific data exchange between different software tools. The project partners of ON-TIME contributed to the improvement and extension of different aspects of the railML model, in particular a new approach for modeling the interlocking aspects has been developed together with the railML consortium and market leading manufacturers of signaling technology in Europe (Siemens, Thales, Alstom).

The simulator HERMES was connected to the algorithms described in the previous sections of this chapter using a service-oriented architecture developed in the project (see Figure 12.5). This allowed for using the simulator as a stand-in for a real railway system and permitted to show how algorithms perform in a closed control loop in practical settings. In addition to that, an HMI of the current product line of project partner Ansaldo STS was connected to the architecture to display train graph and train describer view based on the current traffic state and RTTP as computed by the Perturbation Management Module. Furthermore, the train speed control/driver module of the HERMES simulator permitted the demonstration of all speed control architectures described in section 12.3.2.

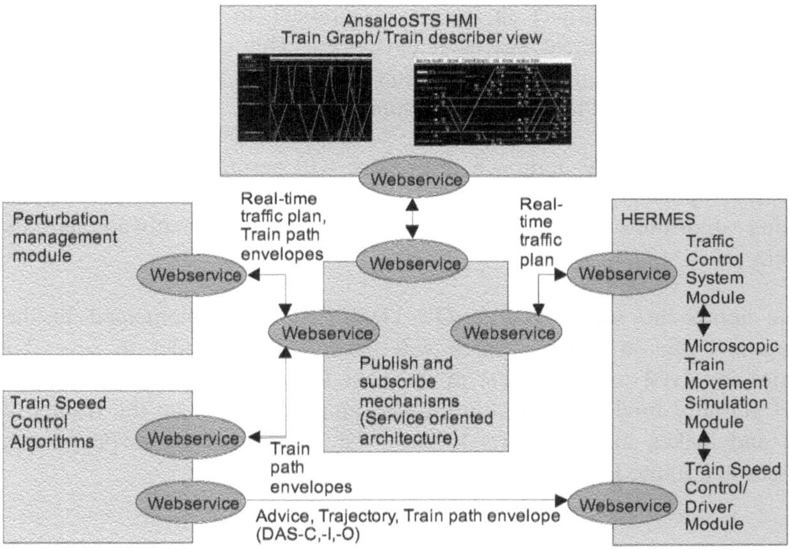

Figure 12.5. *Flexible service-oriented architecture used to demonstrate*

In order to test and demonstrate this framework a high amount of effort had to be put into the automatic operation and interaction of the different modules which went far beyond the typical scope of a research project. This concerned in particular the interaction between the conflict detection and resolution algorithms, the automatic route setting module to execute the RTTP and the simulator such that no deadlocks occur in operation e.g. due to lost messages.

Simulation results showed that current track conflict detection and resolution algorithms are able to deliver robust solutions of the RTTP in real time if applied in a rolling horizon optimization cycle [QUA 15, QUA 13]. It is however still very difficult to estimate the potential of these algorithms concerning the reduction of delays or the avoidance of unplanned stops compared to real railway operation as the simulator is only able to compare different algorithms with each other. In order to obtain reliable estimates about the effect of optimization algorithms applied to real railway systems, the behavior of all the different human actors in the railway system (signalers, dispatchers and train drivers) would need to be modeled in a simulator which was beyond the scope of the ON-TIME project and must be part of future projects.

12.5. Conclusions

In the ON-TIME project an integrated approach to timetabling and real-time railway traffic management has been developed. This chapter presented findings from the functional requirements definition phase which have been implemented in different prototypes and tested in simulations.

Using the ON-TIME approach in practice will permit the railways to react on future developments in the field of decision support systems more quickly by replacing algorithms independently from the underlying hardware and its long lifecycle.

The approaches developed in ON-TIME will be implemented in the next generation TMSs in Europe which will be called for tender in several of the countries involved in the project in the next few years. The CAPACITY4RAIL project partially funded by the European Commission took up the results of ON-TIME and makes further efforts towards the automation of traffic management processes in a future European railway network.

12.6. Acknowledgments

The ON-TIME (Optimal Networks for Train Integration Management across Europe) project was co-funded by the European Commission within the 7th Framework Programme (2007–2013), Grant agreement FP7-SCP01-GA-2011-285243.

12.7. Bibliography

[ALB 12] ALBRECHT T., BINDER A., GASSEL C., "Applications of real-time speed control in rail-bound public transport systems", *IET Proceedings on Intelligent Transport Systems*, 2012

[CHE 10] CHEN L., SCHMID F., DASIGI M. *et al.*, "Real-time train rescheduling in junction areas", *Proceedings of the IMechE: Part F – Journal of Rail and Rapid Transit*, vol. 224, no. 6, pp. 547–557, 2010.

[COR 13] CORMAN F., MENG L., "A review of online dynamic models and algorithms for railway traffic control", *Proceedings of the 1st IEEE International Conference on Intelligent Railway Transportation*, Beijing, 2013.

[DOL 13] DOLLEVOET T., CORMAN F., D'ARIANO A. *et al.*, "An iterative optimization framework for delay management and train scheduling", Working paper, available at: http://repub.eur.nl/res/pub/32416/, 2013.

[EUR 11] EUROPEAN COMMISSION, Roadmap to a Single European Transport Area – Towards a competitive and resource efficient transport system, White Paper, available at: http://eur-lex.europa.eu/LexUriServ/LexUriServ.do?uri=CELEX:52011DC0144:EN:NOT, 2011.

[GOL 13] GOLIGHTLY D., SANDBLAD B., DADASHI N. *et al.*, "A sociotechnical comparison of automated train traffic control between GB and Sweden", in DADASHI N., SCOTT A. *et al.* (eds), *Rail Human Factors – Supporting Reliability, Safety and Cost Reduction*, CRC Press, 2013.

[GOV 13] GOVERDE R.M.P., HANSEN I.A., "Performance indicators for railway timetables", *Proceedings of the 1st IEEE International Conference on Intelligent Railway Transportation*, Beijing, 2013.

[JAE 13] JAEKEL B., ALBRECHT T., "Interfacing conflict resolution and advisory systems", *Proceedings of the 3rd International Conference on Models and Technologies for Intelligent Transportation Systems*, Dresden, 2013.

[LAG 11] LAGOS M., "CATO offers energy savings", *Railway Gazette International*, pp. 50–52, 2011.

[MON 09] MONTIGEL M., "Operations control system in the Lötschberg base tunnel", *RTR – European Rail Technology Review*, no. 2, pp. 43–44, 2009.

[ONT 13] ON-TIME, Official project website, available at: www.ontime-project.eu, 2013.

[PEL 13] PELLEGRINI P., MARLIÈRE G., RODRIGUEZ J., "A mixed-integer linear program for the real-time railway traffic management problem modelling track circuits", *Proceedings of the 5th International Seminar on Railway Operations Modelling and Analysis*, Copenhagen, 2013.

[QUA 13] QUAGLIETTA E., CORMAN F., GOVERDE R., "Impact of a stochastic and dynamic setting on the stability of railway dispatching solutions", *Proceedings of the 16th International IEEE Annual Conference on Intelligent Transportation Systems*, The Hague, 2013

[QUA 15] QUAGLIETTA E., GOVERDE R.M.P., ALBRECHT T. *et al.*, "Optimal management of railway perturbations by means of an integrated support system for real-time traffic control", *Proceedings of the 6th International Seminar on Railway Operations Modelling and Analysis*, Tokyo, 2015.

[RAI 13] RAILML, Official website, available at: www.railML.org, 2013.

[ROB 13] ROBERTS C., CHEN L., "Optimal networks for train integration management across Europe", *European Railway Review*, vol. 19, no. 3, pp. 48–52, 2013.

13

A Multi-lane Capacity Model Designed for Variable Speed Limit Applications

Recent research has investigated the effects of variable speed limits (VSLs) on traffic flow processes. The findings show that VSLs profoundly impact lane flow distribution (LFD) and increase the use of the slow lane by reducing the speed difference between the slow lane and the overtaking lanes. Finally, VSLs reduce the capacity drop phenomenon on multi-lane sections and significantly increase the proportion of flow on the slow lane. These results highlight the major impact of LFD on the capacity of multi-lane sections.

The paper presents a new analytical model that estimates the capacity of multi-lane sections and reproduces the capacity drop caused by LFD characteristics. It can be applied to estimate the capacity of motorway sections with two or three lanes under common traffic conditions; it can also estimate the impact of VSL on the capacity of multi-lane sections.

The multi-lane capacity model is coupled with a dynamic mesoscopic traffic flow model initially designed for network applications: the meso-LWR model. The resulting dynamic traffic flow model is relevant for motorway network applications and VSL evaluation. The dynamic model is implemented on a French motorway site. The results show the ability of the model to reproduce actual traffic situations (capacity and congestion period) without VSL, including the amplitude of capacity drop phenomenon caused by the under-use of the slow lane. It also shows the ability of the model to estimate the effect of VSL on future traffic flow conditions.

13.1. Background

During the last decades, travel demand has steadily increased driven by economic development and growing needs for mobility. Today travel demand near big cities exceeds the total capacity of major infrastructures. Daily congestion occurs that hinders local economic development and causes pollutant and noise emissions

---

Chapter written by Aurélien DURET.

for the surrounding environment, and discomfort and stressful situations for drivers. Congestion reduction has become an important goal for traffic managers.

Traffic flows work similarly to water running down a plug hole: when the demand is low, the network is free flowing and when the demand exceeds the capacity, the network overflows and congestion occurs. Traffic control strategies can help traffic managers to improve traffic conditions on the network. The objective is to adjust the travel demand and the network supply in order to minimize the presence of congestion on the network. Traffic control strategies consist of controlling demand (through pricing and dynamic assignment), controlling the network supply (traffic lights, ramp-metering and high occupancy vehicle lanes) or changing traffic flow characteristics (free-flow speed and overtaking ban for trucks).

In this context, variable speed limit (VSL) strategies have been used on major French motorways since 2000. The initial goal was to improve the flow on major motorways by diminishing the speed on motorway sections. The benefits of VSL have been observed and measured for several years, but the reason for the capacity increase was unknown. Duret *et al.* [DUR 12] have recently proposed a physical mechanism that explains the efficiency of VSL. The authors have measured flow distribution on a multi-lane motorway and have observed that lane flow distribution (LFD) respects a recurring pattern. When the total flow is low, most vehicles drive on the slowest lane(s); the fastest lane(s) receive a higher proportion of the flow as the total flow increases, and when the demand approaches capacity, the slowest lane receives less than 20% of the total flow: this measurement corresponds to under-use of the slowest lane.

The authors show that VSL strategy modifies the LFD pattern, homogenizes lane use, reduces under-use of the slow lane (far right lane in France) and increases the overall capacity of the section. Weikl *et al.* [WEI 13] have recently observed the same homogenizing effect of a VSL system on German motorways.

Dynamic traffic flow models can be used to demonstrate the benefits of VSL strategies on the capacity of multi-lane sections. Classically, microsimulation tools are preferred to perform real field evaluations, mainly because of their precise representation of traffic flow (individual trajectories). However, these tools present many weaknesses for VSL applications on congested motorway networks:

– they are hardly calibrated, because they are often based on dozens of behavioral parameters;

– they are time consuming, because the computational time is proportional (1) to the length of the network and (2) to the number of vehicles on the networks during the simulation, and congested multi-lane motorway sections can easily contain thousands of vehicles;

– they require multiple replications (to account for stochastic effects) that complicate the analysis of the results;

– most importantly, the ability of the lane-changing components to replicate LFD patterns observed on multi-lane sections can be questioned, as demonstrated by Duret [DUR 10].

Consequently, microscopic simulation tools are not the most effective solution to model traffic flow at a large network scale. Macroscopic simulation tools could be an alternative. However, some difficulties may appear for network applications, specifically for dynamic traffic assignment where the flow has to be distinguished by destination.

Recently, Leclercq and Becarie [LEC 12] have proposed a mesoscopic resolution of the well-known LWR model. This is an event-based resolution that enables traffic flow modeling at links. The mesoscopic LWR model calculates the individual entry/exit time of vehicles at link extremities (start/end). This solution has many advantages; in particular, the calculation time is very low, and it only depends on the number of nodes crossed by the vehicles of the network. It does not depend on the length of the network or the number of actual vehicles on the network during the simulation. Moreover, mesoscopic resolution is fully consistent with earlier Eulerian and Lagrangian resolution methods of the LWR model.

This model is a good compromise between microscopic and macroscopic simulation tools for motorway network applications. The main limitation of the LWR model is that the capacity drop phenomenon is not represented, and this weakness remains in its mesoscopic formulation. Links are considered homogeneous and it does not take into account LFD characteristics. Consequently, it does not model the capacity drop causes through the under-use of the slowest lane.

This chapter proposes to associate the meso-LWR model with the analytical multi-lane capacity (MLC) model that takes into account capacity drops caused by LFD. The chapter is organized as follows. Section 13.2 presents the analytical MLC model. Section 13.3 briefly presents the meso-LWR model coupled with the MLC model. Section 13.4 presents an application on a French motorway and estimates the benefits induced by the deployment of a VSL strategy. Section 13.5 concludes on the application of the meso-LWR model to evaluate VSL strategies, summarizes its strengths and weaknesses and opens up the discussion on further possible research.

13.2. MLC model

13.2.1. *Field measurements*

Since the beginning of traffic flow theory, most research has been focused on the fundamental variable of traffic flow: the flow $\boldsymbol{Q}$, the speed $\boldsymbol{V}$ and the density $\boldsymbol{K}$, and the resulting traffic flow characteristics such as shockwave propagation. Cassidy and Bertini [CAS 99] and Daganzo [DAG 02a,DAG 02b] analyzed the characteristics of multi-lane sections and noticed the specificities of each lane. Carter *et al.* [CAR 99] and Banks [BAN 06] demonstrated a link between LFD and the capacity of a multi-lane section. Wu [WU 06] studied the LFD on German and US motorways and demonstrated that LFD highly depends on the total volume on section. More recently, Duret *et al.* [DUR 12] observed and measured similar recurring patterns that describe LFD under free-flow conditions on two-lane and three-lane sections, as illustrated in Figure 13.1.

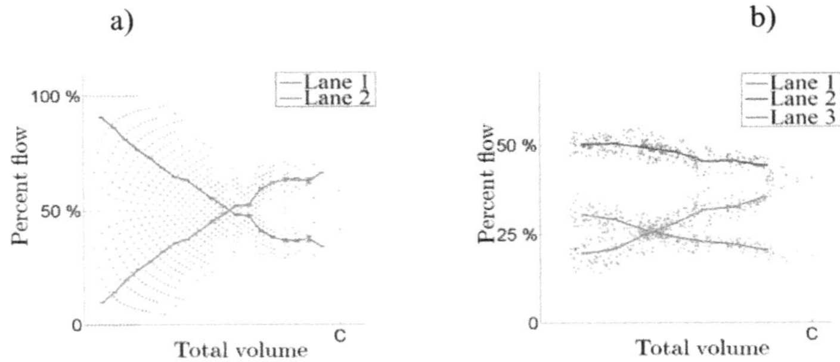

Figure 13.1. *a) LFD on two-lane motorway, France; b) LFD on three-lane motorway, France. For a color version of this figure, please see www.iste.co.uk/jacob/traffic.zip*

Figure 13.1(a) illustrates the classical LFD measured on a two-lane French motorway (A6). It represents the flow percentage on each lane versus the total volume on the section. When the total volume is low, most vehicles stay in the right-most lane (red). Then, the proportion of vehicles on the left-most (overtaking) lanes increases as the total volume increases. When the total volume reaches 2,500 vph, both lanes receive a similar proportion of vehicles. When the total volume exceeds 2,500 vph, most vehicles drive on the left-most lane (green).

Figure 13.1(b) illustrates the classical LFD measured on a three-lane motorway (A7). Here, the process is slightly different. When the total volume is low, vehicles are equally distributed on the two right-most lanes (red and blue). As the total volume increases, the right-most (red) lane receives a lower proportion of vehicles, meanwhile the left-most (green) lane receives a higher proportion of vehicles, and the middle lane receives a stable proportion of the total volume (around 45%).

In both figures, one can observe that when the total volume reaches its highest value C (capacity), vehicles are unequally distributed over lanes as follows:

– on the two-lane highway, the right-most lane and the left-most lane receive 40 and 60% of the total volume, respectively;

– on the three-lane highway, the right-most, the middle and the left-most lanes receive 20, 40 and 40% of the total volume, respectively

Regardless of the number of lanes on the section, the LFD can explain lane-use heterogeneity and the under-use of the slowest lane.

Here, a simple multi-lane capacity model is proposed to estimate the capacity of a two-lane section based on LFD characteristics. The model is simple and has only one parameter. It can easily be extended to three-lane sections.

13.2.2. *Presentation of the model*

Let us consider a section with two lanes where the traffic flow equilibrium of lanes 1 (slow) and 2 (fast) is described by two triangular fundamental diagrams. The fundamental diagram on the lane i is defined by three parameters: the free-flow speed u_i, the maximum wave speed w, the maximum jam density kx and the capacity Cx_i of the lane i is:

$$Cx_i = \frac{u_i . w . kx}{u_i + w} \qquad [13.1]$$

For the sake of simplicity, w and kx are supposed to keep the same value on lanes 1 and 2. Note that this assumption can easily be relaxed in the formulas that are proposed throughout the chapter. Classically, the capacity of a two-lane section is considered as the sum of the capacity of individual lanes. However, empirical evidence shows that the LFD of a two-lane section follows a recurring pattern.

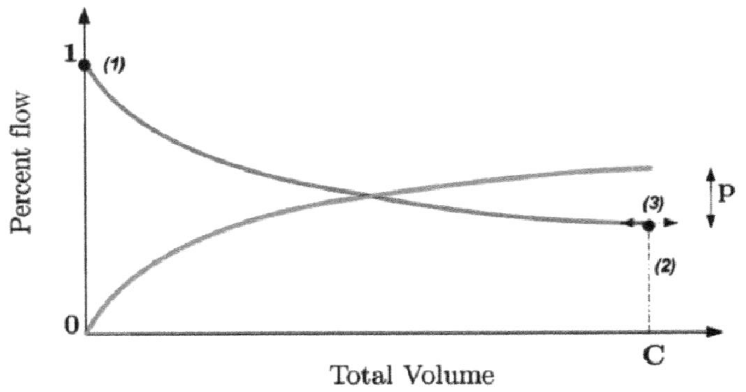

Figure 13.2. *LFD versus total demand on a two-lane section. For a color version of this figure, please see www.iste.co.uk/jacob/traffic.zip*

P1 (*P2*) denotes the percentage of flow on lane 1 [13.2]. Here, we propose a simple quadratic formula to model the variation of P1 versus the total volume *Q*.

$$P1(Q) = a + b.Q + c.Q^2 \qquad [13.2]$$

Here, parameters a, b and c can easily be estimated based on real field data, considering the three following conditions:

$$P1(0) = 1$$

$$P1(C) = (1-p)/2 \qquad [13.3]$$

$$dP1/dQ\,|_C = 0$$

System [13.3] has a unique solution and one can conclude that:

$$P1(Q) = 1 - \frac{1+p}{C}.Q + \frac{1+p}{C^2}.Q^2 \qquad [13.4]$$

$$P2(Q) = 0 + \frac{1+p}{C}.Q - \frac{1+p}{C^2}.Q^2$$

As the total volume increases, lane 2 first reaches its capacity Cx_2. Once lane 2 reaches its capacity, lane-changing activity increases and the entire section quickly switches to congestion. Based on this process, one can then estimate the total capacity $\boldsymbol{C}$ of the section by:

$$C = Cx_2 . \frac{2}{(1+p)} = u_2 . \frac{w.kx}{u_2 + w} . \frac{2}{(1+p)} \qquad [13.5]$$

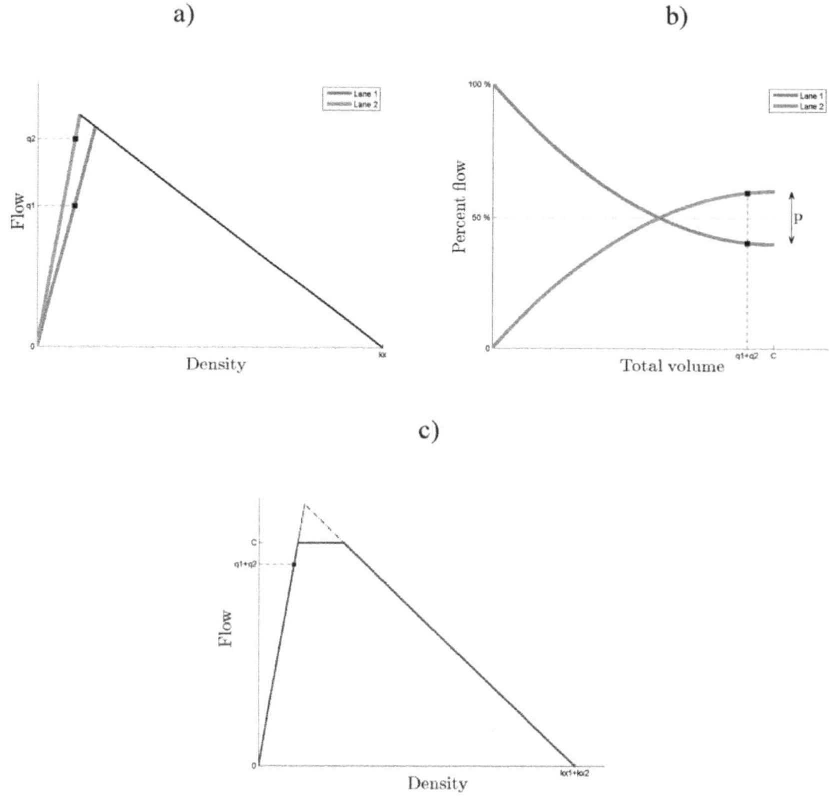

Figure 13.3. *a) Fundamental diagrams on lane 1 and lane 2; b) LFD on a 2-lane section; c) multi-lane capacity model for a 2-lane section. For a color version of this figure, please see www.iste.co.uk/jacob/traffic.zip*

Finally, the MLC model combines the characteristics of the fundamental diagram and the characteristics of the LFD to estimate the whole capacity of the motorway section. It only requires four parameters: u, w and kx that are related to the fundamental diagram, and p that is related to the LFD. All these parameters have a physical meaning, are easy to measure and calibrate.

It should be noted that the capacity of a 2-lane section is the result of a competition between the free-flow speed on lane 2 (that decreases capacity) and the flow percent difference (that increases capacity). A sensitivity analysis is proposed in Figure 13.4 to evaluate the combined effects of both variables on capacity.

Figure 13.4(a) shows the variation of the capacity drop versus p. The capacity drop caused by LFD is null when p is null. Then, it increases with p capacity and it can exceed 20% when p reaches 30% (commonly observed on French motorways). Figure 13.4(b) shows the variations of the capacity on a 2-lane motorway. It clearly illustrates that, for a given free-flow speed, there are strong variations in capacity according to flow percent difference between lanes. For example, at 110 km/h, the capacity is approximately 3,300 vph when the flow percent difference equals 30%, and theoretically reaches very high values (~4,300 vph) if lane use is optimal (p=0%).

Please note that the three-lane capacity model can easily be proposed, based on the same reasoning. For the sake of brevity, it is not presented in this paper.

a) b)

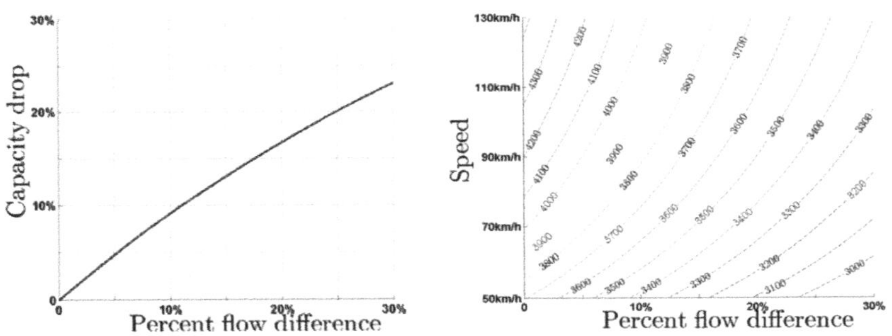

Figure 13.4. *a) Capacity-drop on 2-lane motorway versus flow percent difference; b) capacity on 2-lane motorway versus flow percent difference and speed on lane 2*

13.2.3. *Applications for variable speed limit strategies*

VSL strategies are usually activated when traffic flow approaches critical demand. The activation of VSL strategies changes the free-flow speed on multi-lane sections. Duret *et al.* [DUR 12] have demonstrated that VSL also modifies the LFD and, consequently, the total capacity of a multi-lane section. Most dynamic traffic flow models are able to take into account the modification of the free-flow speed, since it is an input parameter of the fundamental diagram. However, existing traffic flow models do not take into account the impact of VSL strategies on the LFD and the capacity of multi-lane sections.

Here, we propose to use the MLC model to represent the impact of VSL (1) on the free-flow speed and (2) on the capacity of a multi-lane section. Let:

– u_2' denote the free-flow speed of the lane when VSL;

– p' denote percent flow difference between lane 1 and lane 2;

– C' denote the capacity of the section.

When the VSL is activated. Table 13.1 summarizes the parameters and the formulae used to calculate the capacity of a 2-lane section.

Parameters	Without VSL	With VSL
Free-flow speed on lane 2	u_2	u_2'
Flow percent difference	p	p'
Maximum wave speed	w	w
Jam density	kx	kx
Capacity of lane 2	$Cx_2 = u_2.\frac{w.kx}{u_2 + w}$	$Cx_2' = u_2'.\frac{w.kx}{u_2' + w}$
Capacity of the section	$C = u_2.\frac{w.kx}{u_2 + w}.\frac{1}{(1+p)}$	$C' = u_2'.\frac{w.kx}{u_2' + w}.\frac{1}{(1+p')}$

Table 13.1. *Parameters and formulae for the MLC model on a 2-lane section*

The MLC model (with and without VSL) is represented in Figure 13.5. The gray lines represent the MLC model without the activation of the VSL strategy. One can observe a high free-flow speed (Figure 13.5(b)) and a high flow percent difference between lanes when the total volume approaches the capacity (Figure 13.5(a)). The resulting capacity of section C is low. On the contrary, when VSL is activated (solid black lines), the free-flow speed of the section is lower, the flow percent difference between lanes when the total volume approaches the capacity is lower and the resulting capacity C' is higher. Overall, the MLC model includes both effects of VSL strategies on traffic flow: decrease in speed and increase in capacity.

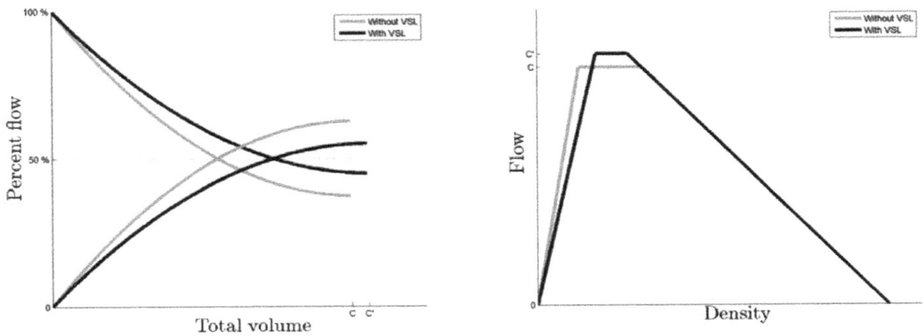

Figure 13.5. *MLC model with and without VSL a) LFD b) fundamental diagram*

Let **g** denote the ratio of capacity with and without the activation of the VSL:

$$g = \underbrace{\frac{u_2'}{u_2} \cdot \frac{w+u_2}{w+u_2'}}_{<1} \cdot \underbrace{\frac{1+p}{1+p'}}_{>1} \qquad [13.7]$$

Here, one can observe that g has two terms:

– the first term (left) stems from the FD and takes into account the speed reduction. It is expected to be lower than 1;

– the second term (right) stems from the LFD and accounts for the homogenization of the flow over lanes. It is expected to be higher than 1.

The objective of the VSL is to achieve a balance that guarantees that g is higher than 1.

13.3. Meso-LWR model and multi-lane capacity model

Traffic flow is the result of the dynamic competition between traffic demand and the capacity of the network. Traffic demand is dynamic: it is highly dependent on the time of the day and can vary significantly within a short period of time (hour). Capacity can also be dynamic, for example during the activation of a VSL strategy. Because of the dynamic nature of traffic flows, it is important to have dynamic traffic flow models that are able to take into account both the dynamics of demand and the dynamics of capacity.

In this section, the meso-LWR model initially proposed by Leclercq and Becarie [LEC 12] is briefly introduced. This model does not account for capacity drop phenomenon and, consequently, we present the combination of the MLC model and meso-LWR model. The resulting dynamic traffic flow model is able to take into account:

– the capacity drops induced by the LFD characteristics of multi-lane sections;

– the impact of VSL on both the free-flow speed and the capacity of multi-lane sections.

13.3.1. *Meso-LWR model*

The meso-LWR, initially proposed by Leclercq and Becarie [LEC 12], has been designed for large-scale networks. It consists of a mesoscopic resolution of the LWR model, fully consistent with prior resolution. Here we summarize the key concepts of the meso-LWR model. It is recommended to read the original paper for a complete description.

Let us consider a link with the length *L* composed of *n* lanes. A triangular diagram defines the traffic equilibrium of the link where *u* is the free-flow speed, *k* is the jam density, and ***w*** is the maximum wave speed.

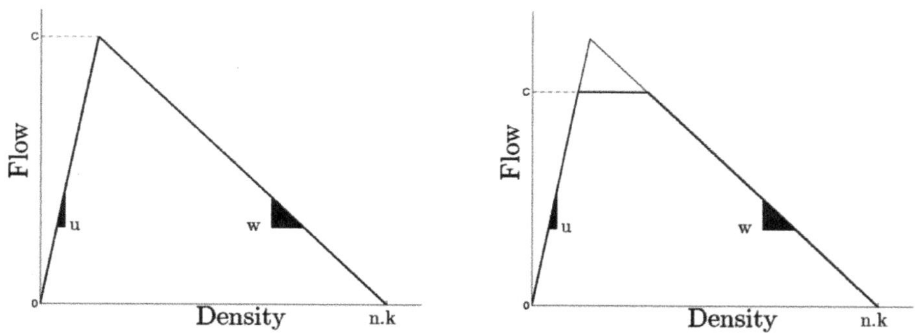

Figure 13.6. *Fundamental diagram a) in its original version and b) coupled with the multi-lane capacity model*

The capacity *C* of the link is equal to *C=n.k.w.u/(u+w)*; see Figure 13.6(a). The authors propose an event-based resolution of the LWR model that consists of calculating the exit time of vehicle i which entered the link at time $t^{in}(i)$. The authors demonstrated that the effective exit time of the vehicle i $t^{out}(i)$ is:

$$t^{out}(i) = \max\left(t^{a}(i), t^{s}(i)\right) \qquad [13.8]$$

where $t^{a}(i)$ is the arrival time at the end of the link and $t^{s}(i)$ the supply time at the end of the link.

$t^{a}(i)$ and $t^{s}(i)$ can then be easily calculated:

$$\begin{cases} t^{a}(i) = t^{in}(i) + L/u \\ t^{s}(i) = t^{out}(i - n.L.k) + L/u \end{cases} \qquad [13.9]$$

See Leclercq and Becarie [LEC 12] for a sketch of proof.

Here, we propose to modify the shape of the FD to account for the capacity drop calculated with the help of the MLC model. The resulting trapezoidal fundamental diagram is illustrated in Figure 13.6(b). MLC do not modify free-flow characteristics and the capacity of the section. The congestion parameters (w and kx) remain the same and, consequently, equations [13.8] and [13.9] still hold when the meso-LWR model is combined with the MLC model.

The LWR-meso model is a link model that needs to be completed with node models to describe traffic flow conditions at merges and diverges. Leclercq and Becarie [LEC 12] propose to combine the link model to the node models proposed in Daganzo [DAG 95] and one can refer to the original paper for a complete description.

13.4. Application

13.4.1. *Presentation of the scenarios*

The meso-LWR model coupled with the MLC model is specifically designed for networks that present multi-lane (motorway) sections. It this section, we apply this model to a section of the A31 motorway located near Thionville (France) with one on-ramp and one off-ramp. The layout of the network is proposed in Figure 13.7. The network is composed of:

– 6 links (4x2-lane links and 2x1-lane links);

– 7 nodes (2 entries, 2 exits, 1 merge, 1 diverge and simple node).

The parameters of each link have been calibrated based on 6-min data from loop detectors (flow and speed per lane). The calibration process is illustrated in

Figures 13.8(a) and (b), and the values of calibrated parameters are summarized in Table 13.2.

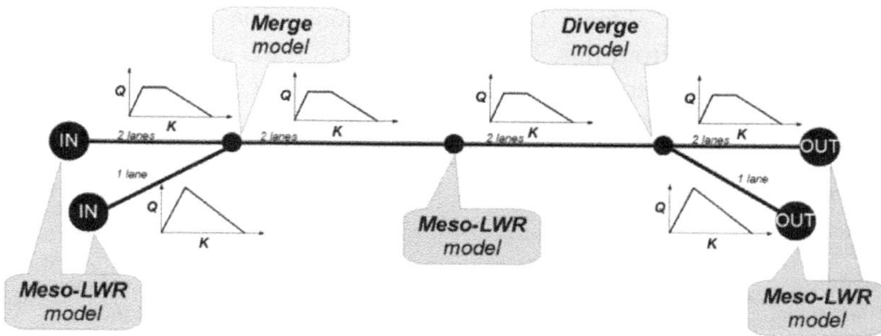

Figure 13.7. *Layout of the A31motorway corridor, Thionville (France)*

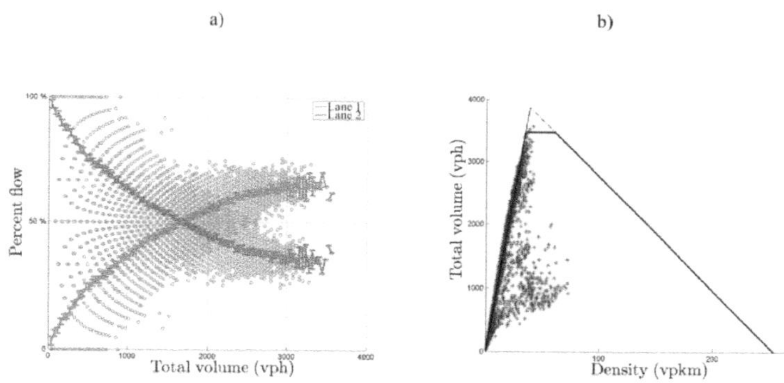

Figure 13.8. *Calibration of link 3: a) estimation of the LFD and b) estimation of the fundamental diagram. For a color version of this figure, please see www.iste.co.uk/jacob/traffic.zip*

	n	u_2 [km/h]	p [%]	C [veh/h]
Link 1	2	118	24	3526
Link 2	2	116	18	3697
Link 3	2	112	25	3476
Link 4	2	118	21	3614
Link 5	1	65	–	1852
Link 6	1	65	–	1852

Table 13.2. *Parameters calibrated with real field data*

13.4.2. *Analytical analysis*

Link 3 presents the lower capacity and it can be explained by its specific configuration (positive slope). Consequently, congestion is expected to occur immediately upstream from link 3. The objective of VSL strategy is to reduce the free-flow speed in order to increase capacity. As explained above, a competition is engaged between free-flow speed on lane 2 and the flow percent per lane when VSL strategy is activated. The VSL strategy has a positive impact on capacity if and only if the flow percent difference p drops below a critical value p^*, as illustrated in Figure 13.9.

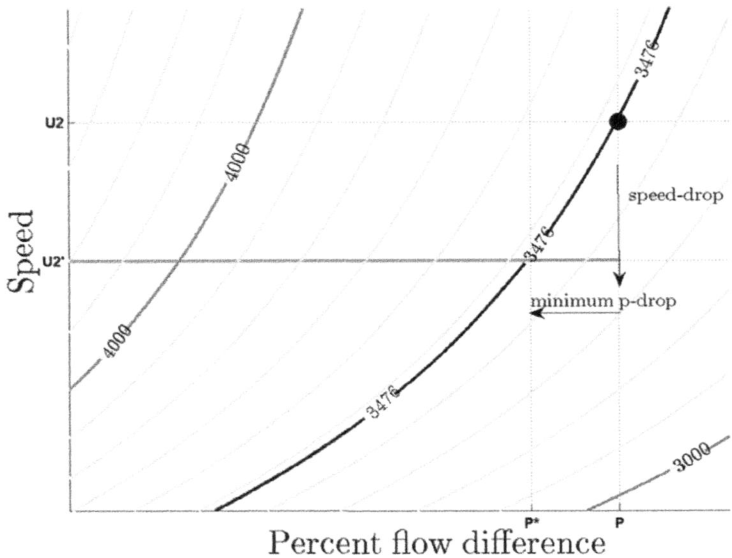

Figure 13.9. *Capacity of 2-lane motorway (vph) versus flow percent difference and speed on lane 2: minimum p-drop for a given speed-drop. For a color version of this figure, please see www.iste.co.uk/jacob/traffic.zip*

In this figure, the black spot represents the situation on link 3 without VSL. When VSL is activated, the speed in lane 2 decreases to 90 km/h, which can be represented by a simple shift downward. Then one can deduce the minimum flow percent difference drop p^* (shifted to the left) that guarantees a higher capacity. In this particular case, p^* is 21%, meaning that if the flow percent between lanes is lower that 21% when VSL is activated, then the VSL strategy increases the capacity on link 3.

13.4.3. *Simulation results*

VSL strategies were applied on the A31 motorway starting from November 2013. In the context of an *ex ante* evaluation of the VSL strategy, assumptions are formulated to model the impact of VSL strategy on traffic flow characteristics. Here, the free-flow speed and flow percent difference are supposed to be modified according to the values presented in Table 13.3 (with single quote). On this basis, a demand scenario has been estimated from vehicle counts measured by loop detectors during a weekday (03/20/2013, 6 am–10 am).

Two simulations have been run. Since the meso-LWR model does not present stochastic effects, it is not necessary to run the same scenario several times. The VSL strategy was not activated during the first simulation. The VSL was activated between 7 am and 9 am (morning peak period) during the second simulation. When VSL is activated, the parameters values are modified, as presented in Table 13.3. Otherwise, the parameters remain the same (see Table 13.2).

	n	u_2' *[km/h]*	p' *[%]*	C' [veh/h]
Link 1	2	90	10	3818
Link 2	2	90	10	3818
Link 3	2	85	15	3616
Link 4	2	90	10	3818
Link 5	1	65	–	1852
Link 6	1	65	–	1852

Table 13.3. *Parameters with the activation of VSL*

Figure 13.10(a) presents the outflow of the corridor during the simulations and Figure 13.10(b) illustrates the travel time along the whole motorway corridor. Without VSL strategy (black solid line), one can obtain a situation commonly observed during the morning rush hours. Before 7.20 am, the observed outflow overlays the outflow demand, no congestion occurs and the travel time remains at a stable value (free-flow travel time). From 7.20 am, the outflow demand exceeds the capacity of the corridor, congestion occurs upstream from link 3 (that presents the lowest capacity of the whole corridor), and the travel time gradually increases. At 8.50 am, the delay reaches 11 min (its maximum value). Then the demand drops below the capacity of the corridor and the travel time starts decreasing. At 9.45 am, the congestion disappears.

With VSL strategy (green solid line), the situation is rather different. Before 7.00 am, the observed outflow overlays the outflow demand, no congestion occurs and the travel time remains at a stable value (free-flow travel time). At 7.00 am, the rush hour starts and VSL is activated. The free-flow speed decreases and the travel time suddenly increases. From 7.25 am, the outflows demand exceeds the (increased) capacity of the whole corridor, congestion occurs upstream from link 3 (still the link that presents the lowest capacity of the whole corridor). The total outflow is now higher because VSL is activated and the travel time gently increases. At 8.10 am, the delay reaches 5 min(its maximum value) and the travel time starts decreasing. At 9.00 am, the VSL strategy is deactivated. At 9.10 am, the congestion disappears.

a) b)

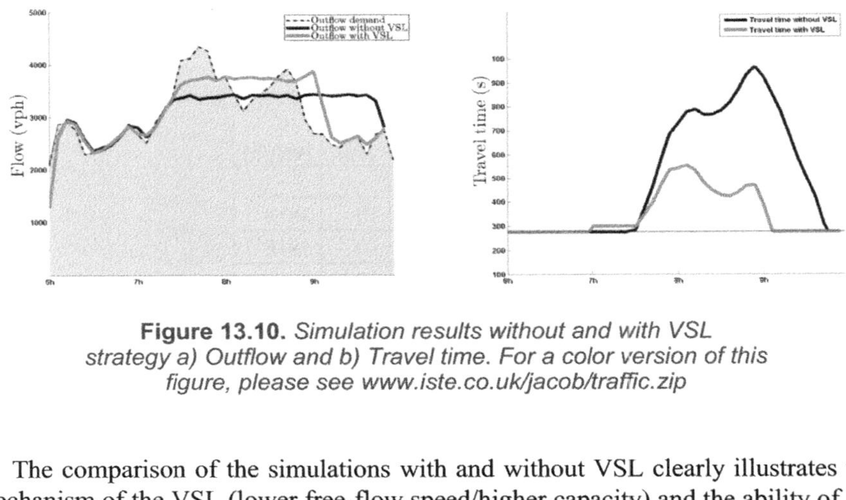

Figure 13.10. *Simulation results without and with VSL strategy a) Outflow and b) Travel time. For a color version of this figure, please see www.iste.co.uk/jacob/traffic.zip*

The comparison of the simulations with and without VSL clearly illustrates the mechanism of the VSL (lower free-flow speed/higher capacity) and the ability of the meso-LWR model to reproduce this mechanism.

	Without VSL	*With VSL*	*Impact*
Start of congestion	7.20 am	7.25 am	
End of congestion	9.45 am	9.10 am	
Congestion duration	2:25	1:45	-27%
Number of vehicles delayed	7421 vehicles	5650 vehicles	-23%
Total time spent	1437 h	872 h	-39%
Total delay	857	292	-65%
Maximum delay	11 min	5 min	-54%

Table 13.4. *Comparison of simulation results, without and with VSL strategy*

This application does not hold for a validation. The MLC model coupled with the meso-LWR model has only been calibrated with real field data without VSL. To properly validate the approach, real field data with and without VSL strategy and under comparable situations are needed. It should become possible soon since the VSL strategy was deployed in the field starting from November 2013.

13.5. Discussion

This chapter has proposed a model that takes into account the impact of LFD characteristics on the amplitude of the capacity drop observed on multi-lane sections. The model for two-lane sections only requires a single parameter, easy to measure and to calibrate. It is also able to reproduce the mechanisms engaged when VSL strategy is activated: the speed reduction that decreases capacity and the homogenization effect that increases capacity.

The MLC model is coupled to the meso-LWR model and the resulting dynamic traffic flow model is now able to reproduce the impact of LFD characteristics (when the model is mono-pipe) and the capacity drop caused by LFD. It can consider a dynamic capacity that depends on the status of VSL strategy.

The dynamic traffic flow model is applied to the A31motorway, near Thionville in France. The results confirm the ability of the model to provide important insights into the impact of VSL strategy on traffic flow conditions. Two scenarios are simulated, with and without VSL strategy. The model also provides measurements of effectiveness that confirm the potential benefits produced by VSL strategy.

However, further research is required to improve and properly validate the MLC model. First, it should be extended to three-lane sections, where similar phenomena have been observed (LFD, recurring patterns, under-use of the slowest lane, capacity drop). Moreover, the MCL model should be properly calibrated and validated on multiple sites. Without VSL strategy, it should be straightforward since the MLC model only relies on four parameters that are directly measurable based on data from loop detectors from any multi-lane section. To study VSL strategy impact, one should have a site with comparable demand scenarios with different VSL status (active/inactive). This should soon be possible since VSL was applied on A31, Thionville, starting from November 2013.

Further research is also needed to improve the performance of the meso-LWR model. Use of the meso-LWR model is relevant on motorway networks. Today, most dynamic traffic flow models are used to test traffic control strategies. Consequently, it is necessary to develop the meso-LWR model along similar lines. The initial version of the model is mono-pipe, FIFO, does not reproduce capacity drop phenomenon. Here, the MLC model is the first extension that enables the meso-LWR model to reproduce the capacity drop caused by LFD characteristics. The results are promising and the next stage will focus on network merges, where: most of congestions are triggered, capacity drops occur (widely referenced in the literature) and ramp-metering strategies can be an efficient control tool.

13.6. Acknowledgments

The author sincerely thanks Professor Ludovic Leclercq for the insightful and fruitful discussions that helped to produce the scientific content of the chapter. The author also thanks the road operator *Direction Interdépartementale des Routes Est* for providing traffic data presented in the chapter.

13.7. Bibliography

[BAN 06] BANKS J.H., "Part I: Traffic flow theory and characteristics: Effect of time gaps and lane flow distributions on freeway bottleneck capacity", *Transportation Research Record: Journal of the Transportation Research Board*, 2006.

[CAR 10] CARLSON R.C., PAPAMICHAIL I., PAPAGEORGIOU M. *et al.*, "Optimal motorway traffic flow control involving variable speed limits and ramp metering", *Transportation Science*, vol. 44, no. 2, pp. 238–253, 2010.

[CAR 99] CARTER M., RAKHA H., AERDE M.V., "Variability of traffic-flow measures across freeway lanes", *Canadian Journal of Civil Engineering*, vol. 26, no. 3, 1999.

[DAG 95] DAGANZO C.F., "Cell transmission model, part II: network traffic", *Transportation Research Part B*, vol. 29, no. 2, pp. 79–93, 1995.

[DAG 02a] DAGANZO C.F., "A behavioral theory of multi-lane traffic flow part i: long homogeneous freeway sections", *Transportation Research Part B: Methodological*, vol. 36, no. 2, 2002.

[DAG 02b] DAGANZO C.F., "A behavioral theory of multi-lane traffic flow part ii: merges and the onset of congestion", *Transportation Research Part B: Methodological*, vol. 36, no. 2, 2002.

[DUR 10] DURET A., Hétérogénéité du trafic autoroutier – identification, quantification, modélisation et impact sur l'écoulement, École Nationale des Travaux Publics de l'État, Doctoral dissertation, 2010.

[DUR 12] DURET A., AHN S., BUISSON CH, "Lane flow distribution on a three-lane freeway: general features and the effects of traffic controls", *Transportation Research Part C: Emerging Technologies*, vol. 24, pp. 157–167, 2012

[LEC 12] LECLERCQ L., BECARIE C., "A mesoscopic model designed for network application", *91st Annual Meeting of the Transportation Research Board*, January 22–26, 2012.

[WEI 13] WEIKL S, BOGENBERGER K., BERTINI R., "Empirical assessment or traffic management effects of variable speed limit system on a German autobahn: before and after", *92nd Annual Meeting of the Transportation Research Board*, January 13–17, 2013.

14

Evaluation Parameters of Re-routing Strategy

This chapter presents the results obtained from the simulation of various dynamic route guidance algorithms. The following conceptual approaches are used as tools for the solution of these problems: traffic monitoring; LOS determination and estimation of congestion conditions; traffic simulation on urban network and estimation of traffic scenarios; the forecast of the system efficiency. Simulation experiments have been carried out for the part of the Rostov-on-Don network using the software AIMSUN. For the procedure assignment, the following features were considered: alternative route determination, computation of the average number of re-routing vehicles, the re-routing period and congestion threshold. In terms of application, it is a part of transport planning and traffic simulation to prepare Rostov-on-Don transport infrastructure for FIFA World Cup 2018.

14.1. Introduction

Efficient use of a city's transport infrastructure is mostly determined by the existence and operation of intelligent transport systems (ITS). The analysis of current traffic conditions and congestion levels of main highways reveals the topical character of scientific research in the field of efficient traffic flow management. That is why the dynamic traffic re-routing from road sections with heavy traffic to the alternative routes is so important. The development of monitoring traffic flow systems and methods of traffic assignment on the road network using simulation models allows us to analyze the problem of dynamic traffic re-routing on a new level.

Chapter written by Vladimir ZYRYANOV and Anastasia FEOFILOVA.

Features of the traffic assignment on the network based on the use of route choice models are given in Sheffi [SHE 85], Mahmassani [MAH 01], Ramming [RAM 02], Barcelo and Casas [BAR 06], Florian *et al.* [FLO 08], Prato [PRA 09], and Sundaram [SUN 10]. The results of studies to find the conditions for starting re-routing traffic flows, the definition of congestion, methods of detection and measurement of congestion are given in Strickland *et al.* [STR 95], Rao and Rao [RAO 12], and Kim and Chang [KIM 12]. Some studies show the experimental data to determine the efficient amount of the re-routing vehicles. So Ben-Elia and Shiftan [BEN 10] determined that the decision for re-routing is to a large extent dependent on the content of information. Chang *et al.* [CHA 02, CHA 05] defined that detour rate grows when the number of lane blockages increases from one to three lanes in the I-95 Corridor between Baltimore and Washington Beltways. Pan *et al.* [PAN 11] suggested proactive vehicle re-routing strategies for congestion avoidance in Brooklyn and Newark. These studies have achieved significant results but the variety of networks, traffic conditions and traffic management requires new data for the current conditions.

The following conceptual approaches are used as tools for these problems' solution in our research: traffic monitoring; LOS determination and estimation of congestion conditions; traffic simulation on urban network and the forecast of the system efficiency. A part of our research is conducted on the basis of the AIMSUN model of Rostov-on-Don CBD network. A set of simulations were carried out using various random seeds for the stochastic results. A database of traffic flow parameters is very important for accurate simulation. It is obviously needed to prove the correspondence of relationship between flow, density and speed for observed and simulation data.

Simulation has been created on the basis of historical data of traffic flow parameters. The main objectives of the research are to estimate the routing strategy and relationship between the shape of macroscopic fundamental diagrams and traffic assignment on the network. In order to get a real model of traffic assignment, one needs to study the features of alternative route determination, calculation of the average number of re-routing vehicles and the re-routing period for the congestion threshold.

For the procedure assignment the following features were considered:

The alternative route determination. In order to calculate the dynamic traffic assignment we defined a set of route choice model parameters. These parameters determine the probability of each available path and model the driver's behavior by randomly selecting an alternative path. This also allows us to adapt simulation for the current traffic network and to predict different traffic flow conditions in some scenarios.

Calculation of the average amount of re-routing traffic. In the implementation of dynamic traffic re-routing it is necessary to ensure the stability of network's elements, which included the recommended alternative route. The volume of traffic from re-routed to alternative paths is determined by the level of service (LOS) of the route section.

Calculation of the re-routing period of time and congestion threshold. It is important to note that the threshold depends on the network layout and type of the roads.

14.2. Simulation framework

To determine the effectiveness of the undertaken decision it would be appropriate to run several scenarios for the development of traffic conditions. Based on driver behavior research in terms of their route choosing we propose the following classification of scenarios:

1) The driver will choose the alternative route.

2) The driver will probably not choose an alternative route.

3) The driver will probably choose an alternate route.

To run these scenarios of drivers' behavior, we used the microsimulation software AIMSUN. The objective of the simulation was to determine such parameters of route choice models at which the "behavior" of vehicles will follow the above-stated scenarios (see Table 14.1).

The next objective of the simulation is the influence of the route choice models' parameters on traffic conditions changing in the network. First, we simulated a simple fragment of the network with an opportunity to choose three different routes (see Figure 14.1), then more complex networks and finally a part of the Rostov-on-Don network. In order to run various scenarios it is important to see how traffic and incidents correlate with route choice. Therefore, we analyze scenarios with incidents on the one route.

Simulation scenarios	Route choice models	Recommended model parameters
1) The driver will choose the alternative route	*Proportional model*	α = [2;3]
	Logit model	θ = (60;100]

	C-Logit model	θ = [60;100]
		β does not influence
		β does not influence
2) The driver will probably not choose an alternative route	*Proportional model*	α = (0;0,5)
	Logit model	θ = (0;10)
	C-Logit model	θ = (0;10)
		β = (0,1; 1)
		β = (0,1; 1)
3) The driver will probably choose an alternate route	*Proportional model*	α = [0,5;2)
	Logit model	θ = [10;60)
	C-Logit model	θ = (10;60)
		β = 0,5
		β = 0,5

Table 14.1. *Implementation of the route choice models and their parameters due to the type of scenarios*

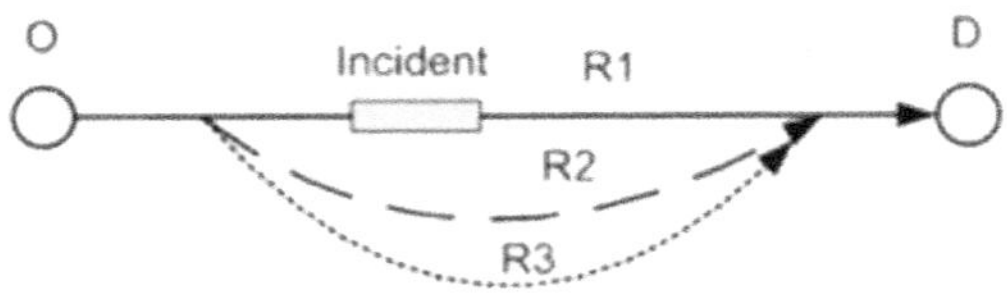

Figure 14.1. *Schematic for traffic routes in a simulated road network*

We found out that the drivers start using an alternate route only when the initial route LOS reaches F. In real conditions, this corresponds to the situation when the driver, himself, sees in front of him congestion and if there is a physical possibility to change the route, he uses an alternative path. Simulating a general driver "obedience" (Scenario 1), we found out that the initial route loading decreases more rapidly. In this case it may lead to LOS D on the alternative path and to LOS E in heavy traffic. Dealing with possible driver routing (Scenarios 2 and 3) we did not observe the LOS increase in the initial route, detecting LOS E and LOS F there (see Figure 14.2). We also found out that in case of heavy traffic the use of an alternative route does not lead to efficient results (there is still a congestion on the main route and traffic conditions on the alternate route R2 are getting worse).

One of the main tasks to be solved is to determine a set of suitable routes. But in the case of on-line traffic re-routing, we need to define the route choice cycle. To study the traffic condition changes in the development of road situations during the calibration stage, we used microsimulation. As a result of the simulation (see Figure 14.3) we found out at that the initial route volume capacity rate (V/C) decreases more rapidly, but in doing so may lead to high V/C of alternative route R2 with LOS D or LOS E in cases of heavy traffic. Reaching $\Delta t \geq 10$ min, we did not observe an increase in the initial route LOS detecting on it the values of F. Optimal distribution of the vehicles on the network occurs at $\Delta t = 3 \div 5$ min according to Ben-Elia and Shiftan [BEN 10], Jizhen *et al.* [JIZ 08], Pan *et al.* [PAN 11].

For more accurate results, we simulated the more complex network site with a total length of 17 km and total traffic volume of 13,000 veh/h (see Figure 14.4). The criteria for MOEs were the average vehicle travel time on the network. The simulation period was 2 hours, the period for actions to recover after incidents – 45 min. The results of the simulation showed that at $\Delta t \in [5;10]$ min we observe a decrease in the vehicle travel time.

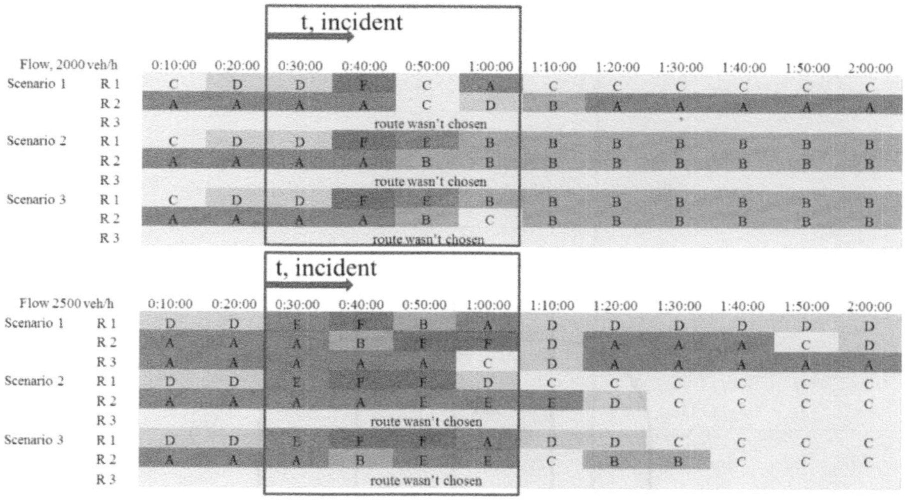

Figure 14.2. *LOS changes in scenarios rank. For a color version of this figure, please see www.iste.co.uk/jacob/traffic.zip*

The simulation results showed that the value Δt in each case depends on the traffic volume and network parameters. If $\Delta t \to$ min, due to the increase of the simulation model reaction, the average vehicle travel time on the network will decrease. On the one hand, too short a re-routing cycle can reduce model operation speed and on the other hand, the excessive traffic recommendations may lead to driver disorientation. A very long cycle will not provide the timely achievement of the dynamic assignment goal, reducing LOS on the network. The recommended value Δt for multipath networks with high traffic capacity is about 10 min.

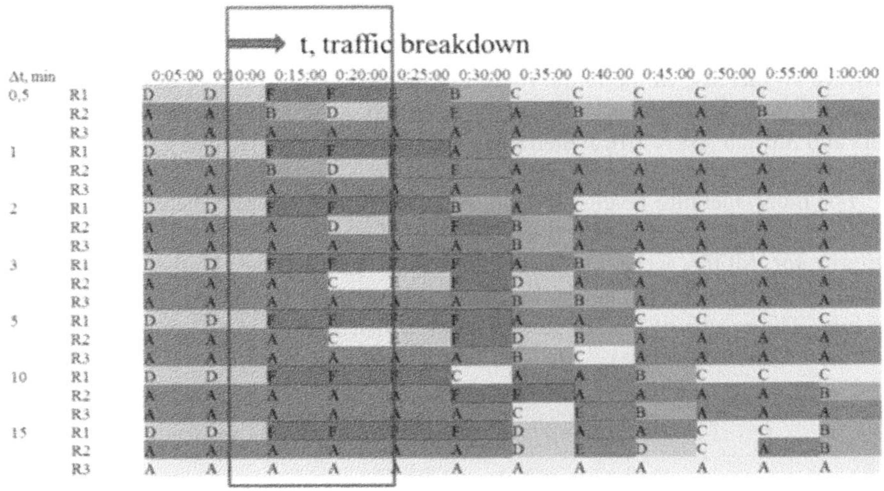

Figure 14.3. *LOS changes depending on the cycle of alternate route calculations. For a color version of this figure, please see www.iste.co.uk/jacob/traffic.zip*

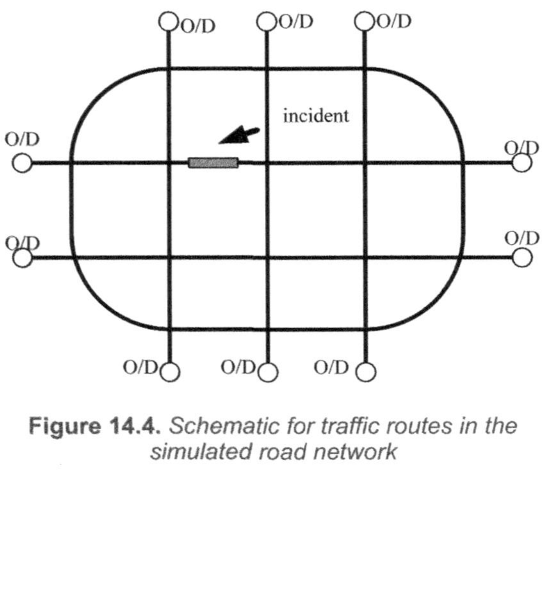

Figure 14.4. *Schematic for traffic routes in the simulated road network*

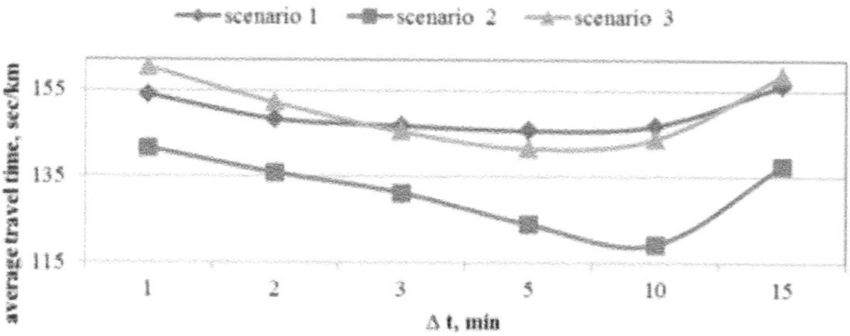

Figure 14.5. *Changes in the average vehicle travel time in the network. For a color version of this figure, please see www.iste.co.uk/jacob/traffic.zip*

14.3. Determination of the dynamic re-routing start based on traffic flow conditions

Most of the research is devoted to finding an alternative path, the influence of traffic information on road users, the analysis of the reactive strategies results for operational traffic flows management. However, the tasks of dynamic traffic flow re-routing in case of incidents and congestions are still topical. Thus, the definition and formalization of such traffic conditions under which their dynamic re-routing is efficient are of particular significance. For many years the study of the relationship between the main characteristics of traffic flow has been one of the main directions of theoretical and experimental research in the theory of traffic flow. The contradiction between the real and calculated data in the basic models and the choice of appropriate models are studied in [DEL 95a, DEL 95b, ZYR 08]. Research and prediction of traffic transition from the free state are reflected in the alternative theory of Kerner's [KER 09] traffic flow. Many traffic simulations using AIMSUN software showed the same type of traffic flow phase diagrams in case of congestions on the study route regardless of their origin.

Traffic flow phase diagrams provide the presented characteristics of traffic flow q_{max}, k_{crit}, k_{jam}. More accurate studying reveals k_{thresh} traffic flow density at which there is a transition to synchronized flow phase. This indicates a change in traffic conditions for the worse and leads to congestion if there is a lack of traffic management. These fundamental dependencies shown in Figure 14.6 allowed us to use k_{thresh} as a result of dynamic traffic flow re-routing. The volume at k_{thresh} is the width of the hysteresis loop which describes $F \to J$ and the reverse re-routing $J \to F$.

We define the width as $\Delta q = q_2(k_{thresh}) - q_1(k_{thresh}) = Qeffect$. Figure 14.7 graphically presents a methodology for determining the threshold parameters of the traffic flow and a part of re-routing vehicles. It also depicts the sustainability of the alternative path that takes the effective volume of re-routing vehicles. Let the alternate path detect traffic flow conditions which can be characterized as $(\bullet)B(k_i;q_i)$. If we are re-routing to an alternate vehicle path Q_{effect}, we will get a new traffic flow condition which can be characterized as $(\bullet)Bnew(k_j;q_i+Q_{effect})$ and meets the free traffic conditions. Consequently, the re-routing Q_{effect} will not cause the rejection of an alternative path (Figure 14.7).

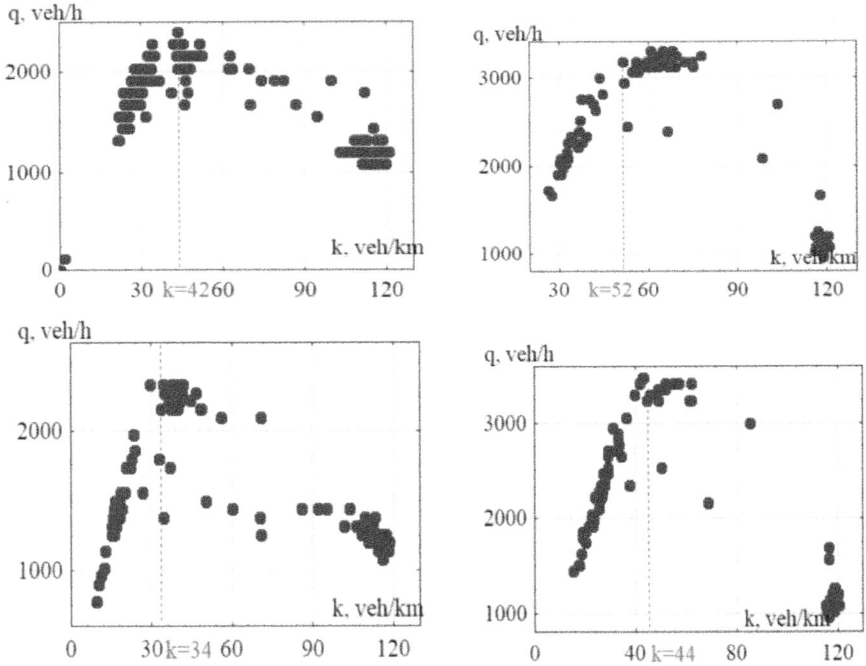

Figure 14.6. *Traffic flow phase diagrams*

Based on the above-mentioned analysis, we found out traffic flow conditions at which their dynamic re-routing is efficient and determined the efficient number of re-routing vehicles (see Table 14.2).

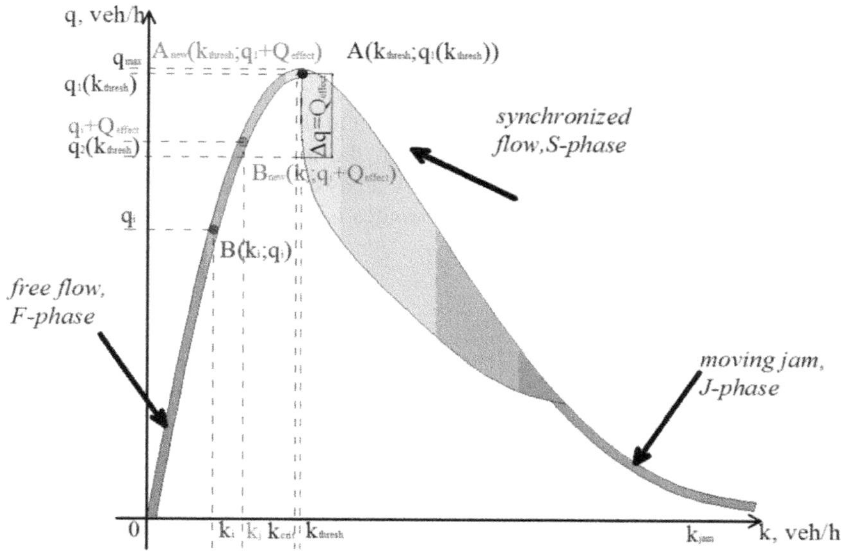

Figure 14.7. *Determination schematic for use condition for dynamic traffic flow re-routing. For a color version of this figure, please see www.iste.co.uk/jacob/traffic.zip*

Speed limit on the route, km/h	Number of route lanes	k_{thresh}, veh/km	Number of re-routing vehicles, %
60	2	42	32
	3	52	35
90	2	34	30
	3	44	36

Table 14.2. *Recommendations for dynamic traffic flow re-routing*

To prove the efficiency of the designated threshold conditions, we used ANOVA analysis on the following series of 36 simulations with the input variable parameters:

1) The number of lanes: 2, 3.

2) Speed limit for each route: 60 km/h, 90 km/h.

3) Detected threshold: LOS D, k_{thresh}, travel time. Travel time is defined as the sum of the average travel time on the route in free traffic flow and the 95th daily observed percentile of historic travel time.

4) The number of dynamically re-routing vehicles: Q_{effect}, $Q_i>Q_{effect}$, $Qi < Q_{effect}$.

The criterion we have chosen is the traffic flow speed, as it closely correlated with other traffic flow characteristics. It also reflects the change in traffic conditions on the initial path, which includes the incident and the alternative path. The aim of simulation was to reduce congestion time on the initial route while ensuring the alternative path sustainability. The ANOVA analysis showed that using k_{thresh} and Q_{effect}, the average flow speed for the two routes is above in comparison with other threshold values (see Figure 14.8).

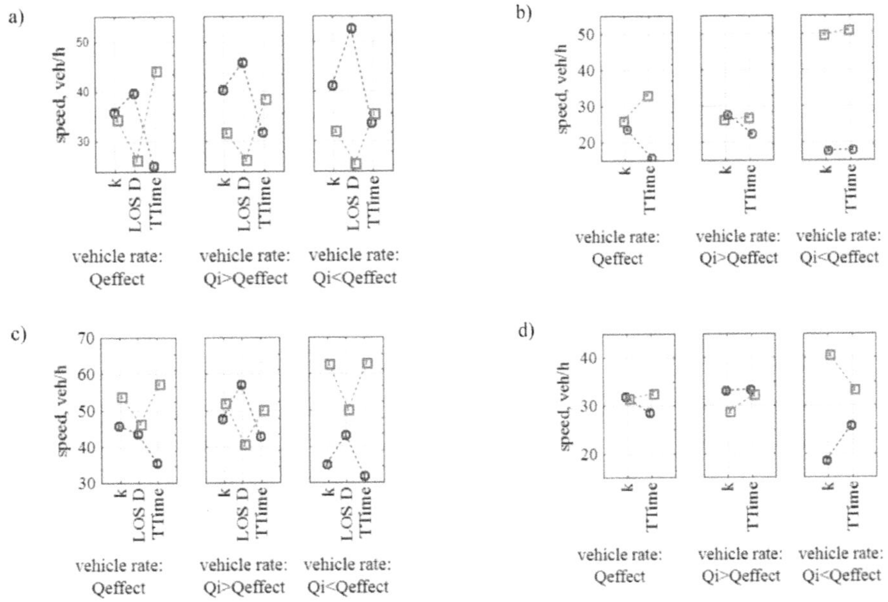

a) number of lanes -2, traffic speed – 60 veh/h
b) number of lanes -3, traffic speed – 60 veh/h
c) number of lanes -2, traffic speed – 90 veh/h
d) number of lanes -3, traffic speed – 90 veh/h

Figure 14.8. *ANOVA analysis graphs. For a color version of this figure, please see www.iste.co.uk/jacob/traffic.zip*

To implement these results was a part of transport planning and traffic simulation to prepare the transport infrastructure of Rostov-on-Don for the FIFA World Cup 2018 (Figure 14.9). As an example of the dynamic traffic flow re-routing, we studied the network area near the stadium functioning as the entrance to the city. For real data, we used incident statistics in the period of 2010–2012 on the Eastern highway of Rostov-on-Don. The statistical data also provided the duration of incident consequence elimination.

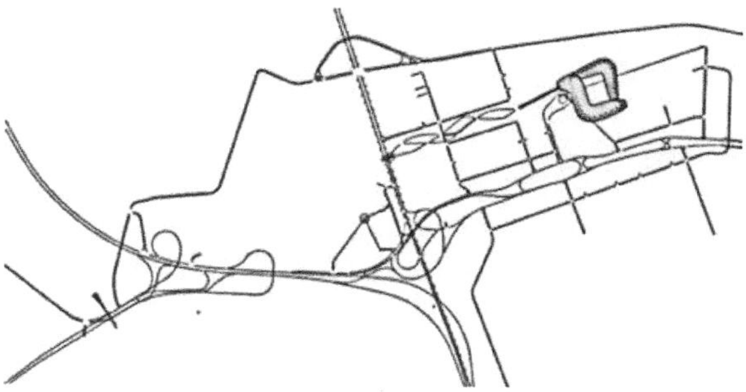

Figure 14.9. *Simulation area*

For qualitative analysis of the strategies, we used data obtained in AIMSUN software with application software (API) for additional data. Constructed space-time diagrams reflect the change in traffic states on the route while performing any impact.

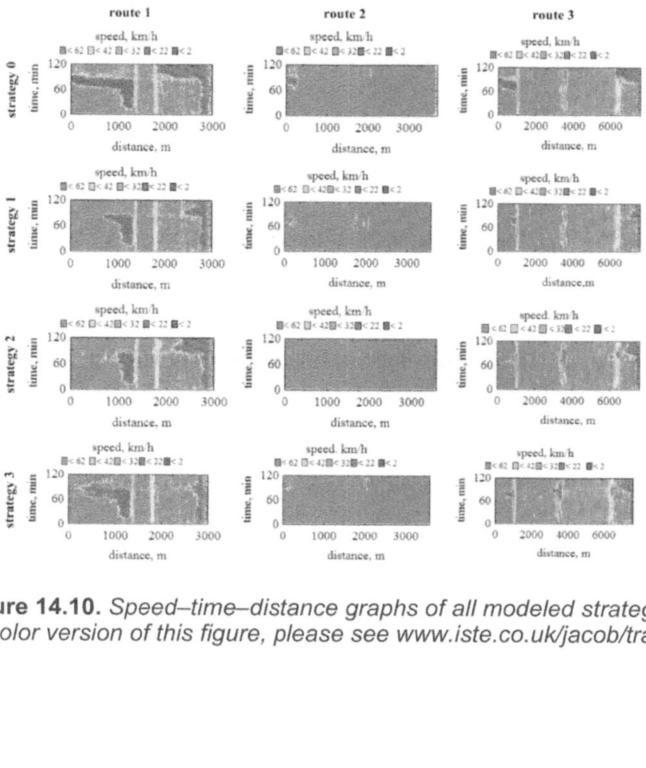

Figure 14.10. *Speed–time–distance graphs of all modeled strategies. For a color version of this figure, please see www.iste.co.uk/jacob/traffic.zip*

Quantitative analysis of the experimental results showed a decrease in the vehicle travel time using strategy 1 on the initial route up to 200% and using strategy 2–3 up to 175% (Figure 14.11).

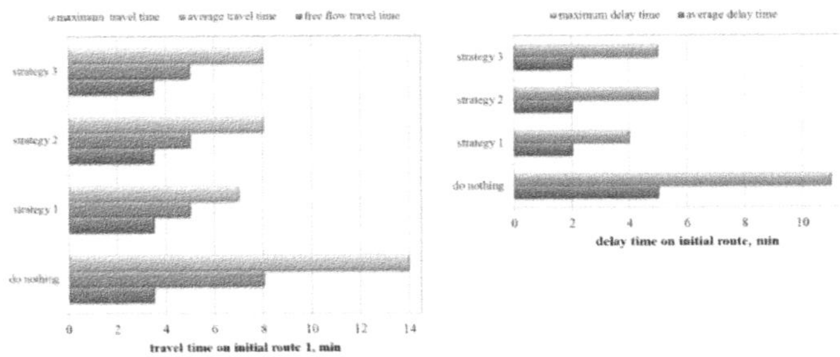

Figure 14.11. *Quantitative analysis of microsimulation's results*

14.4. Conclusion

This chapter presents research aimed at finding and adapting models of routing traffic to real conditions. By calibrating the model parameters, we have determined that the forecast changes in the traffic situation should be under three scenarios: when the driver chooses the recommended route, the driver will probably choose the route and the driver probably will not choose the route.

We also defined the conditions for the start of the dynamic re-routing of traffic flows, and found that the conditions were determined by the parameters of the network and the speed limits. Our proposed method of evaluating the conditions to begin redirection can also determine the effective portion of the re-routing traffic. At the same time, we evaluated whether stable conditions should be maintained on the alternate route, or whether additional routes should be defined to reduce the duration of congestion on the original route.

Further research will continue for the following tasks: in the process of transport planning for predicting the traffic situation in the development of transport infrastructure and to examine how redistributed traffic flows according to the new network; to forecast the traffic situation in the event of incidents; for the improvement of methods and algorithms for dynamic traffic re-routing.

14.5. References

[BAR 06] BARCELO J., CASAS J., "Stochastic heuristic dynamic assignment based on AIMSUN microscopic traffic simulator", *85th Transportation Research Board*, Annual Meeting, 2006.

[BAR 10] BARCELO J. (ed.), *Fundamentals of Traffic Simulation*, Springer, p. 440, 2010.

[BEN 10] BEN-ELIA E., SHIFTAN Y., "Which road do I take? A learning-based model of route-choice behavior with real-time information", *Transportation Research Part A*, vol. 44, pp. 249–264, 2010.

[CHA 02] CHANG G.-L., FEI X., POINT-DU-JOUR J.Y., Interrelations between variable message signs and detour operations in the I-95 corridor, Final report, p. 21, 2002.

[CHA 05] CHANG G.-L., FEI X., POINT-DU-JOUR J.Y., "Empirical analysis on influence of traffic information by divergence rate using observed traffic data", *Proceedings of the Eastern Asia Society for Transportation Studies*, vol. 5, pp. 1484–1495, 2005.

[DAG 08] DAGANZO C.F., GEROLIMINIS N., "An analytical approximation for the macroscopic fundamental diagram of urban traffic", *Transportation Research Part B*, vol. 42, pp. 771–781, 2008.

[DEL 95a] DEL CASTILIO J.M., BENITEZ F.G., "On the functional form of the speed-density relationship – I: General theory", *Transportation Research Part B: Methodological*, vol. 29B, no. 5 , pp. 373–389, 1995.

[DEL 95b] DEL CASTILIO J.M., BENITEZ F.G., "On the functional form of the speed-density relationship – II: Empirical investigation", *Transportation Research Part B: Methodological* , vol. 29B, no. 5 , pp. 391–406, 1995.

[FLO 08] FLORIAN M., MAHUT M., TREMBLAY N., "Application of a simulation-based dynamic traf-fic assignment model", *European Journal of Operational Research*, vol. 189, no. 3 pp. 1381–1392, 2008.

[JIZ 08] JIZHEN G., CHANGQING Z., XUELIANG Z. *et al.*, "VMS release of traffic guide information in Beijing olympics", *Journal of Transportation Systems Engineering and Information Technology*, vol. 8, no. 6, pp. 115–120, 2008.

[KER 09] KERNER B.S., *Introduction to Modern Traffic Flow Theory and Control: The Long Road to Three-Phase Traffic Theory. The Long Road to Three-Phase Traffic Theory*, 2009.

[KIM 12] KIM W., CHANG G.-L., Review and enhancement of CHART operations to maximize the benefit of incident response and management, report, University of Maryland, Department of Civil and Environmental Engineering, Maryland, 2012.

[LER 09] LERNER N., SINGER J., ROBINSON E. *et al.*, Driver use of enroute real-time travel time information, Final Report, 2009.

[MAH 87] MAHMASSANI H.S., WILLIAMS J., HERMAN R., "Performance of urban traffic networks", *Proceedings of the 10th International Symposium on Transportation and Traffic Theory*, pp. 1–20, 1987.

[MAH 01] MAHMASSANI H.S., "Dynamic network traffic assignment and simulation methodology for advanced system management applications", *Networks and Spatial Economics*, vol. 1, no. 3, pp. 267–292, 2001.

[PAN 11] PAN J., KHAN M.A., POPAY I.S., *et al.*, Proactive vehicle re-routing strategies for congestion avoidance, report, Department of Computer Science, 2011.

[PRA 09] PRATO C., "Route choice modeling: past, present and future research directions", *Journal of Choice Modelling*, vol. 2, no. 1, pp. 65–100, 2009.

[RAM 02] RAMMING M., Network knowledge and route choice, PhD Thesis, Massachusetts Institute of Technology, Department of Civil and Environmental Engineering, 2002.

[RAO 12] RAO A.M., RAO K., "Measuring urban traffic congestion – a review", *International Journal for Traffic and Transport Engineering*, vol. 2, no. 4, pp. 286–305, 2012.

[RUP 03] RUPERT B., WRIGHT J., PRETORIUS P. *et al.*, Traveler information systems in Europe, available at: www.international.fhwa.dot.gov, 2003.

[SHE 85] SHEFFI Y., *Urban Transportation Networks: Equilibrium Analysis with Mathematical Programming Methods*, Prentice-Hall, Englewood Cliffs, USA, 1985.

[STR 95] STRICKLAND, SHELDON G, BERMAN W., Congestion control and demand management, public roads on-line, available at www.tfhrc.gov/pubrds/winter95/p95wi1.htm, 1995.

[SUN 10] SUNDARAM S. *et al.*, "Simulation-based dynamic traffic assignment for short-term planning applications", *Simulation Modelling Practice and Theory*, vol. 19, no. 1, 2010.

[ZYR 08] ZYRYANOV V., "Simulation of impact of components of ITS on congested traffic states", *Proceedings of 7th European Congress on Intelligent Transport Systems*, Geneva, pp. 1–5, 2008.

PART 4

Travel Information

15

Pre-Trip Road Information Impact Assessment: A Literature Review

Although pre-trip road information has existed for years, the evolution of technologies today allows road travelers to plan a trip in a more timely and accurate way. However, although both the quantity and quality of information have considerably improved, the impacts of such information on the traveler's mode, route and departure time choices have not often been assessed. As a result, pre-trip road information impacts are not precisely known, and road operators do not know precisely to what extent the provision of pre-trip information is useful for road users and traffic management. A literature review of the studies and surveys has, therefore, been carried out so as to determine the impact of pre-trip road information. This chapter presents the results of this review, and also its limits: each study and survey often relates to the specific information content and to a specific trip purpose, and study results are dependent on existing broadcasting media at the time of the study. The chapter concludes with the possibility of dividing travelers into pre-trip changers and non-changers.

15.1. Introduction

Although pre-trip road information has existed for years, the recent evolution of technologies allows travelers to plan their trip in a more timely and accurate way. Indeed, many public or private traffic information websites have been developed during the last 20 years, and are regularly enhanced. Information on real-time traffic conditions has improved thanks to the deployment of specific sensors or the ability to follow part of the traffic. Predictive models have also been developed in order to provide knowledge of the traffic conditions in advance.

Chapter written by Sylvain BELLOCHE, Charlotte PIERREFEU and Caroline SORAND.

However, though both the quality and quantity of information have considerably improved, there is little knowledge about the impacts of such information on a traveler's mode, route and departure time choices: a few impact assessment studies or surveys have been carried out on this issue, but most of the time, these studies or surveys focus on part of the information content, or on a specific trip purpose, such as commuting trips. Therefore, information service providers do not know, precisely, to what extent the information that they provide is useful to road users, or how much the traffic demand is distributed in time and space thanks to the pre-trip road information.

In order to determine the impacts of pre-trip road information on the travelers' choices, a literature review has been carried out. Although, the objective of this review was to assess pre-trip road information impacts, this review first analyzes the context of the studies and surveys so as to identify the information broadcast content that was assessed and the available broadcasting media at the time of the study. This issue is presented in the next section. The second part of the analysis is to determine the drivers' characteristics, and also the determining factors that encourage travelers whether to change their mind or not. The last section then presents the pre-trip road information impacts on different aspects of travelers' travel plans. The chapter concludes with the possibility of splitting up the community of road information users into changers and non-changers and discusses the limits of the results of the review.

15.2. Pre-trip road information content and broadcasting media

The different assessment studies show that pre-trip road information groups together with the different categories of information. The Easy Way programme has suggested the following classification for pre-trip road information [EAS 12]:

– event information; this predictive or real-time information takes into account the expected or unexpected events such as roadworks, traffic jams, incidents or demonstrations, and includes parking information;

– traffic conditions; this information is either predictive or real-time;

– speed limit information; however, speed limit information is mostly on-trip information and not pre-trip;

– travel time information; this information is either predictive or real-time;

– weather information;

– multimodal (or co-modal) information; this information refers to transportation means other than the road and includes information on park-and-ride facilities.

Most pre-trip road information assessment studies only deal with either event information, traffic conditions or travel time information. According to Jou *et al.* [JOU 97], road users are more interested in the traffic condition information for their professional-related trips, whereas, they are more interested in getting parking information for their personal trips. The importance of the traffic condition information for road users is confirmed by the evaluation of 511 in Virginia [VIR 04], in the impact assessment of the Strasbourg traffic management centre [ZEL 09] and by Vaughn *et al.* [VAU 95]. Travel time information appears in the third position in the Jou *et al.*, [JOU 97] survey, and also as the most sought-after information in a survey in Rennes [DOR 07]. Moreover, travel time information is considered to be directly linked to the changes in travel plans, especially for professional-related trips [KHA 08]. Event information, and notably incidents, or weather information may be of a higher importance locally for trip planners, such as for leisure trips in Yellowstone [KAL 04].

Assessment studies and survey results are also dependent on when these studies and surveys were carried out. Indeed, broadcasting media have considerably evolved over the last few years, and media that were fashionable 20 years ago may now be out-of-date. Two surveys carried out in Tsirimpa *et al.* [TSI 00] and in Khattak *et al.* [KHA 06] show that radio use for pre-trip road information has definitely declined between these two dates, whereas, television and internet use has definitely increased. The main broadcasting media are – or were – the following:

– computer (Internet): broadcasting capacities have considerably increased over the last 10 years. This broadcasting media allows highly detailed information, and may be considered as one of the most powerful pre-trip road information broadcasting media;

– smartphone and tablet computer: both have appeared a few years ago and are now quite widespread, but still with increasing broadcasting capacities. They may be used for either pre-trip or on-trip: determining the pre-trip road information impact when it is delivered with either a smartphone or a tablet may be rather difficult;

– telephone: there are two ways to get information with a telephone, the first one by calling a phone number such as 511, the second one is by receiving texts. Both ways can be used for either pre-trip or on-trip;

– radio: although, there are still a lot of radio programs that broadcast traffic information, this media was certainly the most prevalent before the deployment of the Internet. Today, this media is mostly seen by users as an on-trip road information media [ZEL 05], and radio information content is limited by the fact that it often delivers a vocal message to the listeners;

– television: television programs or reports may give road information to the listeners. This media is almost always used pre-trip;

– navigation devices: on-board navigation systems may be used in pre-trip to get traffic information;

– paper: even if this broadcasting media may appear old-fashioned, it is still in use such as for information and advice for specific major events.

As indicated in a few studies, someone consulting the Internet will be more likely to make changes in his travel plans than if he has used another broadcasting media [YIM 00, KHA 08]. Yim suggests that this is due to the fact that information broadcast on the Internet is perceived to be of better quality and is more reliable. However, even in the most recent studies, the Internet is not often the most consulted media to get pre-trip road information. A 2011-study [MED 11] reports that only 13% of those polled use the Internet to plan a trip, against 29% who use the radio. Another 2011 study based on logbooks and interviews confirms that radio is more widely used than Internet for pre-trip information. However, if personalized real-time information is available by a navigation device, most users will prefer this source of information to others [HOO 12].

Moreover, the frequency of the consultation of the pre-trip road information is often low: in North Carolina, about half of all the travelers have never consulted road information while planning a trip [KHA 08], and only 31% of Lyon residents have consulted the traffic information website of the urban area at least once [COR 04]. Table 15.1 gives the results of a survey [COR 04] on the frequency of the consultation of the traffic information website of the Lyon urban area CORALY (http://www.coraly.com).

Frequency of the website consultation	Local people (%)	Non-local people (%)
At least once a day	5	4
4–5 times a week	3	0
1–3 times a week	4	2
2–3 times a month	5	4
Less than 2–3 times per month	14	7
Never	69	83

Table 15.1. *Frequency of the consultation of CORALY (Lyon urban area traffic information website) [COR 04]*

A study based on logbooks also concludes that for only 4.1% of the trips, a source of information has been consulted [TSI 05]. Hoogendoorn *et al.*, [HOO 12] found out that for 16% of the trips, pre-trip road information has been consulted according to logbooks if people did not have the access to real-time information given with a navigation device. If a navigation device delivering real-time information is available, pre-trip road information is checked for 38% of the trips. Consultation of pre-trip road information is, therefore, quite irregular.

15.3. Determining factors for user choice

There is a consensus in the literature on the fact that the users' reactions toward information is highly dependent on the conditions in which they receive this information and its perceived quality. Viswanathan [VIS 01] sums up this idea by saying that "information use depends on a range of personal, travel-related and contextual factors." This point is crucial while assessing pre-trip road information impacts, since there is no direct link between broadcast information and its effects. However, the literature studied provides some explanations on the factors that play a part in the decision-making process of a trip planning.

15.3.1. *Trip purpose*

The trip's purpose is a determining factor since the user does not require the same information content for a professional-related trip as for a personal-related trip (see Chapter 2), and since the reaction to information is not the same. Indeed, for a commuting trip, information users are less prone to changing the trip characteristics since it is a regular trip and since users often know the road network they use [JOU 97]. Paradoxically, for this trip purpose, users are more receptive to information [TSI 05]. The explanation of this paradox may be that since users know the network, they can easily adapt the trip characteristics but only in the case of a major event or very bad traffic conditions. Indeed, Van der Horst [VAN 04] indicates that for a normal and regular trip, the pro-active decision process is not engaged, unless there are specific or unforeseen circumstances.

15.3.2. *Information perceived quality and reliability*

When information is provided to road users, its quality and reliability determine whether these road users take it into account or not. Indeed, Zhang and Levinson [ZHA 08] show that inaccurate information may tend to discredit it in road users' minds. Also, according to Jou [JOU 01], road drivers are able to determine information accuracy very quickly and, therefore, may adapt to their behavior accordingly. Moreover, according to a survey in New York, the more information is reliable, the more people will take it into account [JOU 01].

15.3.3. *Personal factors*

Several personal or subjective determining factors occur in the decision process while planning a trip. In a review concerning travelers' characteristics, Abdel-Aty and Abdalla [ABD 04] note that these factors may be linked to demographic, behavioral or even "technology-interest-related" aspects. Jou *et al.* [JOU 97] also indicates that the traveler's age influences the choice. The list below sums up the main personal or subjective determining factors identified in the review:

– traveler's age: this factor seems to have two consequences: first, the older a traveler is, the less he will seek information [ETT 04, ZHA 08]. Second, once informed, the older a traveler is, the less he will change route or departure time [JOU 01].

– traveler's level of education: Tsirimpa *et al.* [TSI 05] show that travelers with a university degree are more inclined to change their trip characteristics.

– traveler's salary: middle or high-income earners are the most likely users of pre-trip road information [ZHA 12]. However, this may be simply due to the fact that the wealthier a traveler is, the more possible it is for him to get or to pay for information.

– traveler's behavior: once information has been taken in, and despite bad traffic conditions, some travelers will prefer not to change their travel plans so as not to be surprised during their trip. On the contrary, some travelers are likely to take other routes, or make changes to their travel plans without knowing the outcome. These are called "experience-seekers" by Shiftan *et al.* [SHI 11]. Men are more inclined to change than women [JOU 01].

15.3.4. *The traveler's memories*

Another determining factor indicated in the studied literature is the traveler's memories. Indeed, previous experiences for a regular trip and the way the traveler remembers it has an influence about how this traveler will take into account pre-trip road information.

According to Sun *et al.* [SUN 05], the use of pre-trip road information services is in fact mainly due to the traveler's recent experiences. Tsirimpa *et al.* [TSI 05] indicates that the more a traveler has experienced delays due to travel time variability the more he will be inclined to change his trip characteristics when he gets pre-trip information.

However, the traveler's memories may disappear or be less present in the long term when changes are made quite often for a regular trip [SUN 05] with the traveler's increased experiences

15.4. Pre-trip road information impacts

As already stated in section 15.2, around or more than half of the travelers do not consult information before making a trip. This is consistent with the results of section 15.3, unless there are specific events or traffic conditions, travelers do not try to obtain information for regular trips.

Therefore, is it worth broadcasting information, and why? For the second part of this question, the road operator's point of view and the traveler's point of view should be both considered. The road operator's point of view is easy to understand, since on a congested network, it is better to spread the demand over time and over less congested roads. By limiting congestion, the road operator is likely to improve both safety and network efficiency.

From the traveler's point of view, several reasons may explain the pre-trip information consultation. Table 15.2 sums up the results of a survey in the Puget Sound region on the main reason for changing travel plans when a traveler gets information [TSI 05]. Even if this survey does not distinguish pre-trip information from on-trip information, it points out that about 40% of travelers do not try to obtain a reduction in journey time when making a change. Indeed, the three main parameters that seem to influence a traveler's choice are: efficiency, safety and comfort of travel.

Reason for changing travel plans	Percentage
Reduce travel time	60.1
Reduce travel time uncertainty	10.5
Get less stressful travel conditions	18.3
Get safer travel conditions	6.5
Other reasons	4.6

Table 15.2. *Benefits sought by travelers when changing travel plans in the Puget Sound region [TSI 05]*

15.4.1. *Overall impact of pre-trip road information*

If pre-trip road information impacts travel plans, these impacts may deal with the following traveler's choice:

– impact on route choice;

– impact on departure time choice;

– impact on modal choice;

– impact on the destination choice or trip status;

– impact on several components of the trip.

Study or survey references	Study or survey methodology	Absence of changes in travel plans (%)
San Francisco [KHA 96]	Stated-preference on a major event	40
San Francisco [KHA 96]	Survey	54
Study [KHA 96]	Stated-preference	15
France [CNI 96]	Intention	80
Taiwan [JOU 97]	Interviews at travelers' home	23–35 (depending on trip purpose)
San Francisco [YIM 00]	Online survey	less than 20
Puget Sound [TSI 05]	Logbooks	67.7
511 Virginia [VIR 04]	Study	51
Basel [RAP 03]	Survey	83
Paris [CGP 04]	Intention	45
England [TRA 05]	Survey	48.7
North Carolina [KHA 08]	Postal questionnaire	21.4

Table 15.3. *Assessment of pre-trip road information impacts: the absence of changes in travel plans*

Several studies and surveys have assessed whether or not information has any impact on travel plans. The results of these studies and surveys are listed in Table 15.3 above.

In stated-preference surveys, people generally tend to overestimate their ability to act. Therefore, if these surveys are put apart, and so is the Basel survey since it was done on a short-lived demonstrator, most of the results approach 50%. In other words, when people get information, about half of them may change their travel plans. However, as indicated by Khattak *et al.* [KHA 99], travel plan changes are not regular. Indeed, in the San Francisco survey, about 36% of those who had changed travel plans did so once a month or less, about 26% did so every two or three weeks, and 38% did so almost once a week.

15.4.2. *Impact on route choice*

Pre-trip road information may have an impact on the route that a traveler had planned to use. This impact might be on part of the route or concern the whole route. However, since very few studies distinguish the two, this chapter does not make a distinction. Table 15.4 below sums up the results of the studies and surveys on the impact of pre-trip road information on the route choice for a trip. The table also highlights if the impact on the route choice is the major impact, in cases where the study or survey also assessed other impacts of pre-trip road information.

From this review, an average of about 20% of the travelers change route once they have taken into account the pre-trip road information. However, this average value hides a few lower and several higher percentages that may be due to local differences such as a grid or non-grid network, or to the survey itself, such as the sample choice or the trip purpose that was effectively surveyed.

The frequency of the route change for a regular trip is, however, quite low. The Los Angeles survey shows that more than half of the travelers that have changed route have done so once or twice a month [VAU 95]. In Korea, the results of the questionnaire show that travelers change route about 3.55 times a month on average [TSI 05]. In conclusion, the frequency of route changes is about a few times a month.

Study or survey references	Study or survey methodology	Impact on route choice (%)	Major impact?
Los Angeles [VAU 95]	Phone survey	44.2 (at least once a month)	Same as the impact on the departure time
San Francisco [KHA 96]	Stated-preference on a major event	17	No
Seattle [JOU 01]	Survey (commuting trips)	11	No
San Francisco [KHA 99]	Survey	16.6	Yes
Taiwan [JOU 97]	Interviews at travelers' home	29–35	Yes
Seattle [YIM 00]	Online survey	about 20	No
San Francisco [YIM 00]	Online survey	about 50	Yes
Washington [SHA 01]	Real-data based simulation	3.3	No
Puget Sound [TSI 05]	Logbooks	about 21	Same as the impact on the departure time
Korea [TSI 05]	Questionnaire to drivers	40	–
Basel [RAP 03]	Survey	7	Same as the impact on the departure time
Paris [CGP 04]	Intention	31	–
England [TRA 05]	Survey	16.8	Same as the impact on the departure time
France [ZEL 05]	Survey	about 20	Same as the impact on the departure time
North Carolina [KHA 08]	Postal questionnaire	34.6	–
Study [ZHA 12]	Stated-preference based simulation	5.7	No

Table 15.4. *Assessment of pre-trip road information impacts: impact on the route choice*

15.4.3. *Impact on departure time*

The literature reviewed provides heterogeneous definitions of what is a real change of departure time. A survey based on logbooks has considered a change of departure time to be if there was a difference of 10 min between the average departure time and the effective departure time, whereas, another survey took a 5-min threshold.

Another comment occurring in the literature is the traveler's room for the maneuver. Since time is limited, travelers cannot differ their departure time by much. A French survey has indicated that the average room for the maneuver is around 20 minutes [CGP 04].

Table 15.5 sums up the results of the studies and surveys about the impact of pre-trip road information on departure time. As for Tables 15.4 and 15.5 also states if the impact on departure time is the major impact, in cases where the study or survey assessed other pre-trip road information impacts.

Study or survey references	Study or survey methodology	Impact on departure time (%)	Major impact?
Los Angeles [VAU 95]	Phone survey	41.1	Same as the impact on route choice
San Francisco [KHA 99]	Stated-preference on a major event	33	Yes
Seattle [JOU 01]	Survey (commuting trips)	13	Yes
San Francisco [KHA 99]	Survey	5.2	No
Taiwan [JOU 97]	Interviews at travelers' home	14–19%	No
Seattle [YIM 00]	Online survey	45.5	Yes
San Francisco [YIM 00]	Online survey	about 40	No
Washington [SHA 01]	Real-data based simulation	11.1	Yes
Puget Sound [TSI 05]	Logbooks	11.3	Same as the impact on route choice
Paris [CGP 04]	Intention	20	–
Basel [RAP 03]	Survey	6	Same as the impact on route choice
England [TRA 05]	Survey	16.8	Same as the impact on route choice
France [ZEL 05]	Survey	15–20	Same as the impact on route choice
Study [ZHA 12]	Stated-preference based simulation	12.1	No

Table 15.5. *Assessment of pre-trip road information impacts: impact on departure time*

From studies and survey results, approximately 20% of the travelers that consulted pre-trip road information changed their departure time. For more than 75% of the studies, this impact is at least equal to or higher than the impact on the route choice. Therefore, as for the impact on the route choice, the change of the departure time is one of the major impacts of pre-trip road information.

As far as the frequency of the changes is concerned, there are very few results. From the Los Angeles survey, it seems that the changes occur from one to six times a month [VAU 95].

15.4.4. *Impact on mode choice*

Pre-trip road information impact on mode choice is not discussed much in the literature reviewed. This may be explained by the fact that multimodal information such as the location of park-and-rides or the bus travel time to the city center is not often given by road information services.

Table 15.6 sums up the results of the studies and surveys on the impact of pre-trip road information on the mode of choice. In this review, the focus has been put on a change from car to other means of transportation.

Study or survey references	Study or survey methodology	Impact on mode choice	Major impact?
San Francisco [KHA 96]	Stated-preference on a major event	2% (23% if multimodal information was given)	No
San Francisco [KHA 99]	Survey	"Some have changed of mode and of route or departure time"	No
Taiwan [JOU 97]	Interviews at travelers' home	10–15 %	No
Seattle [YIM 00]	Online survey	2–3 %	Yes
Puget Sound [TSI 05]	Logbooks	1.4%	No
Basel [RAP 03]	Survey	4%	No
England [TRA 05]	Survey	5%	No
France [ZEL 05]	Survey	Less than 5%	No
Study [ZHA 12]	Stated-preference based simulation	27.7% (including walk, bike and scooter)	Yes

Table 15.6. *Assessment of pre-trip road information impacts: impact on the mode choice*

In Asian studies and surveys, modal shift is very high. This is not the case elsewhere, with modal shift of about 2–3% after the consultation of pre-trip information.

However, as indicated in the San Francisco survey [KHA 99], the impact is likely to be more significant when multimodal information is broadcast to travelers, such as in the Basel survey or in the Transport Direct (England) survey. Nonetheless, and despite the possibility to increase the impact on the modal shift, this impact is minor when compared to the impacts on route choice and departure time.

15.4.5. *Other impacts*

Other impacts of pre-trip road information concern the change of destination and putting off the trip. For both impacts, information in the literature reviewed is seldom, undoubtedly because such impacts are themselves rare.

The change of destination may occur for specific trip purposes, such as shopping-related trips: there are very few consequences while changing from one shopping mall to another one. A survey in Austin shows that 25% of the shoppers changed destination when bad traffic conditions or major events were reported on their route [KRA 00]. However, the fact that the survey was stated-preference based may have biased the results. Indeed, in other surveys, the changes are less numerous: in the Transport Direct survey, 3.4% changed their trip [TRA 05].

The literature also shows that putting off a trip altogether is also rare, except for leisure trips [JOU 97].

15.4.6. *Coupled impacts*

Since assessment studies and surveys focus on only one component of a trip, the issue of coupled impacts is not often discussed in the literature reviewed. The only coupled impacts described deal with route and departure time changes. Indeed, Jou [JOU 01] and Khattak *et al.* [KHA 95] agree that there is a correlation between the change of route and the change of departure time: travelers changing their route are likely to change the departure time, and travelers that are not prone to changing their route are not prone to changing the departure time either.

15.5. Conclusions and discussion

Although pre-trip road information is now widespread thanks to different (but efficient) broadcasting media, just under half of the travelers seem to be consumers of such information. The more regular the trip is, the less the travelers consult information before travelling. Those who consult information do not act in the same way: their reaction depends on personal factors, on the trip's purpose and on the traveler's memories of past experiences. Moreover, consulting pre-trip information results in no change in the behavior for about half of the travelers. If changes in travel plans occur, the primary reason for these changes may also differ from one traveler to another: some travelers are sensitive to gain in travel time, whereas others are more sensitive to experience less stressful travel conditions. The major changes in behavior concern route choice and departure time: on an average, about 20% of the travelers who consulted the information changed their route or departure time, and did so several times a month. However, locally, these changes may be more significant. The impact of pre-trip road information on the traveler's modal choice is less significant: only a very low percentage of the travelers change their mode after consulting information. Nonetheless, this change is likely to be more important if multimodal information is provided to travelers. Other impacts such as the change of destination and putting off a trip have also been observed, but there is very little literature on these aspects.

From these results, it seems that the splitting of travelers into changers and non-changers suggested by Haselkorn *et al.*, is a relevant way of classifying travelers' reactions to pre-trip road information [KHA 95]. Haselkorn *et al.*, also split changers into route changers, pre-trip time changers and route and time changers. However, this classification does not take into account modal shifts, or trip destination changes.

These results and especially the average results indicated in this paper provide initial answers to the question of the impact of pre-trip road information broadcasting. Nonetheless, they are limited by the different surveys and studies methodologies undertaken. Moreover, in a study or survey, it is sometimes difficult to distinguish pre-trip road information impacts from on-trip road information impacts, since some mix both. Thus, in order to have more precise knowledge of the reasons that make a traveler effectively change his travel plans and the consequences of information broadcasting on traffic management, another study is required, focusing solely on pre-trip road information.

15.6. Bibliography

[ABD 04] ABDEL-ATY M., ABDALLA F., "Understanding drivers' route choice under long-term pretrip and short-term en-route traffic information using generalized estimating equations", *Journal of Transportation Engineering*, pp. 777–786, 2004.

[CNI 96] CNIR, Bison-futé: 20 ans au service du conducteur, 1996.

[CGP 04] CGPC – CONSEIL GÉNÉRAL DES PONTS ET CHAUSSÉES., Évaluation socio-économique des systèmes d'exploitation de la route en milieu urbain, in Rapport du groupe de travail, 2004

[COR 04] CORALY & MV2, Suivi barométrique 2001-2004, 2004.

[DOR 07] DOR BREIZH, Enquête sur le dispositif Dor Breizh, Report, 2007.

[EAS 12] EASY WAY, Traveller Information Services Reference Document: TIS Deployment Guideline Annex, 2012.

[ETT 04] ETTEMA D., TIMMERMANS H.J.P., ARENZTE T.A., "Modelling perception updating of travel times in the context of departure time choice under ATIS", *Journal of Intelligent Transportation Systems*, vol. 8, pp. 33–43, 2004.

[HOO 12] HOOGENDOORN S., RAMOS G.M., FREJINGER E. *et al.*, "Commuters and traffic information: a revealed preference study on route choice behavior", *TRAIL Research School*, p. 10, 2012.

[JOU 01] JOU R.C., "Modeling the impact of pre-trip information on commuter departure time and route choice", *Transportation Research Part B*, vol. 35, pp. 887–902, 2001.

[JOU 97] JOU R.C., HU T.Y., LIN C.W., "Empirical results from Taiwan and their implications for Advanced Traveler Pretrip Information Systems", *Transportation Research Record*, vol. 1607, pp. 126–133, 1997.

[KAL 04] KALINOWSKI A., EIDSWICK J., "Greater Yellowstone regional traveler weather information system", *Western Transportation Institute*, p. 93, 2004.

[KHA 95] KHATTAK A.J., SHOFER J.L., KOPPELMAN F.S., "Effect of traffic information on commuter's propensity to change route and departure time", *Journal of Advanced Transportation*, vol. 29, no. 2, pp. 193–212, 1995.

[KHA 96] KHATTAK A., POLYDOROPOULOU A., BEN-AKIVA M., "Modeling revealed and stated pretrip travel response to advanced traveler information systems", *Transportation Research Record*, vol. 1537, pp. 46–54, 1996.

[KHA 99] KHATTAK A.J, YIM Y., STALKER L., "Does traveler information influence commuter and noncommuter behaviour?" *Transportation Research Record*, vol. 1694, pp. 48–58, 1999.

[KHA 08] KHATTAK A.J., PAN X., WILLIAMS B., ROUPHAIL N., FAN Y., "Traveler information delivery mechanisms: impacts on consumer behavior", *Transportation Research Board*, vol. 2069, pp. 77–84, 2008.

[KRA 00] KRAAN M., MAHMASSANI H.S., HUYNH N., "Traveler responses to advanced traveler information systems for shopping trips", *Transportation Research Record*, vol. 1725, pp. 116–123, 2000.

[MED 11] MEDIAMOBILE, Mediamobile sonde l'attitude des français face au trafic, available at: http://blog.v-traffic.com, 2011.

[RAP 03] RAPP P., BAUDEZ G., EGELER C., "Multimodal travel information service for trinational regional transport, Final report, Trans-3 project", *European Commission*, p. 29, 2003.

[SHA 01] SHAH V.P., WUNDERLICH K., LARKIN J., "Time management impacts of pretrip advanced traveler information systems: findings from a Washington, D.C., case study", *Transportation Research Record*, vol. 1774, pp. 36–43, 2001.

[SHI 11] SHIFTAN Y., BEKHOR S., ALBERT G., "Route choice behaviour with pre-trip travel time information", *IET Intelligent Transport System*, vol. 5, no. 3, pp. 183–189, 2011.

[SUN 05] SUN Z., ARENTZE T., TIMMERMANS H., "Modeling the impact of travel information on activity-travel rescheduling decisions under multiple uncertain events: distributed myopic decision heuristics", *Transportation Research Record*, vol. 1926, pp. 79–87, 2005.

[TRA 05] TRANSPORT AND TRAVEL RESEARCH LTD., Transport Direct evaluation online survey analysis, *report*, 2005.

[TSI 05] TSIRIMPA A., POLYDOROPOULOU A., ANTONIOU C., "Modelling the impact of advance traveller information systems on travellers' behaviour: Puget Sound region case study", in *European Transport Conference*, p. 15, 2005.

[VAN 04] VAN DER HORST R., "Use of travel information and effects on location choice for recreational trips", *European Transport Conference*, Strasbourg, 2004.

[VAU 95] VAUGHN K.M., REDDY P., ABDEL-ATY M.A. *et al.*, *Transportation Research Record*, vol. 1516, pp. 61–69, 1995.

[VIR 04] VIRGINIA DEPARTMENT OF TRANSPORTATION, 511 Virginia Evaluation, 2004.

[VIS 01] VISWANATHAN K., GOULIAS K.G., "Travel behavior implications of information and communications technology in Puget Sound Region", *Transportation Research Record*, vol. 1752, pp. 157–165, 2001.

[YIM 00] YIM Y., TravInfo Field Operational Test evaluation: information service providers customer survey, California Partners for Advanced Transit and Highways (PATH), report UC Berkeley, 2000.

[ZEL 05] ZELT, Contribution à l'évaluation du SDIR : enquêtes auprès des usagers - réseaux VRU de Nantes et Bordeaux - Autoroute A20 (entre Vierzon et Brive), 2005.

[ZEL 09] ZELT, Réseau routier Gutenberg – Système d'aide à la gestion du trafic et d'information des usagers: Évaluation des impacts, report, 2009

[ZHA 08] ZHANG L., LEVINSON D., "Determinants of route choice and value of traveler information: a field experiment", *Transportation Research Record*, vol. 2086, pp. 81–92, 2008.

[ZHA 12] ZHANG Y., YUN M., YANG X., "Who will use pretrip traveler information and how will they respond? Preliminary study in Zhongshan China", *91th TRB Annual Meeting*, p. 17, 2012.

16

Transferability Study on Full-scale Implementation of Real-time Passenger Information

The University of Maribor is carrying out testing and feasibility studies on real-time passenger information (RTPI) in the city of Maribor. A methodology for bus stop hierarchy is performed to set RTPI panel locations. During a 3-month testing period, accuracy of the real-time passenger information on bus stops and perceived waiting time were analyzed. Additional cost benefit analysis and the possibilities for full scale implementation are elaborated. The results of the pilot within EU South-East European (SEE) project ATTAC (attractive public transport for accessible cities) are presented as lessons learned in a project output-mobility toolbox. A mobility toolbox is a collection of good strategies/tools in the form of policy recommendation for providing seamless travel to local citizens and commuters like using public transport in South-East European cities.

16.1. Introduction

Cities nowadays face increasing mobility needs, intensified sub-urbanization, escalating traffic flows and congestion. This leads to poor accessibility and emissions that negatively affect the environment. A major reason for congestion is the ever-growing use of private cars, which results in air pollution and increases travel times. Public transport (PT) is, therefore, seen as a good alternative; today's efforts should be directed towards increased use of group transport modes of traveling. Travelers will keep using public transport and non-users will start using public transport only if public transport will be attractive and user-friendly. One way of promoting greater use of public transport is by providing reliable and easily-accessible real-time passenger information (RTPI). It is widely accepted that RTPI

Chapter written by Mitja KLEMENČIČ, Vlasta RODOŠEK, Marko ČELAN and Marjan LEP.

offers significant benefits to PT users, present and potential, by providing a wide spectrum of information. In this respect, RTPI systems improve public transport services and help trip makers change their travel behavior towards more sustainable transport modes.

Mid-size cities, where local public transport is operated by buses only, very often face the problem: where should the limited financial recourses to increase ridership be put? Small and mid-size cities have some common characteristics: (1) the headways of departures are greater than acceptable few minutes, (2) operators do not have expensive and extensive integrated ticketing, tracking and information systems installed, (3) there are no significant priority schemes at signalized crossings or dedicated lines for busses, and (4) timetables are adapted to demand and vary a lot between peak and off-peak periods.

RTPI, as an option to improve friendliness of PT services, is a well-known solution in bigger urban agglomerations. The challenge of its implementation in small or mid-sized cities consists of several tasks, such as:

– to find an autonomous ("small scale") solution;

– the cost of implementing RTPI should not exceed the limited budget; and

– there must be some measurable benefit of implementing it.

16.2. RTPI testing in Maribor

The city of Maribor has approximately 100,000 inhabitants. The local public transport is operated by one municipality owned operator. The system includes 42 busses. There are 19 lines with headways of 15–25 min on the busiest routes and up to 60 min on the routes with lower demand. Approximately 400 bus stops are served. Overall, 2.5 million bus-km per year is offered to 3.8 million travelers. The operating speed of the bus lines is below 15km/h. Only 12% of all the trips in the city are made by public transport. The punctuality and reliability of the service is not measured by the contractor and no *bonus-malus* system for the quality of service is implemented. Users often claim to be confused, waiting for the unreliable service.

Implementing priority schemes or dedicated lines for busses is a long-term and expensive solution, so the municipality has started to implement RTPI as a measure to improve the information for the regular and potential users.

The main purpose of the testing period is to predict the possibilities of implementing a low-cost RTPI system in Maribor. One two-side LCD display has been installed at one of the busiest bus stops ("City Center"). A seven-line display shows real-time information of bus arrivals for four city bus lines permanently, and timetable information for other city bus lines exchange information with weather prognosis and the date/time. A real-time arrival prediction is based on the current position of the buses, determined by GPS and map matching technology and the historical data of the bus runs. Fifteen buses have been equipped with tracking devices for obtaining the current position of the buses. Adopted software has been developed for data transmission between components of the RTPI system and calculating the prediction of bus arrivals to the bus stop.

Basic architecture (Figure 16.1) of the system consists of the mobile units installed at the buses whose main task is to send the current position of the buses to the control center (server), where bus arrival time prediction is calculated according to the developed algorithm and sent to the LCD display in order to offer passengers and potential passengers traveling out of the city center the real-time information about bus arrivals.

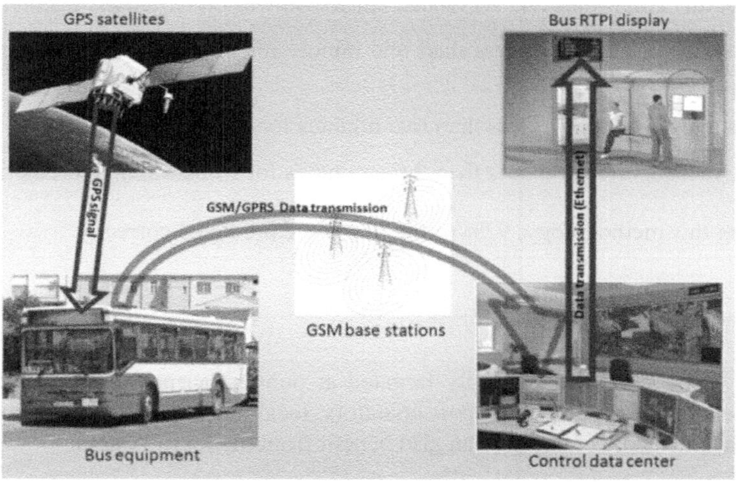

Figure 16.1. *Architecture of the test RTPI system in Maribor*

Cost Benefit Analysis (CBA) for RTPI systems in Maribor has various limitations that impose constraints on CBA. There are currently no appropriate data on key parameters required to perform a comprehensive CBA. The analysis presented below is only a demonstration and does not constitute a full scale of the cost and benefit evaluation. Benefits of the RTPI system such as increasing willingness to pay, decreasing external costs and greater use of the PT and costs of unreliable information are not considered. This analysis only focuses on the potential benefits from users' perceived waiting time and foresees costs of implementation and maintenance of an RTPI system in Maribor.

The RTPI system in Maribor was tested over 3 months (February–April 2013). The quality of the RTPI system was assessed from two aspects, reliability of the RTPI system from a technical point of view and accuracy of the bus arrival prediction. The "reliability" errors were considered from the following aspects:

– organizational errors; abnormal or unacceptable delays of bus runs from the start bus stop, cancelled bus runs and runs without mobile units;

– technical errors; e.g. inappropriate data transferring between units of the RTPI system and software (SW) errors in case when the data from the mobile units were not considered (correctly);

– human factor errors caused by drivers by inputting the data into to on-board units manually.

The accuracy of the predictions depends upon many factors including traffic flow, weather conditions and accidents. An arrival was classified as "accurate," if:

– the margin of error is less than one minute for predictions of five-minutes or less;

– the margin of error is less than two minutes for 10-min predictions;

– the margin of error is less than three minutes for fifteen-minute predictions.

Using this methodology, 92% of the runs were predicted correctly.

16.3. Benefits of the RTPI system

The RTPI system is primarily targeted at public transport users and potential users; nevertheless, public transport operators, local authorities and even road users as well as the whole of society can also benefit from the RTPI system.

16.3.1. *Benefits for passengers*

Each person using public transport or contemplating its use can benefit from the RTPI system. Real-time information can make journeys more efficient by minimizing travel and waiting times and also help to build confidence. Real-time information sources, such as websites and mobile applications, can have a direct impact upon the reduced passenger waiting times and help to increase the choice for traveling. Although stop signage with the next-arrival information does not directly reduce the waiting times, since passengers have to be at the stop to know this information, it reduces anxiety and may provide a perceived benefit of less safety and security risks. Real-time information is most valuable to passengers in the departure time decisions, i.e. what time to depart to a stop. By knowing the next-bus or train arrival information, passengers may be able to make better use of their time or seek alternate modes of transportation (e.g. if the wait time is too long). Other likely passenger benefits of the RTPI systems are as follows: increased willingness to pay, more efficient traveling through better use of waiting time, positive psychological effects (reduced uncertainty, increased feeling of personal security, creation of a general sense of trust in the PT system, easier use), better overall image of the system and greater passenger satisfaction.

16.3.2. *Benefits for public transport operators and public transport authorities*

Public transport companies and public transport authorities achieve a better image when providing real-time information and when using innovative information systems in their communication with the users, resulting in higher customer satisfaction [INS 10]. The possibility of higher income might be a long-term benefit when the measures result in higher demand for public transport. Effective use of RTPI technology can help reduce bus journey times in a number of obvious ways. In addition to the RTPI system, benefits can be achieved by enabling technologies for tracking and identification of public transport vehicles. The monitoring capability of enabling technologies together with the analysis of the data allows the agency/operator to make better long-term decisions about service planning (fleet management, schedule planning/efficiency, operating performance, incident response). Enabling technologies for tracking and vehicle identification can alsobe "accessory" by the implementation of other IT systems such as voice announcement system inside vehicle and bus traffic signal priority system,allowing late running buses to force a sequence change in the signals which enables them to pass the junction freely.

16.3.3. *Benefits for society*

By improving the information on the public transport system, public transport can develop as a real alternative to private car use. An increased use of public transport can reduce vehicle kilometers and congestion, which results in a total value of time, reduced vehicle O&M costs, reduced emissions and costs of environmental mitigation.

For the calculation of benefits of the RTPI system, the reduction of perceived waiting time was taken into account. Other costs were not considered, although they might be more important.

16.4. Cost benefit analysis and RTPI system

Cost benefit analysis produces information about the relative magnitude of benefits and costs that accrue (over time) to society as a result of any given action. The purpose of such an analysis is to provideinformation about project or program performance, relative to a baseline set of conditions where the project or program has not been implemented [PUG 10]. General methods for CBA according to PSRC are:

– *Comparison against a baseline.* – Alternatives generated and measured to the same base (situation).

– *Measuring costs and benefits.* – For transportation investments benefits of travel time savings and reduction in other cost (accidents costs, Vehicle operating and ownership costs, travel time unreliability savings, facility operating cost, facility capital costs, vehicle emissions costs) should be considered. When the perceived costs of a trip are reduced, consumer surplus increases. Additionally, the willingness to pay and the user's perception of their costs of travel are relevant to calculating the benefits of the improvements.

– *Estimating streams of annual benefits and costs.* – Use of models to analyze more than one analysis year both with the project and without the project

– *Discounting benefits and costs to present values.* – Future benefits and costs are converted into present value terms by applying an appropriate discount rate.

– *Unit cost inputs and other parameters.* – A number of key assumptions are made in order to develop estimates of the user benefits from transportation projects and programs.

16.4.1. *Cost benefit analysis of RTPI system*

Various studies concerning CBA have been done, but evaluation results are usually related only to the qualitative benefits of RTPI system or are based on the assumption of increasing of public transport use and consequently on the increase of public transport carrier revenue. Project ATTAC's best practice study reveals that the average ridership increase, after implementing RTPI, was 3%. Costs of RTPI systems are mostly divided into the costs of the project preparation and implementation of the RTPI system (capital costs) and operating costs.

Welde *et al.* [WEL 11] have presumed four hypotheses of benefits of the RTPI systems and categorized four types of costs.

Benefits of the RTPI system	Cost of the RTPI system
RTPI will reduce bus dwell time.	Investment costs
RTPI will reduce passengers' wait time.	Reinvestment costs
RTPI will increase passenger satisfaction.	Maintenance and operating costs
RTPI will increase the total number of passengers	Marginal cost of public funds

Table 16.1. *Benefits and costs of RTPI system by Welde et al. [WEL 11]*

Overall assumptions used in the estimation were:

– time frame for analysis: 25 years;

– expected life time of investments: 12 years;

– discount rate: 4.5%;

– average value added tax (VAT) rate: 6%;

– marginal cost of public funds: 20%;

– benefit per trip when RTPI is available: $ 0.70.

According to the above assumptions CBA analysis showed that RTPI provides significant value for money.

Cham *et al.* [CHA 06] mentioned that effective application of cost–benefit analysis (CBA) for the assessment of an RTPI system investment requires the identification and collection of all potential costs and benefits to all relevant stakeholder groups. The analyst needs to collect these data through agency interviews, passenger surveys and related sources. Investment in RTPI systems requires investment in the "customer facing" technologies that the public can access (e.g. websites, phone applications, electronic information displays, terminals) and also in "enabling" technologies that are required for RTPI systems, but which may be implemented independently of the RTPI systems (e.g. Automatic Vehicle Location (AVL), Automatic Vehicle Identification (AVI), Automated Passenger Counting (APC)). Stemming from the fact that PT agencies and carriers also do invest in the enabling technologies for other purposes, the question arises of whether the CBA should include the costs and benefits of the enabling technologies or not. If they decide to include all technologies, then it is only appropriate to include the benefits of all technologies, not just those relating to real-time information to passengers. If the purpose of the analysis is solely to evaluate the return on (the incremental) investment in real-time information systems, then the CBA should be limited to costs and benefits relating to the incremental investment. In every case it is very important that the current situation of the enabling technologies is already in function. If the enabling technologies are already in function, then it is advisable to consider only incremental investments and benefits of RTPI system, if the enabling technologies are not yet in function than all costs and benefits should be considered, including possibility to use enabling technologies by implementation of other ITS system such as bus priority and voice announcement inside the vehicles.

[CAU 03] assumed that the provision of RTPI accrues a 0.5–1.5% increase in revenue to the operators, emanating from increase in passenger numbers. The result of analysis showed that on the basis of the relatively low benefits, the operators could pay for the RTPI system within 3–12 years depending on the benefits realized.

16.4.2. *Perceived users waiting time in Maribor*

Perceived users' waiting times were measured by methods of observation and interviews. The point of observation allowedthem to havea good view of the bus stop. The tasks of the observer were to record information about:

– time of passenger arrivals to the bus stop;

– time of bus arrival to the bus stop;

– time of interviews – how long have you already waited for the bus?

– passenger purpose of the trip – is your trip time-bound?

Results of the analysis showed that the passengers arrive to the bus stop at an average of 7.5 min before bus arrivals and 7 min before foreseen bus arrivals according to the timetable [HOJ 13]. At the bus stops and bus lines with headways of less than 20 minutes, the average waiting time for a bus is 5.9 min and at the bus stops with headways, of bus runs between 20 and 30 min, it is 8.7 min.

The perceived waiting time for a bus is 1.35 times higher than the actual waiting time. At the bus stops with headways of bus runs between 20 and 30 minutes the perceived waiting time is 3.45 min, i.e. 1.47 times higher than the actual waiting time and at the bus stops with headways under 20 min the perceived waiting time is 1.3 minutes, which is 1.23 times higher than actual waiting time. From this we can conclude that real-time information about bus arrivals/departures is even more important at the bus stops where the headway of the bus services is higher.

16.4.3. *Steps towards full-scale RTPI implementation*

Three options of RTPI implementation were investigated with the goal of helping the decision makers in reallocation of the budget intended to support public transport improvements:

– implementation of an RTPI system at the 10 bus stops according to the passenger rate;

– implementation of an RTPI system in the frame of city PT at bus stops where benefits of perceived waiting time are monetized higher as lifecycle costs of additional costs (display, operation costs, project planning);

– implementation of an RTPI system with the availability of real-time information for all types of PT subsystem operating in the city of Maribor, available in different information media (panels, webpages, mobile phones), with real-time information at stops where benefits of perceived waiting times are monetized higher than lifecycle costs of additional costs (display, operation costs, project planning).

Costs were structured into capital costs and operating costs. The following capital costs were taken into consideration:

– project planning, design and project management. These costs must include the cost of agency staff time invested in concept development;

– equipment (hard- and software) and design of websites, internet- and mobile-phone trip planning tools;

– hardware (e.g. displays, servers, conduit, On-board equipment for PT vehicles tracking and identification) installation costs and parts.

The following operating costs were calculated:

– operation costs (hardware maintenance, software license and operation, marketing and communication, costs for operations staff, cost of electricity required to power the systems, cost of data transfer between units of the system, etc.);

– life expectancy: the life expectancy (in years) of hardware costs;

– specific issues for interchanges (i.e. new buses, new server location, etc.).

– disposals.

16.4.3.1. *Step 1*

The RTPI system will be implemented at the 10 most occupied passenger bus stops where electricity and internet are available without extra costs. The total number of boardings at these 10 stops is 1.35 million per year which represents approximately 36% of all boardings on the buses in the frame of the city PT in Maribor.

Capital costs to equip 10 bus stops with real-time information are estimated at 130,000 €. Costs of project planning, design and project management are estimated at 30,000 € and costs of hardware (HW) and SW with installation are estimated on 100,000 €. Operation costs, included power consumption, costs for operation staff, costs of data transfer and SW license are estimated to 10% of capital costs per year. Lifetime of LED/LCD displays installed at bus stops and equipment on bus stop is estimated on 10 years and lifetime of servers is estimated on 5 years. Additional cost could be related in case of PT vehicle changes, reconstruction of bus stops and change in bus run itineraries. These costs are estimated to 2% of capital costs per year. Disposal costs are not taken into consideration by this analysis.

Project planning, design and project management	30,000 €
Hardware and SW (included installation)	100,000 €
Operation costs	130,000 €
Specific costs for interchanges	26,000 €
Lifecycle costs (10 years)	286,000 €

Table 16.2. *Foreseen costs of RTPI implementation at 10 bus stops for 10 years period*

The benefits of the RTPI system in Maribor are considered only by decreasing the perceived waiting time. From the observation and interviews, it was found that the passenger waiting for a bus in the frame of the city's PT in Maribor was on average 7.5 min and their perceived waiting time was above 2.7 min. If we consider 1.35 million entries at the chosen bus stops, the total perceived waiting time on a yearly level could be decreased by about 60,000 h. If we consider that the value of time by traveling by PT is about 35% of the value according to the average salary in Slovenia [LEP 04] than the value of one hour is 2.55 €. Taking into account this value of time than the total benefit as a result of RTPI system in Maribor is estimated to be 153,000 € per year.

16.4.3.2. *Step 2*

Step 2 foresees the implementation of the RTPI system at those bus stops in Maribor where monetized benefits of perceived waiting time are higher as lifecycle costs of additional costs. Additional costs are considered as an upgrading of an RTPI system from option 1. The bus stops where availability of electricity and intranet is limited will be provided by solar power. Additional costs by implementation of the RTPI system at one bus stop are estimated 13,200 € in at 10-year period which is equal to the 5,176 h of perceived waiting time or 115,022 entries, respectively.

Project planning, design and project management	1,000 €
Hardware (included installation)	5,000 €
Operation costs	6,000 €
Specific costs for interchanges	1,200 €
Lifecycle costs per additional bus stop (10 years)	13,200 €

Table 16.3. *Additional costs of RTPI system implementation at one bus stop for 10 years period*

The entire cost of implementation of the RTPI system at 83 bus stops is estimated to be 1.353 million €. The total number of entries to the buses at these stops is 3.7 million per year. The benefit – monetized value of perceived waiting time – for a 10-year period is estimated to be 4.25 million €.

Project planning, design and project management	100,000 €
Hardware and SW (included installation)	515,000 €
Operation costs	615,000 €
Specific costs for interchanges	123,000 €
Lifecycle costs (10 years)	1,353,000 €

Table 16.4. *Foreseen costs of RTPI implementation at 83 bus stops for 10 years period*

16.4.3.3. *Step 3*

Step 3 foresees the implementation of an RTPI system with the available real-time information for all types of PT subsystems operating in the city of Maribor, accessible in various information media (panels, webpages, mobile phones), with real-time information at stops where benefits of perceived waiting time are monetized higher than lifecycle costs of additional costs (display, operation costs, project planning). This calculation assumes that the intercity bus and railway transport systems already have information of the current position of buses and trains. Additional costs for integration of city bus PT, intercity bus PT of several carriers and railway transport are related to the SW for unification between data of different systems. Costs increase also for SW and HW by establishing webpages where real-time information is available. The website can be adapted for usage with mobile phones. For the purpose of intercity bus transport, installation of additional 10 displays is predicted. At the main bus stop a panel with real-time information about train departures/arrival from/to main railway station in Maribor is foreseen and at the railway station real-time information about bus services is available.

The entire costs of full-scale RTPI system, implementation are estimated to be nearly 2 million €. The total foreseen monetary benefit is estimated to be over 5.5 million € in a 10-year period.

Project planning, design and project management	200,000 €
Hardware and SW (included installations, webpage design, mobile phone adaption, data unification interface)	685,000 €
Operation costs	885,000 €
Specific costs for interchanges	177,000 €
Lifecycle costs (10 years)	1,947,000 €

Table 16.5. *Foreseen costs of RTPI system full scale implementation for 10 years period*

16.5. Mobility toolbox as transferability tool

Lessons learned from the testing period are part of the main ATTAC project output– the Mobility Toolbox. The ATTAC Mobility Toolbox (MT) is a collection of good strategies and tools for making SEE cities and regions effective nodes of transnational accessibility.

The main purpose of the MT is to be a useful knowledge provider and planning support tool for local policy makers within local SUMP's (Sustainable Urban Mobility Plan). The aim is to provide and use a transferable toolkit for creating local and regional SUMPs with focus on public transport measures gained from ATTAC experience. The first version of MT feeds the local SUMP and then in general is an input for local administration to design the SUMP.

The scheme below represents the MT concept. The MT concept is structured to tackle all phases of the SUMP. It represents the structure of the elaborated SUMP with hierarchy of strategic, tactical and operational planning.

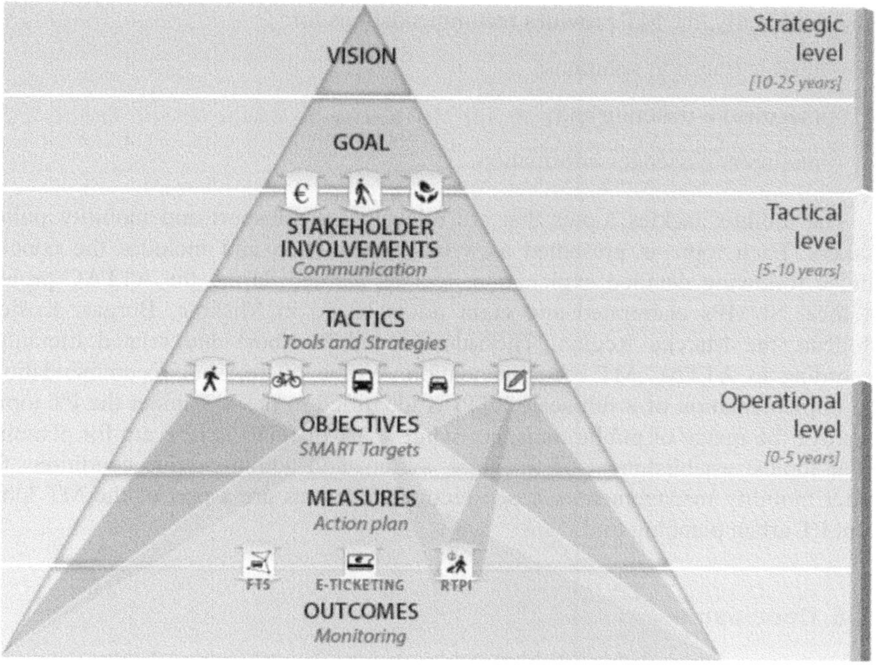

Figure 16.2. *Mobility Toolbox concept [ATA 13]*

Economic, social and environmental/health goal setting represents general aims of the cities transport development on strategic level. As realization of all those goals has limited resources, stakeholders on tactical level decide on priorities. Different tools and strategies help decision makers to choose the right scenario of priorities. According to the selected scenario objectives, *smart* targets and measures can be set. Experience showed that it is better to manage measures in packages. The MT will tackle the transport related challenges in 5 pillars:

– walking;

– cycling;

– public transport;

– individual car + freight and;

– planning.

Additionally, the MT provides recommendations for:

– flexible transport solutions;

– integrated e-ticketing and;

– intelligent passenger information.

The toolbox tackles topics that are relevant for transport and mobility policy makers. Each topic is presented as a recommendation and includes the concise advice following detailed explanation, lessons learned during the ATTAC project (8 local SUMPs elaborated and eight pilots tested in Miskolc, Burgas, Košice, Modena, the Marche Region, Thessaloniki and Maribor) and related literature, according to ATTAC MT recommendation. Within 60 relevant recommendations the implementation of a full-scale RTPI is identified in the MT under the PT topic: improve the image of public transport, while CBA of RTPI is relevant for planning pillar topics: establishment of financial, system and administrative conditions for better mobility management. Chosen recommendations are a part of the MT long-term PT urban planning tool.

16.6. Conclusion

This chapter and underlying pilot project should help designers of transport policy in small and mid-size cities to understand the benefits and the costs of RTPI. The applied method for CBA evaluates the RTPI system in Maribor for the next 10 years. The method is limited to the costs of RTPI in 10 years and benefits on reduced perceived waiting time, which represent only a small part of the PT performance evaluation. To evaluate PT investments with a cost benefit analysis, parameters for benefits of travel time savings and reduction of other costs (accidents costs, vehicle operating and ownership costs, travel time unreliability savings, facility operating cost, facility capital costs, vehicle emission costs etc.) should be considered, while on the cost side, other planned technologies for higher PT quality should also be considered.

16.7. Acknowledgments

Research was conducted in the frame of ATTAC project, partially financed by the European Union within the framework of South East Europe Transnational Cooperation Programme.

16.8. Bibliography

[ATT 11] ATTAC PROJECT, Task Force 3 – Best practice report on intelligent Passenger information, University of Maribor, 2011.

[ATT 13] ATTAC PROJECT, WP 3 Creating a SEE Urban Mobility Toolbox for the introduction of public/collective transport solutions; Mobility toolbox – concept, University of Maribor, 2013.

[CAU 03] CAULFIELD B., O'MAHONEY M., Real time passenger information: the benefits and costs, Association for European Transport, 2003.

[CHA 06] CHAM L., DARIDO G., JACKSON D. *et al.*, Real-time bus arrival information systems return-on-investment study, Final Report, National Technical Information Service, Federal Transit Administration, Washington D.C., USA August, 2006.

[HOJ 13] HOJSKI D., Občutenićasičakanjapotnikovnaavtobusemestnega JPP v Mariboru, Project work, Maribor, 2013.

[INS 10] INSTITUTE FOR TRANSPORT STUDIES, Innovative information systems for public transport, University of Natural Resources and Applied Life Sciences (BOKU), Civitas policy advice notes, Vienna, 2010.

[LEP 04] LEP M., MESAREC B., BOŽIČNIK S. *et al.*, *Analiza eksternih stroškov prometa*, Faculty of Civil Engineering, University of Maribor, Maribor, 2004.

[PUG 10] PUGET SOUND REGIONAL COUNCIL INFORMATION CENTER, Benefit-Cost Analysis: General Methods and Approach, report, Seattle, 2010.

[WEL 11] WELDE M., FOSS L., TVEIT Ø., "Evaluating the impacts of real time passenger information and bus signal priority in trondheim", *18th ITS World Congress*, Orlando, 2011.

17

Excess Commuting and Commuting Economy: Peak and Off-Peak Variation in Travel Efficiency Measures

In recent years, there has been an increase in the level of attention being paid to the empirical difference between observed travel patterns and those necessitated by the distribution of jobs and housing in urban areas. In the academic literature, this issue has been investigated fairly extensively within the context of the excess commuting and commuting economy frameworks in various city-regions. Within these frameworks, one area that has received considerably less attention is the case of off-peak travel, which is used as a proxy for non-work travel. Accordingly, the current research specifically addresses this period using the city-region of Dublin, Ireland as a case study. The approach uses data from an urban traffic simulation model to determine the minimum, maximum and random travel costs for the study area, which are compared with observed costs. The results show that non-work travel is associated with more efficient travel behavior driven by the intermixing of land use arrangements associated with these trip types and the transport network. They also show that there are only slight improvements in the efficiency of off-peak travel over the time horizon but considerable improvement during the peak period as a result of the extent of job decentralization.

17.1. Introduction

In recent years, the excess commuting framework, which is characterized by the deviation of observed commuting cost from a theoretical minimum commuting cost [HOR 02a, MA 06], has been utilized extensively to measure the efficiency of urban commuting patterns and the morphology of jobs-housing characteristics [MUR 11, ZHO 14]. Numerous scholars from a range of academic disciplines have used the excess commuting concept to aid the understanding of commuting efficiency in

Chapter written by Enda MURPHY.

various cities [FRO 98, HOR 02a, NIE 06, MUR 09]. In particular, the framework has been used repeatedly as a useful proxy for either regional job-housing balance or imbalance [HOR 02a, HOR 03] but also as a means for assessing the efficiency of regional commuting patterns more generally [MUR 12].

The current chapter applies the excess commuting and recently developed commuting economy framework to off-peak travel in Dublin, Ireland. It uses data from a Dublin traffic simulation model to assess the relative efficiency of trip making in the peak and off-peak periods. Given the skepticism surrounding the application of the excess commuting framework to all but specific types of trips, the chapter provides a rationale for application of the framework to trips in the off-peak period. In this regard, the chapter argues that analysis of aggregate off-peak trip making provides useful indicators of the differing nature of peak and off-peak trips. Applied within the context of the excess commuting and commuting economy frameworks, it is possible to identify the differences in the nature of trip-making for the two periods, which is useful for understanding the nature of travel change for the two periods especially when a temporal analysis is undertaken and the data is disaggregated by the mode of transport (as is the case for this study).

The objectives of the chapter are twofold. First, a comparison of the "efficiency" of peak and off-peak trip making is offered within the excess commuting and commuting economy frameworks. "Efficiency" is defined narrowly here, specifically within the context of the excess commuting and commuting economy framework. Moreover, it is also accepted that differences emerging in the results for the two traffic periods may not, in reality, be inefficient (e.g. trips that are chained in the off-peak, or trips for comparison goods, etc.). Second, the chapter applies the analysis to the data for two time periods and for different modes of transport to investigate the changes in the travel behavior within the context of dynamic changes in the urban structure of Dublin. Neither of these core objectives has been addressed in the related literature and, as such, the current exploration provides additional understanding about the changing dynamics of peak and off-peak commuting over time and within the context of the efficiency indicators used in the excess commuting and commuting economy framework.

17.2. Excess commuting, commuting economy and off-peak travel

Attempts to measure the efficiency of travel and land use arrangements have taken different approaches over the last few decades. However, there has been a relatively fruitful and consistent literature utilizing the excess commuting and commuting economy frameworks (see [HAM 82, WHI 88, HOR 02, MUR 11]). The concept of excess commuting (EC) emerged from the economics literature in the early 1980s when Hamilton [HAM 82] sought to test the robustness of the monocentric urban model for estimating urban commuting. Since then, the concept has evolved quite a bit. Subsequent work by White [WHI 88] utilized a more appropriate linear programming approach for calculating the average minimum commute while Horner [HOR 02a] introduced additional concepts such as capacity utilization. Ma and Bannister [MA 06] provide a comprehensive review of the evolution of the framework as well as the range of methodological and contextual extensions to it from the early 1980s until 2006. The reader is referred to that paper for further details. However, since that review, there have been some further developments including the development of the commuting economy concept [MUR 11] as well as the utilization of the framework to investigate attainable reductions in metropolitan average trip length [O'KE 08]. All of these initiatives have brought new dimensions to what is now a fairly strong conceptual framework for understanding travel behavior within the context of land use change.

Excess Commuting is a measure of the deviation of an urban area's observed average travel cost from a theoretical minimum average travel cost. It has been described as the surplus travel cost resulting from the fact that the actual geography of travel deviates from the pattern that minimizes total journey cost [HAM 82, WHI 88, FRO 98, HOR 02b, MUR 09]. The minimum commute is taken to be the optimal allocation of individuals from origin to destination in a manner that minimizes overall travel cost (e.g. travel distance or time). Effectively, it allocates individuals to job locations assuming that they travel to the closest possible workplace destination on average. It has been applied extensively to commuting where the difference between the actual commute and the minimum commute can be thought of as excessive in the sense that it is not required by the fixed physical separation of origins and destinations. It is expressed as a percentage of the actual commute as follows:

$$EC=\left(\frac{T_{act}-T_{min}}{T_{act}}\right)*100 \qquad [17.1]$$

where EC is the excess commute, T_{act} is the observed commute, and T_{min} is the theoretical minimum commute. In an extension to the excess commuting concept, Horner [HOR 02a] used the transportation problem of linear programing (TPLP) to maximize the cost of allocating workers from residences to workplaces, i.e. the inverse of the minimization objective. He interpreted the maximum commute as a worst case scenario where every worker in the study area travels to the furthest possible workplace. Given a fixed distribution of origins (residences) and destinations (workplaces) in a city-region, the maximum commute (T_{max}) places a cap on the average distance that can be traveled. Horner used the addition of T_{max} to develop a measure for what he refers to as capacity utilization (C_u) – the percentage of travel cost capacity of a city-region being consumed by daily trip making:

$$C_u = \left(\frac{T_{act} - T_{min}}{T_{max} - T_{min}} \right) * 100 \qquad [17.2]$$

If we consider T_{max} – T_{min} as a finite scale determined by the distribution of origins and destinations, then the relative location of the actual average travel cost, T_{act}, on the scale indicates the amount of available travel resources being used by urban trip making. As T_{act} – T_{min} increases relative to the total range of travel available, T_{max} – T_{min}, trip making becomes less efficient because more of the available distance or time resources in the study area are being consumed. Together, T_{min} and T_{max} represent the range of commuting possibilities that a given distribution of trip origins (residences) and destinations (workplaces) permits, i.e. the "commuting capacity". The magnitude of this range, i.e. (T_{min}-T_{max}), is a function of the land-use geography of a city region. Consider a city region where all of the trip origin zones (residences) are located in one area and destinations (workplaces) in another (Figure 17.1(a)). In this case, T_{min} and T_{max} will be quite similar and hence the urban commuting range will be relatively small; the opposite will be true of a city region where the distribution of trip origins and destinations is inter-mixed (Figure 17.2(b)). In this situation, considerable scope exists for T_{min} to be low and for T_{max} to be high. If in a city region, the value of (T_{min}-T_{max}) increases through time, this suggests that land-uses in that city are becoming increasingly inter-mixed (see [MA 06]).

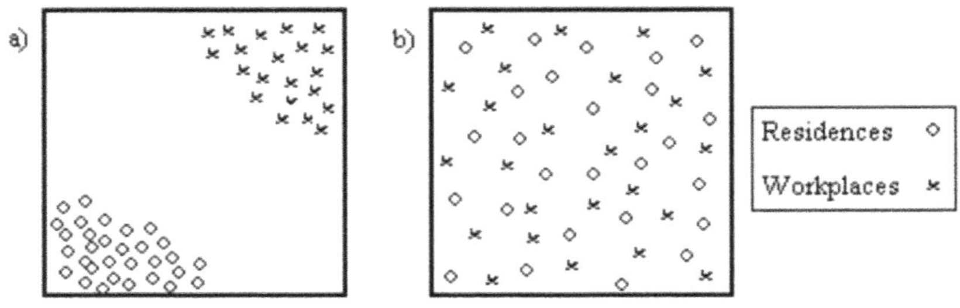

Figure 17.1. *Schematic of hypothetical city-regions. For a color version of this figure, please see www.iste.co.uk/jacob/traffic.zip*

A further extension has been provided recently by Charron [CHA 07] and Murphy and Killen [MUR 11] who introduce the notion of random commuting (T_{rand}). T_{rand} is assumed to be the average commuting cost that is achieved in a city region if individuals are assumed to travel randomly between their place of origin and destination. To be more specific, T_{rand} relates solely to a particular type of travel behavior – that is, one where no disutility is associated with the separation between origin and destination. Put another way, travel cost is considered to be irrelevant under such behavior. The further T_{act} departs from T_{rand}, the greater the role that separation is playing in determining travel behavior whether it is as a disincentive (where $T_{act} < T_{rand}$) or indeed as an incentive (where $T_{act} > T_{rand}$). Murphy and Killen [MUR 11] argue that T_{rand} is the more appropriate upper limit on the commuting cost and suggest that this should be utilized as the *de facto* benchmark for understanding and evaluating the efficiency of travel behavior. They suggest two alternative indicators, which measure the extent to which the behavior is moving from or towards random behavior and, thus, becoming more or less efficient – commuting economy and normalized commuting economy. Using the same logic as that underlying EC in [17.1], commuting economy is given by:

$$C_e = \left(1 - \frac{T_{act}}{T_{rand}}\right) * 100 \qquad [17.3]$$

C_e gives the extent to which T_{act} is falling below (positive) or above (negative) T_{rand}, i.e. the degree to which the behavior as expressed by T_{act} is becoming either more or less random. This represents the extent to which individuals are economizing on commuting costs. To go further, T_{min} and T_{max} represent the greatest extent to which T_{act} can depart from T_{rand} in either direction. Where T_{act} is less than T_{rand}, as we would expect, the statistic:

$$NC_e = \left(\frac{T_{rand} - T_{act}}{T_{rand} - T_{min}}\right) * 100 \qquad [17.4]$$

represents the extent to which T_{act} is below T_{rand} relative to the theoretical extent to which this could happen as determined by land-use geography i.e. T_{min}. Thus, this is considered to be a normalized commuting economy (NC_e) indicator. Together the values, T_{min}, T_{max} and T_{rand} represent what Murphy and Killen [MUR 11] refer to as the urban travel scale. Its value lies in its ability to offer a more complete framework against which values of T_{act} can be measured and interpreted. Within the context of this paper, all of the foregoing measures are calculated for the study area of Dublin, Ireland. The contribution of this chapter is to calculate these measures for the peak and off-peak period. Research in this area to date has applied these concepts and produced indicators exclusively to assess the morning and evening commute. Of course, this is largely because of the difficulties associated with acquiring suitable data for the off-peak but is also due to a range of potential difficulties with applying such a framework to non-work trips. Horner and O'Kelly [HOR 07] have pointed to some of these which include the discretionary nature of non-work travel in that it cannot be modeled as a relatively fixed distribution of origins and destinations to the same extent as the commute, its diversity in terms of the trip being undertaken, and the problem of trip chaining for non-work travel.

17.3. Data and methods

17.3.1. *Dublin*

Prior to the late 1990s, Dublin was best characterized as having an essentially monocentric urban structure where the majority of employment was located in the city center with scattered employment at the periphery [MUR 04]. After the Celtic Tiger economic boom beginning in the 1990s, the morphology of Dublin's built environment witnessed dramatic changes. This period of prosperity was characterized by industrial restructuring and in particular, a move towards more service-based industry away from heavy manufacturing [WIL 02]. A substantial proportion of this new service-based employment has located in the suburbs while many older manufacturing industries have also relocated there. This has led to a major restructuring in the development of office and industrial space away from the center towards "edge city" locations at the periphery (see [MAC 10]). Simultaneously, the rapid population increases fed a rapid acceleration in housing development and much of that development has occurred in a highly dispersed manner at the periphery of the region and beyond [WIL 02]. The failure to provide adequate public transport infrastructure to facilitate emerging employment centers at the periphery has led to increased demand for car-based commuting with its associated negative environmental consequences for the region. A monocentric city, with the majority of employment and service functions located in the core and with the majority of residences located in outlying areas, generates a primarily radial pattern of trip making with significant flows along a limited number of major radial routes. By way of contrast, a city where residential and non-residential functions are more interspersed (i.e. polycentric cities), as is now the case in Dublin, generates a more diffuse pattern of trip making. Specifically, these land use changes have created geographies of travel that have become increasingly complex and dispersed characterized by greater numbers of inter-suburban, cross-city and reverse commutes with a greater reliance on the private car to serve those trips. Between 1991 and 2002, the proportion of individuals driving to work increased in all counties of the Greater Dublin Area (GDA) with the greatest increases, in relative terms, occurring in the outlying counties [MUR 09]. Over the same period, the proportion of public transport trips has decreased despite significant public transport investment.

17.3.2. *Data*

The study area for the current analysis comprises the Greater Dublin Area (GDA), which consists of seven local authority administrative districts. The data for the study was derived from a Dublin Transportation Office (DTO) traffic simulation model for the 2001 peak (8–9 am) and off-peak periods (2–3 pm) and a comparable

DTO model for the same periods in 1991. The total number of trips recorded for 2001 was 327,001 while the corresponding figure for 1991 was 234,834. The model has two main components: the Highway Model (primarily car based) and the PT Model (public transport based). In the PT model both bus and rail trips are modeled and interchanges between PT modes are permitted. The model is based upon a 463 zonal sub-division of the Dublin Region and is derived primarily from the Irish District Electoral Division (DED) system. The model components have been independently validated to within five percent of actual travel patterns using goodness of fit statistics [WSP 03]. The off-peak data relates solely to the entire suite of non-work travel which includes retail, recreation, leisure and entertainment trips, among others. However, it does not provide an individual breakdown of non-work trips by purpose which, quite obviously, is a limitation of the analysis.

Two types of journey-to-work flow matrices were available for the peak and off-peak period: private transport trips and public transport trips. Thus, the data was such that, unusually for such studies, comparisons could be made by time of day as well as between 1991 and 2001. The key limitation of the 1991 data relates to the off-peak period data where only data for bus trips are available. Thus, we use these in this study as a proxy for public transport trips overall, although, the dataset is not completely comparable with the 2001 off-peak data. However, given that bus trips amounted for 83% of all off-peak public transport trips in the Dublin region in 1991 we feel that using the bus data as a proxy for public transport overall, while not ideal, can be justified.

Euclidean (ED) and road network (ND) distances were available for each mode of transport and they were taken as a proxy for the cost of physical separation between zonal units. Intra-zonal travel distances were estimated by assuming that each zonal unit is approximately circular in shape (see [FRO 98]). In a similar manner to other studies, this study excluded those trips originating and destined for locations outside the study boundary (see [FRO 98, HOR 02a, MUR 09]).

17.3.3. *Formulations*

From the preceding discussion, it should be clear that a necessary prerequisite for attaining values associated with the efficiency indicators described in 1–4 is the calculation of T_{min}, T_{max}, and T_{rand} and T_{act} – the values on the urban travel scale (see [MUR 11]). For T_{min}, the TPLP was used to determine the assignment of the trips from the origin to the destination that minimized mean commuting cost. The objective function and constraints of the TPLP are given by:

$$\text{Min: } Z = \frac{1}{N} \sum_{i=1}^{m} \sum_{j=1}^{n} c_{ij} X_{ij} \qquad [17.5]$$

$$\text{s.t. } \sum_{i=1}^{n} Xij = Dj \ \forall j = 1,\ldots,m \qquad [17.6]$$

$$\sum_{j=1}^{m} Xij = Oi \ \forall i = 1,\ldots,n \qquad [17.7]$$

$$Xij \geq 0 \ \forall i, j \qquad [17.8]$$

where, m = number of origins; n = number of destinations; Oi = trips beginning at zone i; Dj = trips destined for zone j; cij = travel cost from zone i to zone j; Xij = number of trips from zone i to zone j, and N = total number of trips. The objective function [17.2] minimizes the average transport costs. Constraint [17.3] ensures that trip demand at each destination zone is satisfied while constraint [17.4] limits the number of trips leaving each origin zone to the number of trips originating there. Constraint [17.5] restricts the decision variables, Xij, to non-negative values. It should be noted that travel costs, cij, may be expressed in terms of any measure of zonal separation, for example travel distance, travel time or indeed a generalized cost measure. For this study, Euclidean distance (ED) and route network distance (ND) were used as measures of zonal separation.

T_{max} was also determined using the TPLP where the objective function is the inverse of the minimization problem discussed previously [17.5] and is given by:

$$\text{Max } Z = \frac{1}{N} \sum_{i=1}^{m} \sum_{j=1}^{n} cijXij \qquad [17.9]$$

The approach for calculating T_{rand} utilized the Markov Chain Monte Carlo (MCMC) hit-and-run algorithm outlined by Murphy and Killen [MUR 11, MUR 62]). It is given by the following:

$$N_T = N! \qquad [17.10]$$

$$\text{s.t. } \sum_{i=1}^{n} Xij = Dj \ \forall i = 1,\ldots,n \qquad [17.11]$$

$$\sum_{j=1}^{m} Xij = Oi \ \forall j = 1,\ldots,m \qquad [17.12]$$

$$Xij \geq 0 \ \forall i, j \qquad [17.13]$$

where N_T is the number of possible commuting configurations in a city; N! is the factorial of the total number of trips in the urban area. Constraints [17.11]–[17.13] are identical to those of the transportation problem and they limit the commuting possibilities to those supported by the fixed distribution of jobs and residences. Finally, T_{act} was calculated from observed trip data and associated costs.

17.4. Results

17.4.1. *Peak and off-peak travel scales*

From the results in Table 17.1, it can be seen that, on average, there has generally been an increase in the travel ranges (T_{min}-T_{max}) during the peak and off-peak period between 1991 and 2001. The one exception to this trend is the decline in the travel range for car (ED). To be more concrete, a general widening of the travel range suggests a greater intermixing of land use arrangements along the lines of that suggested in Figure 17.1(a). The general decline in the values of T_{min} (for private transport) and increase in the values of T_{max} support this assertion. The decline in the values of T_{min} is worth noting because it suggests that origin and destination opportunities have moved closer together. However, the results show a significant dichotomy between peak and off-peak changes over the time horizon. Over the ten year period, the average range increased by 4.2 kms for the peak but only by 0.95 kms for the off-peak period. This is an interesting result because it suggests quite different land use changes for travel activity associated with the peak and off-peak periods. It implies that there has been much greater inter-mixing of home-work (peak) land uses over the period which has been aided by significant decentralization of employment in the study area [MUR 12]. On the other hand, there has only been a slight increase in the range for the off-peak period implying less intermixing of land uses associated with off-peak (non-work) travel activity (i.e. retail, leisure, recreation etc.). This may be related to the fact that land uses associated with off-peak travel activity (i.e. non-work travel) are generally more dispersed throughout a city-region in the first instance so the possibilities for further inter-mixing are more limited than for land use associated with travel during the peak period (i.e. jobs-housing). Thus, in 1991 Dublin exhibited urban spatial structure characteristics that were largely monocentric in nature with relatively centralized employment and strong radial flows from the center outwards (see [MUR 04]). On the other hand, retail (convenience shopping), recreation, leisure and entertainment among others were much more dispersed and decentralized (see [PAR 99]). By virtue of this arrangement, there was considerably less scope for a decline in T_{min} for off-peak land uses when compared with the peak period.

Overall, the result of these land use changes – where origins and destinations have increased in proximity – has been a decline in average journey distances over the study period. To illustrate, T_{act} has declined on average by 1.5 km across all indicators (ED and ND) for the peak and off-peak periods indicating that the greater intermixing of land use functions has broadly had an impact on reducing the average journey distance travelled. However, the greater inter-mixing of land uses associated with travel during the peak period (i.e. jobs-housing) has facilitated more reductions in average distance travelled over the time horizon when compared with the off-peak period. The results show that mean T_{act} values (for ED and ND) for the period declined by 1.6 km between 1991 and 2001 while the corresponding decrease for the off-peak period was 1.5 km.

Turning attention to Table 17.2, it provides the results associated with the various travel efficiency indicators used in the literature. For the purposes of this analysis, we focus only on the commuting economy (C_e) and normalized commuting economy (NC_e) indicators; it has already been argued elsewhere [MUR 11] that these are more appropriate indicators of travel efficiency because they more effectively capture how behavior is responding to land use change than indicators such as excess commuting (EC) or capacity utilization (C_u). Nevertheless, EC and C_u are provided for reference purposes. Looking at the results (Table 17.2), it is evident that whether measured in terms of C_e or NC_e, the general trend is for an increase in the efficiency[1] of travel patterns over the period – and the trend is consistent across the peak and off-peak periods. The NC_e indicator shows that, in overall terms, actual travel was 15.4% further away from random travel in 2001 compared with 1991. Indeed, this trend towards increasingly non-random behavior is also supported by the C_e indicator.

However, the interesting trend is between the peak and off-peak period. For the peak period, actual travel was 23.2% further away from random travel in 2001 than in 1991 but for the off-peak the corresponding figure was 7.6%. The obvious implication here is that peak travel behavior has reacted to changes in the spatial structure of home-work land uses. In particular, the decentralization of jobs has brought destinations closer to origins (as evidenced by T_{min}) and this has allowed for more efficient home–work travel. On the other hand, land use changes associated with off-peak travel activity have clearly not been as dramatic. The results imply that while travel economies are being achieved during the off-peak when compared with 1991 they are considerably less pronounced than for the peak period.

1 The term "efficiency" is referred to specifically within the context of the commuting economy framework whereby a move away from random behavior denotes greater "efficiency" and vice versa.

		Priv. Tran. (ED) 1991	*Priv. Tran. (ED) 2001*	*Priv. Tran. (ND) 1991*	*Priv. Tran. (ND) 2001*	*Pub. Tran. (ED) 1991*	*Pub. Tran. (ED) 2001*	*Pub. Tran. (ND) 1991*	*Pub. Tran. (ND) 2001*
T_{max}	P	14.7	15.9	19.7	24.3	10.2	11.6	14.0	16.5
	OP	15.3	14.4	22.0	20.6	10.9*	11.4	14.3*	16.1
T_{rand}	P	11.3	11.9	15.1	16.9	8.7	9.0	10.4	13.0
	OP	11.9	10.5	16.0	13.4	8.5*	8.4	10.6*	12.5
T_{act}	P	9.6	7.6	12.2	10.2	8.0	6.5	9.9	9.2
T_{min}	P	2.9	2.1	4.5	2.2	4.5	2.8	5.0	3.7
T_{min}-T_{max}	P	11.8	13.8	15.2	22.1	5.7	8.8	9.0	12.8
	OP	14.0	13.7	18.1	19.8	9.9	11.3	13.3	14.3

*Bus data used as a proxy for all public transport trips

Table 17.1. *Peak and off-peak travel scales*

		Priv. Tran. (ED) 1991	*Priv. Tran. (ED) 2001*	*Priv. Tran. (ND) 1991*	*Priv. Tran. (ND) 2001*	*Pub. Trans. (ED) 1991*	*Pub. Trans. (ED) 2001*	*Pub. Trans. (ND) 1991*	*Pub. Trans. (ND) 2001*
EC	P	69.8	72.4	63.2	79.2	43.6	56.3	49.5	59.9
	OP	86.5	89.4	70.2	90.0	85.7*	81.7	87.1*	79.8
C_u	P	56.8	39.8	50.7	36.0	58.9	42.0	51.1	42.9
	OP	59.3	43.4	50.8	36.2	61.4*	47.6	54.0*	49.7
C_e	P	15.0	36.1	19.2	39.7	8.1	30.2	4.8	29.2
	OP	36.1	37.1	18.1	40.3	20.0*	27.8	26.4*	28.8
NC_e	P	20.2	38.4	27.4	45.6	16.7	40.3	9.3	40.9
	OP	40.6	39.7	24.0	42.9	25.3*	32.9	29.2*	33.6

Table 17.2. *Peak and off-peak travel efficiency measures*

	Mean Priv. Tran. Range (T_{min}-T_{max})	*Change (1991-2001)*	*Mean Pub. Tran. Range* (T_{min}-T_{max})	*Change (1991-2001)*
P	1991 – 13.5 2001 – 18.0	+4.5	1991 – 7.4 2001 – 10.8	+3.4
OP	1991 – 16.5 2001 – 16.8	+0.3	1991 – 11.6 2001 – 12.8	+1.2

*Bus data used as a proxy for all public transport trips

Table 17.3. *Changes in travel ranges for the peak and off-peak period, 1991–2001*

17.4.2. *Peak and off-peak mode analysis*

At the outset of the modal analysis, it is prudent to point to one limitation enforced by the data. That is, the 1991 off-peak data uses only bus trips as a proxy for overall public transport costs while all other costs are calculated from a combined origin-destination trip flow and cost matrix of bus and rail data. This means that off-peak costs are likely to be underestimates because they exclude rail trips which have higher average journey distances in the Dublin region than bus trips (see [MUR 12]).

Table 17.3 shows changes in the mean travel ranges (which include both ED and ND measures) for the 1991 and 2001 peak and off-peak periods for public and private transport. The data shows that the travel ranges increased for both private and public transport between 1991 and 2001 suggesting greater intermixing of land uses over the period (see Figure 17.1(a)). On average, the overall travel range increased by 2.4 kms for private transport and by 2.2 km for public transport indicating a slightly greater intermixing of functions for the users of land uses being accessed by private transport. Moreover, for private transport, the range has increased by 4.5 km and 0.3 km for the peak and off-peak periods, respectively, between 1991 and 2001. By way of comparison, the corresponding range increases for public transport for the peak and off-peak are 3.4 and 1.2 km, respectively, over the time horizon.

Overall, this provides further evidence of a greater juxtaposition of jobs and housing during the peak period for private transport users reinforcing the implication that home-work land use arrangements have been to the advantage of private transport users. However, the same trend is not seen for the off-peak period. Here, the increase in travel range is considerably greater for users of public transport indicating that the intermixing of land use arrangements associated with off-peak travel activities as well as coverage improvements that have been made to the public transport network have been beneficial for users. Broadly speaking, these land use changes have facilitated reductions in the average actual travel (T_{act}) for both private and public transport trips between 1991 and 2001 but the greater reduction in T_{act} for private transport demonstrates that this mode has benefitted more from land use restructuring.

		Mean Changes in C_e and NC_e – Priv. Tran.	*Change (1991–2001)*	*Mean Changes in C_e and NC_e – Pub. Tran.*	*Change (1991–2001)*
C_e	P	1991 – 17.1	+20.8	1991 – 6.5	+23.2
		2001 – 37.9		2001 – 29.7	
	OP	1991 – 27.1	+11.6	1991 – 23.2	+5.1
		2001 – 38.7		2001 – 28.3	
NC_e		1991 – 23.8	+16.2	1991 – 13.0	+27.6
	P	2001 – 42.0		2001 – 40.6	
		1991 – 32.3	+9.0	1991 – 27.3	+ 6.1
	OP	2001 – 41.3		2001 – 33.4	

Table 17.4. *Mean changes in C_e and NC_e for the peak and off-peak period, 1991–2001*

Turning to the travel efficiency measures, Table 17.4 provides a breakdown of changes between 1991 and 2001. It is notable that whether 1991 or 2001 is considered in isolation the indicators are higher for private transport than for public transport. Thus, a key result to highlight is that private transport trip making is considerably further from random than public transport for all results irrespective of peak or off-peak or the measure of zonal separation being considered (see Table 17.2). For example, looking at the mean private and public transport C_e indicators (Table 17.4) for 2001 shows that private transport users were 37.9 and 38.7% away from behaving as random travelers during the peak and off-peak periods, respectively; the corresponding values for public transport were 29.7 and 28.3% respectively. The same trend holds for the NC_e indicator. Thus, it is possible to conclude that private transport users economize more on their travel costs than public transport users. Of course, there is little doubt that this is related to the relative density of the private and public transport network, the latter of which has a fixed and relatively inflexible network thereby ensuring that trip making is considerably less direct on average from origin to destination.

Perhaps more interesting is the trends between 1991 and 2001 for public and private transport. While private transport users generally have travel costs that are further away from random than public transport users, Table 17.4 shows that for the peak period the trend over the time horizon has been for public transport users to gain greater efficiency than their public transport counterparts. The results show that public transport users moved 2.4% (23.2–20.8) further away from random travel than private transport users when the C_e indicator is considered; when the normalized indicator is considered (NC_e) the extent of the shift is even more pronounced (11.2 per cent). By way of contrast, the opposite trend emerges when the off-peak results are consulted. Private transport users moved 6.5% (11.6–2.5) further away from random travel behavior than public transport users; once again, this trend also holds for the NC_e indicator (3.9% despite the magnitude of the shift being slightly less).

Some implications can be derived from these results. The decentralization of jobs over the period has certainly assisted public transport to serve more destinations along its route network and thereby facilitate more direct and less random home to work travel. However, the same cannot be said for off-peak travel. It seems that land use changes over the period have indeed facilitated less random off-peak travel for public transport users but only marginally so when compared to private transport. This supports the assertion of [MUR 12] that while the public transport network has reacted well to changes in the land use arrangements of home to work, it has not reacted very well to changes in the nature of off-peak travel through, for example, route and service changes to facilitate non-work trip making.

17.5. Conclusions and limitations

The first broad conclusion that can be drawn from the foregoing analysis is that the excess commuting and economy frameworks do provide valuable insights into the differences between peak and off-peak commuting. Table 17.4 demonstrates that in 1991 off-peak trips were significantly further away from random travel across both modes indicating considerably more efficient off-peak trips. However, there has been quite a dramatic change in this trend over the ten year period to 2000: both peak and off-peak trip making has moved further away from random behavior but the gap between the peak and off-peak differential has narrowed considerably. In other words, peak hour trip making has gained considerably in its efficiency while off-peak trip making has gained only marginally to the point that travelers during both periods are economizing on trips in a roughly similar manner. This is a very interesting result because it suggests that the general pattern of off-peak trip making has remained similar over the time horizon; on the other hand, travel behavior has reacted to land use rearrangements associated with the journey to work (peak period) in such a manner as to become considerably more efficient over the period.

Ultimately what this suggests is that the decentralization of jobs in the Dublin region has made trip making, on average, more distance efficient. However, it says nothing about the extent to which this decentralization has improved or disimproved the role of public transport. In Dublin, these land use changes and the concomitant increase in efficiency of trip making has come hand in hand with a decline in the role of public transport in serving the journey to work over the time horizon. And of course this is the conundrum associated with such efficiency measures – on the one hand they point towards greater individual behavioral efficiencies being achieved as a result of land use change but say little, on the other, about the role of these changes (i.e. land use rearrangements) in undermining the possibility for supporting collective travel (i.e. via public transport). In this regard, the results for Dublin show that despite the changes being associated with less random behavior for public transport users, there has nevertheless been a decline in the proportional use of public transport and a corresponding increase in the use of the private car over the period under consideration (see [MUR 09]). In this sense, the use of the concept of efficiency is probably a little misleading and perhaps alternative terminology should be adopted to avoid misinterpretation.

Beyond these caveats, the excess commuting and commuting economy frameworks are highly useful in that they provide the possibility for scholars and policy makers to gain a deeper understanding of the nature of land use change during different periods of the day, across variable time horizons and across modes and, in particular, how behavior is reacting to these changes through observed travel patterns. It is precisely because of analyzing changes in behavior through these frameworks that conclusions can be drawn about the nature of land use change over time, across modes and travel periods. In this regard, there is plenty of potential for further study. There has not yet been any serious analysis of off-peak travel using this framework and future research should focus on applying the framework to specific non-work trips (e.g. shopping, recreation, entertainment, etc.). This would allow for more concrete conclusion to be drawn about the nature of off-peak travel and how it is changing over time and in comparison to travel in the peak period.

17.6. Bibliography

[CHA 07] CHARRON M., "From excess commuting to commuting possibilities: more extension to the concept of excess commuting", *Environment and Planning A*, vol. 39, pp. 1238–1254, 2007.

[FRO 98] FROST M., LINNEKER B., SPENCE N., "Excess or wasteful commuting in a selection of British cities", *Transportation Research A*, vol. 32, pp. 529–538, 1998.

[HAM 82] HAMILTON B.W., "Wasteful commuting", *Journal of Political Economy*, vol. 90, no. 5, pp. 1035–1053, 1982.

[HOR 02a] HORNER M.W., "Extensions to the concept of excess commuting", *Environment and Planning A*, vol. 34, no. 3, pp. 543–66, 2002.

[HOR 02b] HORNER M.W., MURRAY A.T., "Excess commuting and the modifiable areal unit problem", *Urban Studies*, vol. 39, pp. 131–139, 2002.

[HOR 03] HORNER M.W., MURRAY A.T., "A multi-objective approach to the regional jobs-housing balance", *Regional Studies*, vol. 37, no. 2, pp. 135–146, 2003.

[HOR 07] HORNER M.W., O'KELLY M., "Is non-work travel excessive?", *Journal of Transport Geography*, vol. 15, pp. 411–416, 2007.

[MA 06] MA K.-R., BANISTER D., "Excess commuting: a critical review", *Transport Reviews*, vol. 26, pp. 749–767, 2006.

[MAC 02] MACLARAN A., KILLEN J.E., "The suburbanisation of office development in Dublin and its transport implications", *Journal of Irish Urban Studies*, vol. 1, no. 1, pp. 21–35, 2002.

[MER 95] MERRIMAN D. *et al.*, "Excess commuting in the Tokyo Metropolitan Area: measurements and policy simulations", *Urban Studies*, vol. 32, no. 1, pp. 69–85, 1995.

[MUR 12] MURPHY E., "Urban spatial location advantage: the dual of the transportation problem and its implications for land-use and transport planning", *Transportation Research Part A: Policy and Practice*, vol. 46, no. 1, pp. 91–101, 2012.

[MUR 04] MURPHY E., "Spatial restructuring and commuting efficiency in Dublin", *Journal of Irish Urban Studies*, vol. 3, no. 2, pp. 25–39, 2004.

[MUR 11] MURPHY E., KILLEN J.E., "Commuting economy: An alternative approach for assessing regional commuting efficiency", *Urban Studies*, vol. 48, no. 6, pp. 1255–1272, 2011.

[MUR 09] MURPHY E., "Excess commuting and modal choice", *Transportation Research Part A: Policy and Practice*, vol. 43, no. 8, pp. 735–743, 2009.

[NIE 06] NIEDZIELSKI M.A., "A spatially disaggregated approach to commuting efficiency", *Urban Studies*, vol. 43, no. 13, pp. 2485–2502, 2006.

[O'KE 08] O'KELLY M.E., NIEDZIELSKI M.A., "Efficient spatial interaction: attainable reductions in metropolitan average trip length", *Journal of Transport Geography*, vol. 16, no. 5, pp. 313–323, 2008.

[PAR 99] PARKER A.J., "Retail trends and the suburbanisation of Dublin's retailing: into the twenty-first century", in KILLEN J.E., MACLARAN A. (eds.), *Dublin: Contemporary Trends and Issues for the Twenty-First Century*, Geographical Society of Ireland, Special Publication 11, Dublin, 1999.

[WHI 88] WHITE M.J., "Urban commuting journeys are not 'Wasteful"', *Journal of Political Economy*, vol. 96, no. 51, pp. 1097–1110, 1988.

[WSP 03] WSP, Audit of DTO transportation model, Final Report. Dublin: Unpublished Internal Report, 2003.

[ZHO 14] ZHOU J., MURPHY E., LONG Y., "Commuting efficiency in the Beijing metropolitan area: An exploration combining smartcard and travel survey data", *Journal of Transport Geography*, vol. 41, pp. 175–183, 2014.

18

Deployment of Interoperable Cross-border Multimodal Traveler Information in Central Europe

Since October 2012, the deployment of interoperable cross-border Multimodal Traveler Information (MTI) has started in Central Europe. Partners from five different countries, namely Austria, Czech Republic, Hungary, Italy and Slovakia have committed themselves to deploying a harmonized cross-border MTI service. Since December 2014, 10 harmonized cross-border MTI services are operative in the different partner countries. The participating partners mainly come from public institutions and transport operators that cooperate to improve their national and regional traveler information systems. This paper presents the applied concept for harmonized traffic data and information exchange for the deployment of cross-border MTI services.

18.1. Introduction

The provision of interoperable and multimodal traveler information services is seen as a major contribution to ensure mobility in and between cities and regions, and additionally they address the challenges of modal split, increasing congestion and greenhouse gas emissions in the transportation sector [EUR 11]. Based on these presumptions various European cities, regions and transport operators have started to provide journey planners for public transport as well as motorized individual transport, in order to serve their citizens and travelers. An important basis for multimodal traveler information services is the underlying digital maps and graphs for the different transport modes (passenger car traffic, public transport, cycling, walking) to enable multimodal routing and interchange information.

Chapter written by Gerhard MENZEL, Martin BÖHM and Katharina ZWICK.

For example, the Vienna Region started to create a multimodal journey planner, available on the web as well as on smart phones (www.anachb.at), which is based on the Austrian wide digital graph system known as the "Graph Integration Platform" (GIP) [MAN 12]. As the name already implies the GIP is a platform that links different digital maps (graphs using various different reference systems) to a common intermodal transport graph according to uniform rules which have been defined in the Austrian GIP Standard [FSV 12]. The GIP can be used as a basis for transport information, transport management and traffic control across all administrative boundaries.

However, systems like the Austrian GIP and similar systems from the neighboring countries of Austria, for instance KIRA, which is the Transport Information System and Database in Hungary [EDI 12b] are still mostly based on regional data and systems that only provide information within the area of the respective cities and regions – therefore there is a need for cross-border (between regional borders as well as across countries) information and solutions for different transport modes.

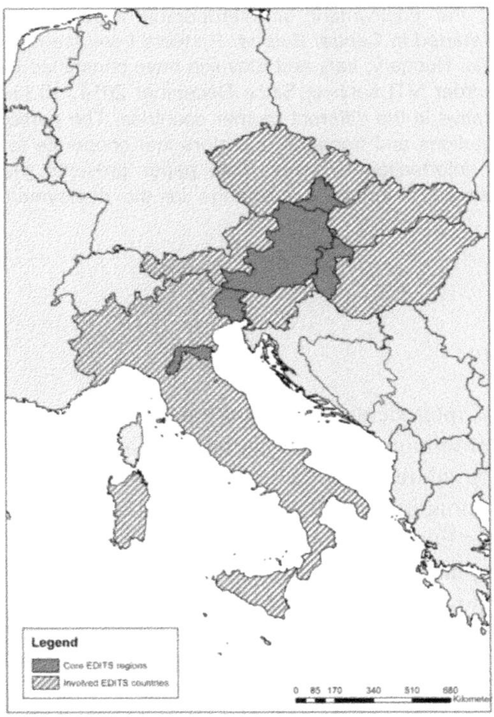

Figure 18.1. *Countries involved*

One of the initiatives currently dealing with this need is the EDITS (European Digital Traffic Infrastructure Network for Intelligent Transport Systems) deployment project – a European Union co-funded project under the umbrella of the Central Europe Programme, which aims to deploy and operate cross-border multimodal travel information based on harmonized traffic data and information gathered on a transnational level [EDI 12a]. The EDITS project took place between July 2012 and December 2014, and involved partners that mainly come from public institutions and transport operators that work together to improve their national and regional traveler information systems within five different central European countries (Austria, Czech Republic, Hungary, Italy, Slovakia) by exchanging traveler information related data.

18.2. The EDITS concept

EDITS focuses on providing interoperable and multimodal traveler and information services based on a harmonized platform for data as well as information exchange. First of all it is important to highlight that the EDITS concept is not based on the principle of setting up a new central multimodal journey planner within the involved EU member states. This is especially due to the fact that a lot of structures and systems have already existed in the national administrations – the challenges now are to improve, update and harmonize the existing systems in a common way to finally ensure the provision of seamless cross-border services to the single traveler.

18.2.1. *The EDITS Multimodal Traveler Information (MTI) service framework – a modular approach*

As the name Multimodal Traveler Information (MTI) service already implies, traveler information shall be deployed for all transport modes. Multimodality in the EDITS context means that the traveler information provided to the end users (pre- and on trip) will support the traveler with different travel options for all transport modes, which are defined in the following transport mode categories:

1) individual transport: covers information for all transport modes that can be used by travelers to move through the traffic network individually, including both motorized individual traffic on the road network (e.g. by motorcycle, car, HGVs) and non-motorized individual traffic (e.g. walking or cycling);

2) public transport: covers information for all transport modes which are publicly available to travel through the traffic network, e.g. by train, subway, bus etc.;

3) intermodal transport: covers information for a combination of more than one specific transport mode, e.g. car and public transport or bike and public transport.

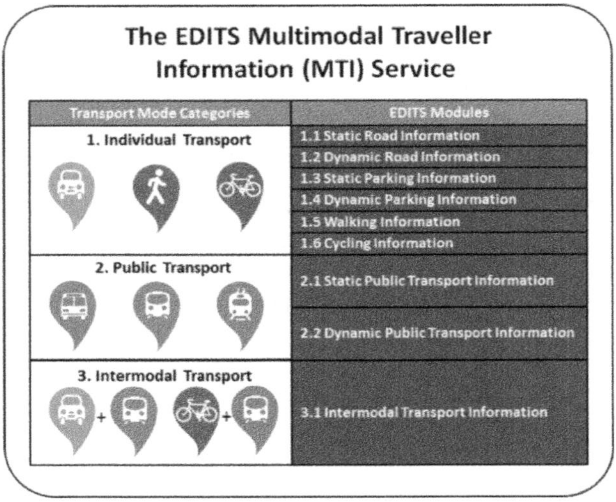

Figure 18.2. *The EDITS MTI service*

Based on the transport mode categories single EDITS Modules are defined, which are all part of the EDITS MTI Service (see Figure 18.3). In other words this means that within EDITS there exists one defined EDITS MTI Service that shall be deployed within Central Europe to support travelers before and during their trip with options and information on all transport modes – however, the involved stakeholders (public administrations, service providers, end-users) within the domain of traveler information provision are usually quite heterogeneous as they come from different organizations with different focus areas. Therefore, this modular approach with the definition of the specific EDITS Modules based on the transport mode categories will ensure the required flexibility when it comes to real deployment based on the existing regional systems and services [EDI 13].

The EDITS Modules can be regarded as a “toolbox” to pick from to define the EDITS MTI end-user services, which are deployed in demonstration pilots within the regions until the end of 2014, depending on the data availability of the participating operators. This means that one region might demonstrate traveler information for end-users coming from EDITS Modules 1.1 (Static Road Information), 1.2 (Dynamic Road Information), 2.1 (Static Public Transport Information) and 2.2 (Dynamic Public Transport Information) while other partners might focus only on 2.1 due to lack of individual transport data and lack of dynamic public transport data. The more the modules are provided within the regionally defined EDITS MTI service, the more the multimodal traffic information can be provided to end-users within the demonstration areas and beyond them – shared through the common data exchange interfaces.

18.2.2. *Exchange of data and information between regions*

The EDITS partners recognized the Austrian GIP (Graph Integration Platform) approach and decided to pursue a system architecture that follows similar principles. Hereby, the development of consistent multimodal transport graphs within the single regions based on common specifications plays an important role as the very basis for the creation and adaptation of partner specific enhanced ITS applications. As shown in Figure 18.1, these regional independent transport graphs, known as the "EDITS-GIPs" are built on existing systems and content databases within the regions.

The local EDITS-GIPs integrate the existing data pools into a common EDITS-GIP format following the specifications that have been developed within the EDITS project. The responsibilities and data ownership always remains within the regions. Hence, the localized regional EDITS-GIPs may look quite different in the regions involved as they are mostly built on existing and proprietary systems (e.g. KIRA in Hungary and the Austrian Graph Integration Platform GIP) and are therefore merely translating the available content into a common form that is understood by all involved parties.

The EDITS-GIP format the interface for the exchange of GIS data between the EDITS source systems (GIP in Austria, KIRA in Hungary, DYNAVIX in Czech Republic, etc.). It covers the GIS data of transport infrastructure (for individual transport and public transport) as well as transport data for individual transport. The EDITS-GIP format consists of data files (.SHP files, .CSV files) and lookup-tables (LUT) that describe value domains for attributes which are common for all EDITS-partners. Based on the mapping of the EDITS source systems to the EDITS-GIP format the regional EDITS-GIPs are set up, which represent the GIS data exchanged among the involved parties.

Since a cross-border MTI end-user service requires more data to be exchanged besides the underlying GIS infrastructure information, the second main component within the EDITS concept is an EDITS-GIP-Interface for commonly agreed data and information exchange between the single regions. The specified EDITS-GIP-Interface will allow the exchange of various kinds of data alongside the GIS data needed for the provision of end-user services (e.g. public transport information, time table data, road related information, etc). These specifications are based on existing European and national standards (like the Austrian GIP Standard, DATEX II, DINO, etc.) and provide the necessary information alongside the GIS data. The specifications will be openly available, in order to foster the deployment of cross-border exchange of data and information within the regions as well cross-border within Europe beyond the project's lifetime.

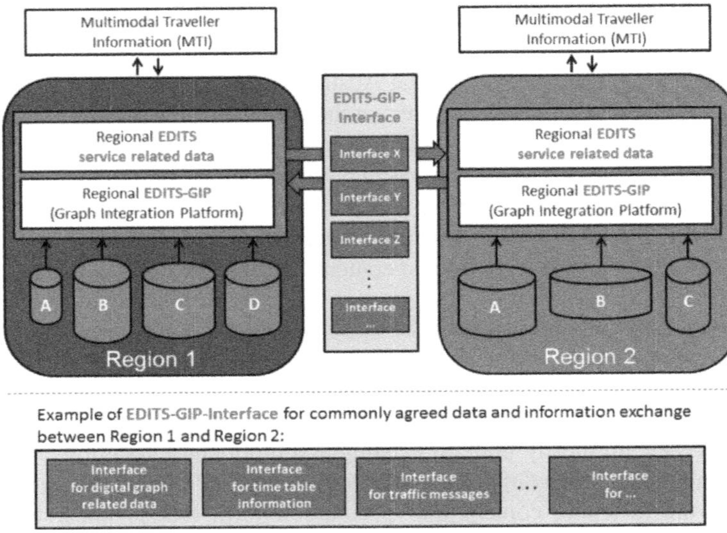

Figure 18.3. *The EDITS concept*

18.2.3. *Provision of multimodal traveler information services – a decentralized approach*

In order to demonstrate the feasibility of the EDITS approach of exchanging data and information, 10 different end-user services were deployed in 2013 and 2014, resulting in demonstrations in the second half of 2014. The eight web-based services and two mobile applications provide interoperable and intermodal pre-trip information and on-trip services to the end-users. Traffic management is also tackled within EDITS, applying the same exchange principle.

An important conceptual paradigm for the EDITS approach is that the responsibilities and provision of the multimodal traveler information services remain the responsibilities of the single regions. This means that the involved parties are enhancing existing systems (when available) through the additional information received from the exchange of data through the regional EDITS-GIP and EDITS-GIP interface within a completely decentralized system. This stands in direct contrast to the possible setup of new centralized EDITS systems that would provide a central MTI end-user service for all involved regions, which is clearly not the approach that has been chosen (compare Figure 18.4 and Figure 18.5).

The end-user services were demonstrated in different pilots among the EDITS partners, within specified corridors where cross-border journey planning is made available. These include show-cases, like for instance door-to-door routing for individual and public transport journey planning from Vienna in Austria to Brno in the Czech Republic, which is currently not available since no cross-border time table data is exchanged among the local public transport operators KORDIS and VOR. The important conceptual aspect is that within EDITS the existing information services (like in the KORDIS-VOR case e.g. the existing journey planning platforms IDS JMK in Czech Republic and VAO in Austria) are enhanced by the exchanged cross-border information through the EDITS interfaces. This means that the existing partner's system will provide the full journey planning and traveler information aspects of the existing services and will in addition be extended by the cross-border components developed within EDITS.

Finally all 10 EDITS services were demonstrated by the end of 2014 within three main demonstration areas:

The CENTROPE demonstration area covered the Vienna Region (including Vienna, Lower Austria and Burgenland), Southern Moravia, the western part of Hungary and the city of Bratislava. Within this demonstration area five web-based end-user services (one based on the service of the Hungarian partners, one based on the Austrian multimodal traveler information service *anachb.at* and three end-user services based on the systems of the Czech partners) as well as two mobile applications (provided by the Czech partners) showed the feasibility of the EDITS concept and fostered the first steps towards a harmonized deployment of traveler information services in Central Europe. The end-user services provided different cross-border traveler information services, including routing for pedestrians, cyclists, motorized vehicles and for public transport.

Within the second demonstration area in Italy the Italian provinces of Modena and Ferrara demonstrated the cross-regional exchange of GIS and public transport data, implemented within a common traveler information platform.

The third demonstration area covered the cross-border triangular area linking the cities of Villach (AT) – Udine (IT) – Ljubljana (SI). Within this demonstration area the motorway operators ASFINAG (Austria), AutoVie Venete (Italy) and DARS (Slovenia) deployed end-user services providing cross-border information about traffic states based on the exchange of traffic messages. In accordance with the European specifications, the EDITS partners within this triangle are early users of a European Standard as a common basis. Therefore, the EDITS partners are taking a leading role in Europe as they demonstrate the exchange of data in the DATEX II format

The end-user services demonstrated traffic information along with routing recommendations in the covered area to drivers planning a trip or looking for route alternatives on their way to their final destination. Information is provided on the basis of the current traffic situation, including weather conditions.

By the end of 2014 the demonstrations showed the feasibility of the approach, and furthermore established a sustainable basis for the harmonized deployment of multimodal cross-border traveler information in Europe taking into account existing structures.

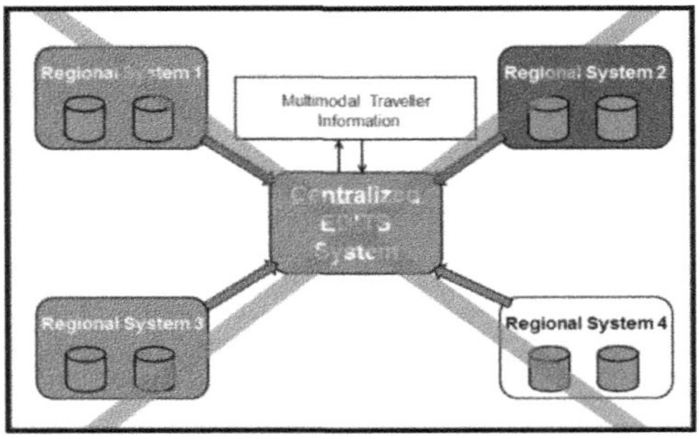

Figure 18.4. *Not the goal: a centralized system and central MTI service*

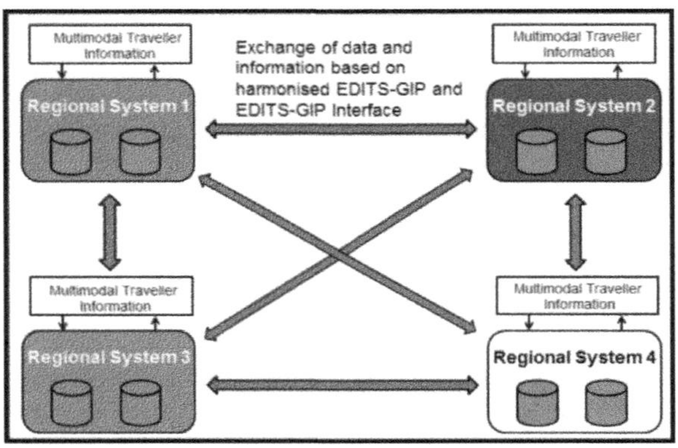

Figure 18.5. *The goal: a decentralized system enabling regionally operated cross-border MTI services*

18.3. Conclusion

EDITS – the European Digital Traffic Infrastructure Network for Intelligent Transport Systems – is a European Union project under the umbrella of the Central Europe Programme, co-financed by the ERDF (European Regional Development

Fund). Bringing together partners from different governmental levels from five different countries within the Central Europe region, EDITS develops common transnational solutions for the exchange of transport related information and the improvement of interoperable and multimodal traveler information services. The fundamental basis necessary to facilitate these services are an interoperable and multimodal EDITS-GIP (Graph-Integration-Platform) and EDITS-GIP-interface, which is a tool to enable the exchange and provision of relevant data and information. This is achieved through the set up or improvement of existing traveler information services and of traffic management solutions. The services developed in this way provide support to EDITS users before and during their journey, within and beyond the regions. Ten end-user services were deployed and demonstrated by the end of 2014 proving the feasibility of the EDITS approach of exchanging data and information and establishing the basis for the next steps – towards the deployment of interoperable cross-border multimodal traveler information in Central Europe.

18.4. Bibliography

[EDI 12a] EDITS PROJECT, EDITS Project Website, Vienna, http://edits-project.eu/, 2012.

[EDI 12b] EDITS PROJECT, Output 3.1.4 Report on State-of-the-Art analysis, Hungarian Transport Administration – Deliverable of the CENTRAL EUROPE EDITS Project Consortium, pp. 7–8, Vienna, available at http://edits-project.eu/, 2012.

[EDI 13] EDITS PROJECT, Output 3.2.3 Report on EDITS Service Definition, Deliverable of the CENTRAL EUROPE EDITS Project Consortium, Vienna, available at http://edits-project.eu/, 2013.

[EUR 11] EUROPEAN COMMISSION, White Paper: Roadmap to a Single European Transport Area – Towards a competitive and resource efficient transport system, COM (2011)144, Brussels, 2011.

[FSV 12] FSV.RVS 05.01.14 Intermodal Transport Reference System for Austria, Standard description GIP (Graphs Integration Platform), Vienna, available at: http://fsv.at/, 2012

[MAN 12] MANDL-MAIR IRMGARD, "GIP.at: The Basis for a Modern Administration of Austria's Transport Routes", *Proceedings of the 19th ITS World Congress*, ERTICO, 2012.

PART 5

Assessment and Impacts

19

Impacts of Cooperative Traffic Systems on Safety, Environment and Travel Times: a Literature Survey

This paper presents the results of the impact assessment study of ten cooperative systems (CS), conducted as a part of the European Cooperative Benefits for Road Authorities project, which aims to provide decision support for road authorities on future CS investments. This literature synthesis is based on 31 preceding studies, including field operational trials and simulation studies that evaluated the impacts of these systems. In total, 10 vehicle-to-infrastructure/vehicle-to-vehicle CS were selected and grouped into three bundles: local dynamic event warnings, in-vehicle speed and signage, and travel information and dynamic route guidance. The results have been synthesized in a matrix, distinguishing between various indicators for safety, traffic efficiency and environmental impacts. This paper is intended to serve as a basis for future projects and a comprehensive standard reference for readers interested in CS impacts.

19.1. Introduction

The term cooperative system (CS) is used to define a specific part of the intelligent transport systems (ITS) that use the technology of communication between vehicles and transport infrastructure to effectively share information. A wide variety of data can be transmitted with the aim of improving safety and efficiency, one step above stand-alone systems. CS can be based on three different types of communication: vehicle to infrastructure (V2I), infrastructure to vehicle

Chapter written by Isabela MOCANU, Philippe NITSCHE and Kerry MALONE.

(I2V) and vehicle to vehicle (V2V). In recent years, several projects have focused on the evaluation of CS, as a step toward their future usage. Field operational tests (FOTs) such as euroFOT [BEN 12], teleFOT [WEL 12] and the most recent DRIVE C2X [FLA 11], were carried out in order to identify the possible challenges and the methods of dealing with CS and, more importantly, to identify the benefits that could be gained from this technology.

19.1.1. *Description of the Cooperative Benefits for Road Authorities (COBRA) project*

The European research project COBRA [FAB 13] aims to aid road authorities in benefiting from changes in the field of CS. This is done by providing them with an insight on the costs and benefits of possible investments, both from a societal and business case perspective. These observations are based on a decision support tool that gives road authorities the opportunity to compare costs and monetized benefits of cooperative services in various contexts. Ten CS were grouped into three bundles and evaluated in terms of their impact on safety, traffic efficiency and environment.

The tool developed in the COBRA project aims to assess the societal costs and benefits of using CS on motorways, and then use those costs and a subset of the (monetized) benefits in the calculations for certain business cases for the road authority. The methodology for achieving the COBRA project goals can be divided into two parts, impact assessment and cost-benefit analysis [FAB 13]. For realizing the decision support tool, it was necessary to determine positive or negative impacts that can be expected with CS implementation and translate those to costs and monetized benefits. Figure 19.2 provides an overview of the relationship between inputs, impact assessment and the cost-benefit analysis in the COBRA tool.

19.1.2. *Outline of the chapter*

Section 19.2 presents the CS which were analyzed in this paper, as well as their classification into the three bundles. Sections 19.3 and 19.4 present the literature, which served as the basis for the evaluation of the systems' impact in terms of safety, environment and travel times and the methodology by which the assessment was performed. The final results are shown in section 19.5, while the conclusions can be found in section 19.6.

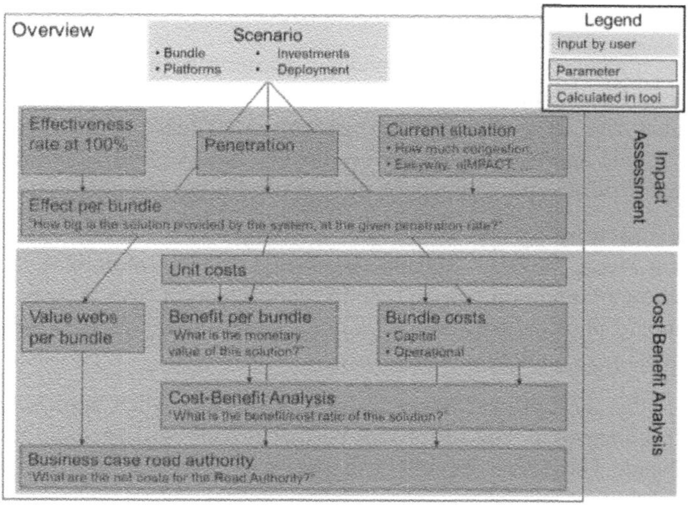

Figure 19.1. *COBRA methodology overview*

19.2. Description of systems and bundles

In total, 10 CS (see Table 19.1) were chosen from an array of technologies and grouped into 3 bundles: local dynamic event warnings, in-vehicle speed and signage, and travel information and dynamic route guidance. The selection of bundles was based on the assumption that, due to technical requirements (wireless vs. cellular communication), implementation costs and legal issues, CS will likely be used in a bundle, rather than as individual systems. Only V2I and I2V technologies were considered. More details can be found in the COBRA Deliverable D3.1 [MOC 12].

Bundle	System
#1: Local dynamic event warnings	Hazardous location notification (HLN)
	Road works warning (RWW)
	Traffic jam ahead warning (TJW)
	Post-crash warning (eCall)
#2: In-vehicle speed and signage	In-vehicle signage (IVS)
	Intelligent speed adaptation (ISA)
	Dynamic speed limits (DSLs)
#3: Travel information and dynamic route guidance	Traffic information and recommended itinerary (TIN)
	Multimodal travel information (MTI)
	Parking information and guidance (PIG)

Table 19.1. *Description of systems and bundles*

The systems are described in more detail below:

– HLN – a safety system that provides a warning notification to the driver, in order to direct his attention to a potential hazardous situation or area. These areas have a higher risk of collisions and incidents and thus require more attention. Examples include approaching an icy or slippery road section, approaching fog or approaching a road section with a poor surface condition.

– RWW – a warning and notification system, which enables faster and safer passing near work zones. Repairs performed on a motorway require the implementation of temporary speed limits, lane changes and merges, as well as traffic signs and physical barriers. A warning system can provide instructions and alerts to drivers for a smoother traffic flow.

– TJW – this system provides a warning to drivers when approaching a road section with congestion. The result will be that drivers will be more aware of the situation ahead. The safety benefit will primarily decrease the risk of rear-end collisions.

– eCall – this safety system is widely known, and its implementation is expected to increase highly in the next years. If a collision is detected by sensors in a vehicle, an automatic call will be placed to emergency services, providing the location of the incident as well as details regarding the vehicle. The system also provides a direct line of voice and data communication so that the emergency call center can talk with those conscious in the vehicle. This system will speed up both the rescue of injured persons making use of the "golden hour", as well as the clearing of the roadway.

– IVS – a notification and warning system that provides information to drivers regarding upcoming road signs, for a better view of situations beyond or outside the line of sight. Drivers receive notifications on road features such as curves, roundabouts, pedestrian crossings, road markings or segregated bus lanes.

– ISA – a well-known safety system that monitors speed limits on all road sections and provides information, warnings or interventions to drivers if that speed limit is exceeded. Three types of ISA systems are as follows:

- informative: the driver is informed about the regulated speed limits of each road section;

- warning: the exceeding of a speed limit is signaled to the driver through various warning methods, i.e. haptic accelerator pedal. The driver will feel a counterforce when trying to press the pedal. This type of warning can be overridden if the driver presses the accelerator pedal harder;

- intervening: when the speed limit is exceeded, the ISA system takes over and limits through the activation of automated braking;

– DSLs – a notification system with the role of informing drivers of the current speed limit of a specific road section. By taking into account influencing factors such as traffic flow, time of day, weather conditions (e.g. fog and ice) and emission levels, this system offers the ability to communicate dynamic maximum speed limits to drivers.

– TIN – an information system that aids drivers through a vehicle navigation system by providing alternative routes around congested locations. Pre-trip, as well as on-trip route choices, can be made; therefore, a distribution of the traffic load can be achieved.

– MTI – a system similar to the better known advanced traveler information system, that has the purpose of aiding drivers by providing information regarding travel times, routing, schedules, etc. on various modes.

– PIG – an information system that provides guidance and routing to drivers, primarily truck drivers that are looking for a parking place. Through various sensors, the number of free spaces is monitored in real-time. When a driver searches for a space, the system selects one of the free spaces and provides routing information to the specific spot.

19.3. Reviewed literature

The literature synthesis was based on 31 preceding projects, FOTs and simulation studies that evaluated the impacts of various CS. The focus was on the motorway network and the level of service was also taken into consideration, as it can influence a CS impact. The Economic Impact Assessment of Stand-alone and Cooperative Intelligent Vehicle Safety Systems (IVSS) in Europe (eIMPACT), Cooperative Systems Deployment Impact Assessment (CODIA) and Easy Way are among the major projects that focus on the impact assessment of stand-alone and CS.

eIMPACT [WIL 08] assessed the effects of 12 intelligent vehicle safety systems in terms of traffic safety and efficiency at various level of penetration. The estimations regarding penetration were achieved by using data sets on fleet composition and mileage, as well as the expected market penetration of CS. The assessment of traffic impacts was done by performing micro-simulations, while the results for the safety impacts were obtained based on relevant accident data (such as number of fatalities and injuries in EU countries).

CODIA [KUL 08] performed an impact assessment of five CS in terms of safety, traffic efficiency and environment. The methodology was similar to the one used in eIMPACT. Micro-simulations, as well as relevant accident data, fleet composition and mileage and expected market penetration, were the basis for obtaining the results. The assessment regarding the environmental effects was performed with an instantaneous emission model describing vehicles' emission behavior.

Easy Way [FEI 11] assessed the effects of seven CS, grouped into two bundles, in terms of safety, congestion, traffic efficiency, fuel consumption and emissions. The methodology used is based on eIMPACT and CODIA, as well as other projects [MAR 08, BAY 07].

COOPERS (Cooperative Networks for Intelligent Road Safety) [KÖL 10] concentrated on the evaluation of innovative telematics applications on the road infrastructure, with a cooperative traffic management between vehicles and infrastructure.

SAFESPOT (Cooperative Vehicles and Road Infrastructure for Road Safety) [WIL 10, AND 10] assessed the impact of certain CS in terms of traffic efficiency through various evaluation methods, such as micro-simulations.

Other projects reviewed include Cooperative Vehicle–Infrastructure Systems (CVIS) [TSE 10], PReVENT (Preventive and Active Safety Applications Integrated Project) [SCH 08], ITS Test Beds [JON 10], Dynamax [STO 11] and others. Individual projects, conference papers, publications and journal articles were also included in the literature review, in an attempt to find the most comprehensible data.

The literature review revealed a lack of impact assessment data, possibly due to the novelty of CS. It was therefore decided to also use data regarding stand-alone systems while making informed assumptions regarding the added benefit that a cooperative technology would bring. No additional simulations or FOTs were performed to validate or support the results.

19.4. Methodology

19.4.1. *Indicators and penetration scenarios*

The impact of the selected CS was assessed in terms of safety, traffic efficiency and environment. In total, 10 indicators were assessed for a 100% penetration of informed vehicles. When different penetration scenarios for a certain impact indicator were found, assumptions were made to extrapolate the data for the 100% penetration scenario.

The potential benefit was compared to the reference situation. This refers to the current situation of the traffic, the level of traffic management system usage and number of motorway kilometers in the country of interest. Figure 19.2 provides a schematic of the methodology of the impact assessment.

The ten indicators used were as follows:

Safety:

– number of injury accidents – number of road accidents that result in personal injuries;

– number of fatalities – number of fatally injured people due to a road accident;

– number of injuries – number of slightly or gravely injured people wounded in a road accident.

Traffic efficiency:

– travel times – measured time difference between departure and arrival times of vehicles, on a specific road segment (measured in s);

– congestion times – time losses due to traffic events, such as congestion (measured in s).

Environment:

– CO_2 – carbon dioxide (measured in g/km);

– NO_x – nitrogen oxides NO (nitrogen oxide) and NO_2 (nitrogen dioxide) (measured in g/km);

– fuel consumption – consumption of fuel (gas or diesel) (measured in l/100 km);

– particles – particulate matter of different chemical compositions suspended in air, generated from the burning of fossil fuels (measured in $\mu g/m^3$);

– noise emissions – traffic noise pollution (measured in dB).

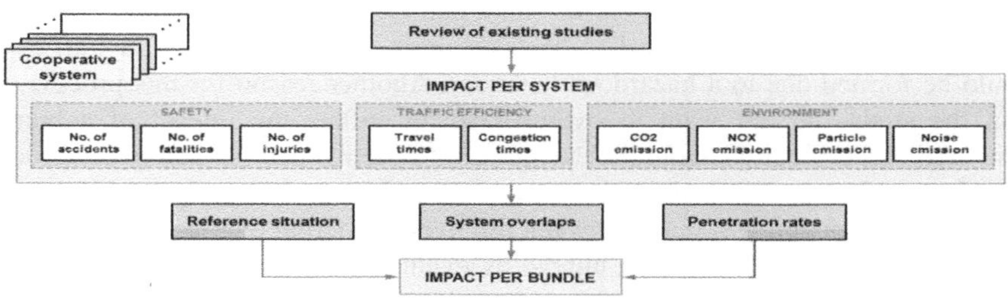

Figure 19.2. *Methodology of the impact assessment*

19.4.2. *Assessment methodology for systems and bundles*

In total, 31 studies were used in the assessment, both FOTs and simulations. Due to the differing nature of these methods of assessment, it was decided that the results of FOTs are more reliable and more accurate.

The literature review revealed multiple studies relevant for some of the systems, which in some cases resulted in having multiple values for each indicator. In order to harmonize the reported data to a final value, a weighting approach was used. The procedure was performed by taking into account several influencing factors such as the method of analysis (FOTs vs. simulations), sample size, study design and input factors. Therefore, when studies were considered similar in their methodology, an averaging function was used to reach the final value. When only one value was found for an indicator, it was either used as it was or informed assumptions were made, based on the respective study's methodology of how that result was achieved. As previously mentioned, it was considered that the results of FOTs are more reliable than those of simulations. In the case when values for the indicators came from studies using both methods of evaluation, a weighting formula was used so that the results of FOTs had more weight.

However, finding relevant results for all impact indicators proved to be challenging. It was decided to also use results from evaluations of stand-alone systems and to make informed assumptions regarding the additional benefit that the cooperative communication would bring. An example is Dynamax [STO 11], a project that evaluated a stand-alone system providing DSLs. Taking into consideration the circumstances of the field operational trials performed in the project, as well as other results found for the same system, final values were reached for the relevant indicators.

Assessing the impact of the three bundles was the next step. No results were found regarding the impacts of CS at a bundle level. Assuming that the systems in a bundle would be implemented on the same platform, the fact that the technologies would influence each other had to be taken into consideration (e.g. a traffic jam could be formed due to a hazardous location). Another reason for this process was that the study did not want to overestimate or inflate the impact of a bundle. Therefore, an evaluation about the extent of the systems' overlap had to be made, in order to perform an analysis of how it would influence the impacts of the whole bundle. A distinction was made between full overlap, partial overlap and no overlap. The analysis was made for each bundle by evaluating how the impact of a system could overlap with another. The following method of estimating the overall impact indicator values of each bundle was chosen:

Bundle 1 – local dynamic event warnings – was comprised of four CS: HLN, RWW, TJW and eCall. Following the assessment performed on each system, it was agreed that the first three might overlap to some extent (e.g. a TJW could be triggered in case of road works ahead). The last system, eCall, is seen as a different technology, as it would be triggered only in the case of an accident. This assessment was also supported by the studies and results obtained in the literature review. Therefore, it was decided that the final impact of the first bundle would be calculated by adding the maximum value for each indicator for the first three systems to the values from eCall.

Impact value = max (HLN, TJW, RWW) + eCall

Bundle 2 – in-vehicle speed and signage – contained three systems: IVS, ISA and DSLs. As all three provide information that involves speed management, it is obvious that an overlap will exist. Therefore, the best method of assessing the impact of this bundle was considered to take the maximum value of each indicator from the results of the three systems.

Impact value = max (IVS, ISA, DSL)

Bundle 3 – travel information and dynamic route guidance – was comprised from three systems: TIN, MTI and PIG. All three have the purpose of providing routing and navigation while maintaining an efficient traffic flow. A lack of data was encountered for operations in congested conditions. Again, the literature review revealed an overlap between the first two technologies, as both deal heavily with route guidance in various traffic circumstances. The last system provides aid in finding truck parking places, and only safety impacts were found for this system, which is not representative for this bundle. Therefore, it was assumed that the safety impacts for this bundle were lower. The final method of bundling was to take the maximum value for each indicator for the first two systems and add the third one.

Impact value = max (TIN, MTI) + PIG

19.5. Results

The results of the impact assessment show that CS have substantial impacts in terms of road safety, traffic efficiency and environment. The study found that they primarily affect road safety, reducing the number of fatalities and injuries. In comparison, the traffic efficiency effects are modest, while the impacts on the environment vary strongly from system to system.

Tables 19.2 and 19.3 present the overall results of the assessment for all indicators at both system and bundle level, as positive or negative percentages. As previously mentioned, the literature review revealed that the effects of CS are influenced by the level of the traffic flow. Therefore, a distinction was made between free flow

(Table 19.2) and a congested situation (Table 19.3). No reliable results were found for the third bundle in the case of congested traffic; therefore, they were not included in the final assessment tables. The methodology described in the previous chapter, as well as in Faber *et al.* [FAB 12], was applied for all functions and bundles. However, in some cases (see values marked with * in Table 19.3), the results found in the literature were not considered reliable by the project consortium and were adjusted based on expert decisions. For example, for the parking and information guidance system, only one study was found that evaluated a road-side parking system [GAR 04]. Based on the data of the study, the values for the overall safety impact were reached using accident statistics from Austria, the UK and the Netherlands. Taking into consideration the characteristics of this system, as well as the expert opinions of the consortium, it was decided that the values reached were too high; therefore, they were downsized. Similar processes were performed for other results that, based on the literature review, were not considered realistic by the project consortium. Therefore, where it was possible, adjustments were made. In the cases where data were insufficient, the consortium preferred not to make assumptions.

Indicators	HLN	RWW	TJW	eCall	*Bundle 1*	IVS	ISA	DSL	*Bundle 2*	TIN	MIT	PIG	*Bundle 3*
Number of fatalities	-4.60	-1.40	-2.00	-2.00	-7.00	-7.20	-5.45	–	-7.00	–	–	-8.55	-4.00*
Number of injuries	-3.70	-2.30	-7.00	-1.00	-8.00	-4.80	-3.64	–	-5.00	–	–	-9.77	-5.00*
Number of injury accidents	-5.30	-2.50	-7.00	–	-7.00	-4.80	-4.60	–	-5.00	–	–	–	-5.00*
Travel time	+2.08	-	+7.70	–	+8.00	+4.23	+2.40	+1.37	+4.00	-11.0	-8.77	–	-11.0
Congestion time	–	–	–	-3.00	-3.00	–	–	–	+6.00*	-6.90	–	–	-7.00
Fuel consumption	–	–	–	-0.55	-1.00	–	-2.00	-4.00	-4.00	0	-11.0	–	-10.0*
Noise	-0.20	–	-0.50	–	-1.00	-0.25	–	–	-4.00*	–	–	–	–
CO_2	-6.00	–	+7.60	0	+2.00	-4.10	-2.30	-4.00	-4.00	-9.14	–	–	-9.00
NOx	-9.88	–	–	0	-10.0	-8.58	-4.00	-5.00	-9.00	-4.90	-6.00	–	-6.00
PM	-5.43	–	+4.70	0	-1.00	-3.75	-1.00	-10.0	-10.0	–	–	–	–

*adjusted values by project consortium

Table 19.2. *Impact assessment results for free traffic flow (given in %)*

Indicators	HLN	RWW	TJW	eCall	*Bundle 1*	IVS	ISA	DSL	*Bundle 2*
Number of fatalities	-4.60	-0.20	–	-2.00	-7.00	-7.20	-5.45	–	-7.00
Number of injuries	-3.70	-0.20	–	-1.00	-5.00	-4.80	-3.64	–	-5.00
Number of injury accidents	-5.30	-0.20	–	–	-5.00	-4.80	-4.60	–	-5.00
Travel time	+3.95	–	0	–	+4.00	+6.07	0	+1.44	+7.00*
Congestion time	-	–	–	–	-3.00*	–	0	–	0
Fuel consumption	–	-	–	–	-1.00*	–	–	-	0*
Noise	-0.20	–	–	–	0	-0.70	–	–	0*
CO_2	+0.25	–	–	–	0	+0.10	–	–	0
NOx	-1.45	–	–	–	-1.00	-1.13	–	–	-1.00
PM	-0.35	–	–	–	-2.00*	-0.93	–	–	-1.00

*values adjusted by project consortium

Table 19.3. *Impact assessment results for congested traffic (given in %)*

19.6. Conclusions and recommendations

The aim of the COBRA project was to develop a support tool for road authorities in order to aid the process of implementing CS. The aim of the impact assessment was to provide impact figures as estimated percentile changes in key indicators for safety, efficiency and environment.

The data found in the literature review were based on FOTs and simulations. Both methods of evaluation have their limitations. Each FOT was carried out in different circumstances (e.g. geographic and regional area, duration of the trial), making it difficult to extrapolate the results to a more general (e.g. country) value. All simulations performed in the studies reviewed differed in input factors (e.g. modeled area and traffic flows), evaluation tools and output results. No additional simulations or FOTs were carried out to validate or support the results.

Based on the assessment performed, several conclusions and recommendations can be made:

– The literature review revealed a lack of impact assessment results, which may be the result of the novelty of CS. More results are expected soon for a number of systems, e.g. DRIVE C2X project.

– Few studies were found which evaluated the additional effects that a cooperative technology could add to an existing roadside system, or the potential differences in impacts between existing roadside implementations and CS implementations. Further research must be undertaken on the difference between roadside and CS (e.g. variable message signs and IVS).

– Most projects which evaluated CS did so by means of simulations. Due to the high variability in input factors and simulation tools, the reliability and accuracy of the results was considered less trustworthy than from FOTs.

– Little research was found regarding how drivers react to CS and the behavioral changes that result. Therefore, it is difficult to evaluate data from different studies, each with their own parameters and scenarios.

– Further attention must be paid towards a more general methodology regarding the extrapolation of results from FOTs carried out in different geographical areas, and segment to generalized figures (e.g. country specific).

– A clear description of the methodology used for further impact assessments should be provided in order to allow for conclusive analyses. Most projects reviewed failed to offer a clear documentation of the methodology used to achieve their respective results.

– Few projects covered the full range of circumstances and influencing factors in which CS would be used. A used scenario for the purpose of an impact assessment should include a wide variety of parameters, e.g. road geometry, traffic flow and weather conditions.

The impact assessment performed in this study not only provided input for the European project COBRA, but also helped gain an insight into the impacts that CS would have. The results of the literature review revealed that these technologies could have positive effects in terms of reducing the number of fatalities and injuries. The more important issue is not the effects themselves, but rather the implementation of steps that need to be taken in order to reach a significant penetration rate of these technologies, so that these positive impacts could actually be reached. The aim of this impact assessment study and other comparable ones (e.g. eIMPACT, CODIA, EasyWay and DRIVE C2X) is to highlight their effects, in order to support the implementation process of CS.

19.7. Acknowledgments

COBRA was funded by the National Road Authorities from Belgium, Switzerland, Germany, the Netherlands, Norway and the UK under the program ENR MOBILITY.

19.8. Bibliography

[AND 10] ANDREONE L., BRIGNOLO R., DAMIANI S. *et al.*, SAFESPOT D8.1.1 Final Report, October 19, 2010.

[BAY 07] BAYLY M., FILDES B., REGAN M.A. *et al.*, TRACE – TRaffic Accident Causation in Europe, D.4.1.1-D6.2 Review of crash effectiveness of Intelligent Transport Systems, November, 2007.

[BEN 12] BENMIMOUN M., PÜTZ A., AUST M.L. *et al.*, European Large Scale Field Operational Tests on In-Vehicle Systems; Deliverable D6.1 Final evaluation results, 2012.

[FAB 13] FABER F., VAN NOORT M., HOPKIN J. *et al.*, Cooperative Benefits for Road Authorities, Methodology framework (Deliverable 2 No. TNO-060-DTM-2013-01482), TNO, The Netherlands, p. 94, 2013.

[FAB 12] FABER F., VAN NOORT M., HOPKIN J. *et al.*, COBRA D2 Methodology framework, COBRA Consortium, 2012.

[FEI 11] FEIJEN M., DE VRIES B., ALKIM T. *et al.*, EasyWay Cooperative Systems Task Force – WP4.1 Business case development, WP4.2 Cost/Benefit Analysis, 2011.

[FLA 11] FLAMENT M., *Introduction to the DRIVE C2X Integrated Project*, overview on the project presented at the ETSI TS ITS Workshop, Venice, February 11, 2011.

[GAR 04] GARBER N.J., TENG H., LU Y., *A* proposed methodology for implementing and evaluating a truck parking information system, Centre for Transportation Studies, University of Virginia, p. 168, 2004.

[JON 10] JONKERS E., FABER F., VAN DER VEEN J. *et al.*, ITS Test Beds, Evaluation tool chain-impact assessment, 2010.

[KÖL 10] KÖLBL R., COOPERS Co-operative Networks for Intelligent Road Safety; D15 Report on common result assessment, June 18, 2010.

[KUL 08] KULMALA R., LEVIAKANGAS P., SIHVOLA N. *et al.*, CODIA final study report (CODIA Deliverable 5), VTT Technical Research Centre of Finland, 2008.

[MAR 08] MARKKULA G., JOHANSSON E., MONTANARI R. *et al.*, AIDE – Adaptive Integrated Driver-Vehicle Interface, D4.1.6 AIDE Final Activity Report, AIDE, June, 2008.

[MOC 12] MOCANU I., NITSCHE P., DEIX S. *et al.*, COBRA D3.1 Impact Assessment, 2012.

[SCH 08] SCHULZE M., MÄKINEN T., IRION J. *et al.*, PReVENT, IP_D15: Final report, Daimler AG, May 7, 2008.

[STO 11] STOELHORST H., SCHREUDER M., POLDERDIJK S. *et al.*, Summary results of Dutch Field Trials with Dynamic Speed Limits (Dynamax), Rijkswaterstaat Centre for Transport and Navigation, The Netherlands, TNO Delft, The Netherlands, 2011.

[TSE 10] TSEGAY S., VERNET G., BURKE R. *et al.*, CVIS D.CVIS.5.2 Test Site Results, CVIS, July 15, 2010.

[WEL 12] WELSH R., MORRIS A., REED S. *et al.*, TeleFOT Field Operational Tests of Aftermarket and Nomadic Devices in Vehicles, Impacts on Safety – Results and Implications, TeleFOT, October 12, 2012.

[WIL 10] WILLEMSEN D., DE KIEVIT M., FABER F. *et al.*, SAFESPOT D 4.6.4 Evaluation of accident impact through simulation, SAFESPOT, September 13, 2010.

[WIL 08] WILMINK I., JANSSEN W., JONKERS E. *et al.*, Socio-economic Impact Assessment of Stand-alone and Co-operative Intelligent Vehicle Safety Systems (IVSS) in Europe (eIMPACT Deliverable D4), TNO, 2008.

20

The Impact of Navigation Support and Traffic Information on Distance-keeping Behavior

Exponential growth of in-vehicle driving systems and their interaction with the driver and other infrastructure elements makes it imminent to focus on the effect and impact of navigation support and traffic information functionalities included in nomadic devices (NDs) to driving behavior. A within-participant design was implemented with eight experienced and common drivers, following preset routes with navigation support and traffic information integrated in the aftermarket device. The combination of speed limit information and navigation support decreased variation significantly (p = 0.12) by 35% when compared to free driving in pure baseline control conditions. A statistically significant decrease ($p < 0.05$) in percentage of time with dangerous car-following behavior with the combination of ADAS and ND functions was found. The analysis indicates that an increase in vehicle headways in the network within safe limits could mean no difference for efficiency.

20.1. Introduction

There are high expectations for in-vehicle and driving assistance systems to contribute to solving traffic-related situations and problems [EUR 10]. Many studies have focused on the impact of using in-vehicle systems with simulators and simulation models in order to identify their effect on traffic. In order to meet the expectations for these systems, on-the-road studies should be conducted in order to provide insight to the potential effects and impacts to driver and traffic efficiency.

Chapter written by Aikaterini TOULIOU, Evangelia GAITANIDOU and Evangelos BEKIARIS.

A general categorization for vehicle systems could divide them into driving support systems (e.g. navigation support, traffic information and speed information) and vehicle support systems (e.g. collision avoidance system, lane departure system, speed control, speed alerts and adaptive cruise control (ACC)) [GOL 01]. These systems, however, would probably be integrated into more complicated hyper-systems for future vehicles.

Mobile phones were one of the first devices to be used in vehicles. As mobile phones became available to people, early studies focused on the effect of talking and messaging while behind the wheel. The primary target was to identify the effect of using mobile phones on safety. Safety is the cornerstone of driving research, and efficiency has been a secondary objective until now.

Navigation support has developed exponentially over the last few decades. Increasingly, more drivers are using dynamic navigation systems with information about potential delays and congestion. Yamashita *et al.* [YAM 04] found that when drivers received accumulated traffic information with a route information sharing system, travel time decreased. If the system usage increased, then travel time decreased even more. Their findings were based on multi-agent simulation in a latent network and radial ring network.

Van Arem *et al.* [VAN 06] examined the traffic flow effects of cooperative ACC. The ACC system under investigation exchanged information with the vehicle ahead so that it could follow this vehicle more closely. The results indicated an improvement of traffic flow stability expressed by fewer shockwaves and smaller SDs of speed, and a slight increase of traffic efficiency was indicated by higher queue discharge flows.

The use of ACC systems could contribute to reducing head–tail accidents, but it could also reduce the traffic throughput, depending on the system settings [MIN 99].

With a 40% ACC equipment rate and one second headway time, [BRO 91] estimated throughput gains at 13%. [VAN 96, MIN 99] found a decrease in average speed as a result of a collapse of speed in the fast lane when ACC with headway times of 1.4 seconds and above were used. Minderhoud and Bovy [MIN 99] performed simulations with headway times as low as 0.8 seconds and concluded that current ACC using a one second headway time could achieve capacity gains of 4%.

A combination of ACC and lane departure warning (LDW) was tested in the study of Alkim *et al.* [ALK 07]. Although LDW was found much less effective than ACC, some of the test drivers reported an interesting positive integration effect. With ACC activated, a slight increase in the variation of lateral position in the driving lane was found. The test drivers, however, claimed that the warning issued by LDW effectively compensated for this and increased their alertness.

Several simulation studies have investigated the potential impact of ACC on traffic flow. The simulation studies used various ACC algorithms, for instance, to find various headway times and applied them in various environments, at various penetration rates using various behavioral models. All these differences strongly influenced the outcomes on traffic capacity and speed; therefore, they make it very difficult to compare these studies and their results. Communication between vehicles and between vehicle and roadside is considered the technology that will make a whole new generation of ADAS possible.

Moving from simulation to real environments – according to a study carried out in both simulated and real environments – as soon as participants in any of the settings became aware of driving performance deterioration as a consequence of the secondary task, they chose to adapt their behavior [BRO 03]. The adoption of lower (i.e. safer) speed, smaller distance to the road shoulder and a longer margin toward vehicles in front was particularly clear in the simulator and field.

Despite the increasing amount of knowledge of driver support systems, it can be concluded that there are no simple, straightforward assumptions on how these systems will influence driving behavior and traffic flow performance. Many effects are still unknown and more research is required [VAN 05].

Advances in diverse areas of new technologies (such as telematics, telecommunications and sensors) have made both driving support systems and in-vehicle communications systems more affordable and, subsequently, more frequent in traffic. New technologies already play a key role in traffic safety, and their interaction with other type of systems (e.g. vehicle–infrastructure communication) is an objective set in near future. The evaluation of existing systems in the market as a potentially unified system would ultimately increase their market penetration rates due to minimized installation costs (i.e. potentially increased cost-effectiveness), but additional harmonization among systems and their interaction with the driver is required. The importance of interactions between these systems and the drivers is still uncertain, and their investigation is of an imminent importance in order to clearly define their interaction and relation. Increasingly, creating more complex and sophisticated systems has a drawback in the inherent difficulty of isolating their effect.

In this study, two aspects of car-following behavior were investigated. Time headway-based summary metrics were estimated with different combinations of in-vehicle functions. Violation of safety margins (i.e. small following distances) could be interpreted as deteriorating driving behavior with potentially negative safety consequences. Time headway was chosen because it is a reliable index of longitudinal behavior fitting better for an efficiency impact assessment than time to collision which is an index, more appropriate for safety analysis.

Two (two-tailed) research hypotheses were investigated as follows:

– H_1: overall time headways will increase with the combination of ADAS and navigation support;

– H_2: time spent with dangerously low time headways will decrease with the combination of ADAS and navigation support provided by a nomadic device (ND).

The effect of the navigation support to distance headway was investigated in order to indirectly address the potential impact to traffic efficiency (from the driver's perspective). Aftermarket devices might affect driver's attention and ADAS could decrease driver's arousal and readiness, as the driver might rely on the system to warn him/her about a dangerous driving situation (e.g. incidents). On the other hand, the navigation system might support safe driving behavior by, for example, controlling lateral behavior and assisting the driver to find the fastest route (both environmentally beneficial and increasing network efficiency).

20.2. Methods

20.2.1. *Participants and recruitment*

Eight participants (five male and three female drivers) drove an instrumented vehicle in each condition. Participants were pre-screened in order to investigate personal characteristics that might affect findings. All participants were experienced and common drivers (>3 years driving experience; average 12,000 km per year; driving on a daily basis). Ages ranged between 30 and 38 years old. They were all healthy volunteers with no eyesight problems or corrected eyesight. Participants gave written consent prior to the study onset. All participants were well-acquainted with new technologies, but were unaware of the aims and objectives of the study. All participants were well-acquainted with smartphones. Their ICT literacy was high and most of them had used a navigation system before, but none of the participants had such a system installed in their own cars. Participants agreed their data to be available as long as anonymity/confidentiality was protected. Participants did not receive reimbursement for their participation in the trials. In addition, it was decided to carry out the tests at specific hours of day and to avoid traffic volume variations. Testing was carried out on specific time intervals (11:00–12:00 and 13:00–15:00). Although routes included urban sections, they did not include city center sections avoiding high congestion, extremely low speeds, still recordings (i.e. collection of recording with zero/null values) and vaguely marked or non-existing lanes.

20.2.2. *Design and procedure*

A within-subject design was adapted with one experimental and two control conditions (Baseline 1: ADAS off; Baseline 2: ADAS on) in Detailed Field Operational Tests (DFOTs). Conditions were counterbalanced. The ADAS included were a LDW and a forward collision warning (FCW) system.

Four different routes including urban, peri-urban and motorway segments were selected for all conditions. Based on previous research experience, it was anticipated that both warnings systems (LDW and FCW) would not perform in all parts of the road, and this was taken into consideration in both data reduction and statistical analysis. The LDW is active when the lanes are well-defined and marked. In addition, the FCW warning is activated above 30 km/h; hence, it might have not optimally performed in urban roads. Participants drove for approximately 40–50 min, depending on the traffic volume for approximately 27.2 km. Routes were selected to reflect daily activities for each driver (e.g. shopping, leisure and travelling).

Approximately 70% of testing was conducted during winter, and the remaining 30% was carried out in spring always with good weather conditions (no rain, snow or fog) to avoid confounding results because of these conditions regardless the fact that warning signals do perform well in light rain conditions. Participants were assigned a unique identification number, and gathered data were renamed and allocated to new codes. Access to real names was granted only to chief investigator. Objective data included log files from the ND and vehicle data (CAN bus) (Figure 20.1). Data were stored in safe place and copied in two different password-protected external hard disks.

Figure 20.1. *In-vehicle data collection*

20.2.3. *Instrumented vehicle*

The Lancia Thesis Emblema is an instrumented vehicle (Figure 20.2) equipped with basic electrical equipment to manage the power supply of all additional equipment (e.g. auxiliary battery, fuses, relays, switches and lamps to configure and monitor the system). An additional electronic unit (gateway) transfers information from the vehicle (CAN buses) and makes it available to a PC in the boot (gas, brake pedals position, longitudinal speed/acceleration, yaw rate, steering angle, lights status, wiper status, external temperature, headway, obstacle distance, sampling rate 100 ms and among others). The vehicle is equipped with a frontal ACC radar which acquires information about the leading vehicle (distance and relative speed); these data are made available on the CAN bus network. The instrumented vehicle has both LDW and collision avoidance systems with the availability of lane and headway data on CAN bus.

The aftermarket ND used was a Samsung Omnius II with Sygic maps installed for the navigation support system. Participants were offered the fastest available route. Navigation support was static. Text files (log files) were directly uploaded and stored to an appointed FTP site. The stored log files included Global Positioning System (GPS) data, certain pre-determined events and acceleration (x, y and z) data.

20.2.4. *Statistical analysis*

Firstly, data were checked for correctness, missing values and erroneous recordings (e.g. impossible values for vehicle parameters). Secondly, data were post-processed in Excel files. Data were both graphical and statistically checked for outliers. Data were filtered in order to create required datasets for further analysis steps. Thirdly, data were analyzed for the whole trip, for certain threshold values for the indicators (e.g. time headway values < 5 s; distance headway values < 50 m) and before and after the activation of warnings (LDW/FCW). Non-parametric tests were used, as sample size was quite small and α was adjusted in pairwise comparisons (Bonferroni adjustment). The Bonferroni adjustment is a correction of p value for the number of performed comparisons (a/n) [ABD 07]. Non-parametric statistical tests are independent from sample distribution and do not assume the data, or population have any characteristic structure or parameters; thus, they can be applied in data with no variance homogeneity. The Friedman's two-way ANOVA by rank test is the non-parametric alternative to the one-way ANOVA with repeated measures. It is used to test for differences between groups when the dependent variable being measured is ordinal. It can also be used for continuous data that has violated the assumptions necessary to run the one-way ANOVA with repeated measures (e.g. data that has marked deviations from normality). The Wilcoxon signed rank test is a non-parametric statistical hypothesis test used when comparing two related samples, matched samples or repeated measurements on a single sample to assess whether their population mean ranks differ (i.e. it is a paired difference test). It can be used as an alternative to the paired Student's T-test, T-test for matched pairs or the T-test for dependent samples when the population cannot be assumed to be normally distributed [CON 71, CON 80].

20.3. Results

Time headway is defined as the distance to the lead vehicle (bumper to bumper) divided by the travel speed of the ego vehicle (i.e. own vehicle) [ÖST 04], and usually large values are discarded.

The addition of the ND and related functions did increase headway but not significantly ($p > 0.05$) in all conditions The average overall increase in time headway when compared to the pure baseline condition (free driving) was 8.85%. For the analysis of central tendency measures, values higher than 5 s were excluded. Increase in headway was found for the condition with ADAS and then the difference pertained with the addition of navigation support.

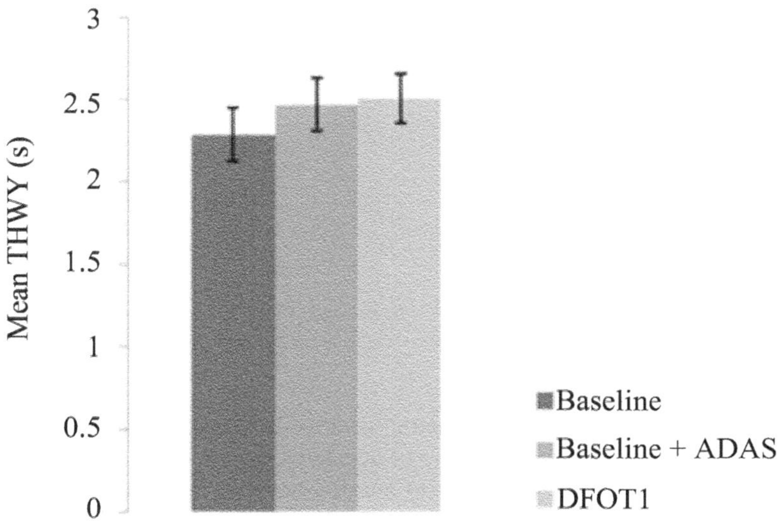

Figure 20.2. *Mean time headway (sec). For a color version of this figure, please see www.iste.co.uk/jacob/traffic.zip*

The next step was to investigate the variations in time headway, as these variations could affect the stability of the network and, subsequently, increase or decrease traffic efficiency (Figure 20.3). SD of time headway was calculated for all DFOTs from where data were derived. Overall, related sample Friedman's two-way ANOVA by rank test was significant ($\chi 2(4) = 15.3$; $p = 0.004$). Pairwise comparisons were carried out with Wilcoxon-signed ranks test (Bonferroni adjustment). As the samples had a mean correlation of $r = 0.379$, then the Bonferroni correction was adjusted for this correlation from 0.005 to 0.012. The correction to Bonferroni adjustment was made in order to avoid making a Type II when trying not to make a Type I error. Stringent significance α values are retained because the number of pairwise comparisons is high, as variance in within subjects design accounts more for similarities than differences.

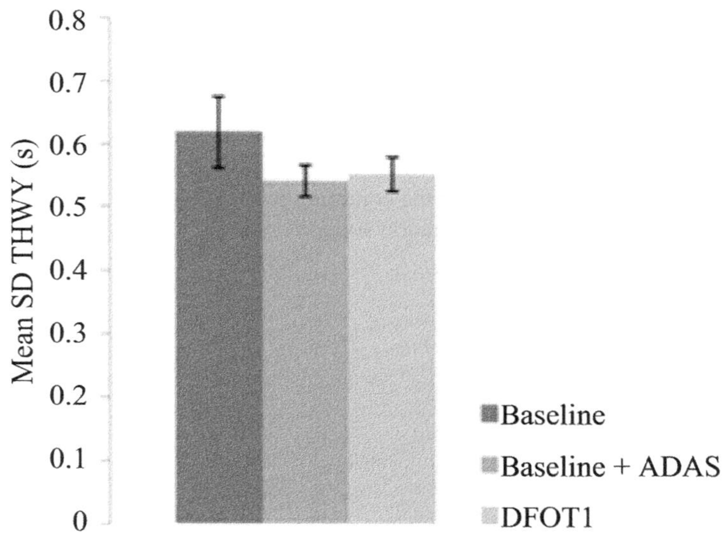

Figure 20.3. *Mean SD for time headway (sec). For a color version of this figure, please see www.iste.co.uk/jacob/traffic.zip*

The minimum time headway metric is directly linked to safety. Extremely low headways increase the probability of accidents that create numerous problems in traffic flow affecting the efficiency of a traffic network. No statistical differences were found for mean minimum time headway among DFOTs ($p > 0.05$); however, the lowest minimum values were observed in the pure baseline condition (0.47 ± 0.09).

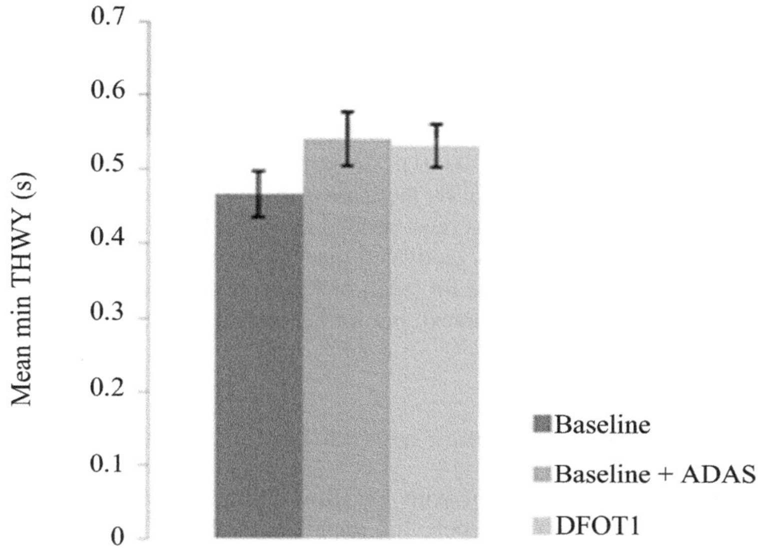

Figure 20.4. *Mean minimum time headway in DFOTs. For a color version of this figure, please see www.iste.co.uk/jacob/traffic.zip*

20.3.1. *Stratification for road types*

The second analysis level included stratification based on different road types. Chosen road types were as follows: urban, peri-urban and rural and motorway.

No differences were found in mean time headway among conditions in DFOTs within the city area. The greatest (but not significant) was reported between the pure baseline condition and DFOT3 where a decrease of 6.43% in mean time headway was found. The integrated functions probably slightly decreased time headway values, but the differences are small within the city area. In addition, all values are above 2.4 seconds and, therefore, are within safety limits.

In peri-urban and rural roads, significant differences were found between DFOT 1 and Baseline 1 ($z = -22.87$). In particular, mean time headway increased by 10.95% with the addition of navigation support. It is hypothesized that users were not so familiarized with these roads; therefore, they paid more attention to the navigator resulting in more "loose" following. If it is assumed that these roads are unfamiliar roads and navigation systems are more useful and used more for these types of roads, then the addition of speed alert functions increases longitudinal control; therefore, it positively affects the stabilization of traffic flow.

Similarly, on motorway segments, significant differences were found, but in this occasion, a significant increase from pure baseline to ADAS condition (14.95%) (Baseline 2) ($z = -19.31$; $p < 0.001$). Activation of FCW warnings was more frequent in motorways, and probably, the increase is the effect of compliance to the warnings. Navigation support was most needed in not so familiar roads, and ADAS warnings were more frequently activated on the motorway. The findings are potentially influenced by the need for "shift of focus" in each context. Drivers seem to prefer to prioritize based on context, but further research is needed on importance of "shift of focus" per road type.

20.3.2. *Percentage of time spent with certain time headway values*

Percentage of time (%) spent with low time headway values were estimated because of its importance for its potential impact on both traffic flow and volume. Drivers in pure baseline condition spent almost three times more driving with dangerously low time headways (8.8%). The addition of ADAS significantly decreases (5%) time spent with dangerous time headway values, but the addition of the other functions did not significantly affect percentage (%) of time spent with very low headways and did not have a negative effect either.

20.4. Discussion

Longitudinal driving behavior shows a great heterogeneity and, therefore, affects traffic flow. Differences in car-following behavior affects to some extent the distribution of vehicles in lanes leading to induced lane changes with more disturbances in traffic flow [KER 04]. On the other hand, low and stabilized headways increase traffic volume and, therefore, increase the efficiency for the traffic networks. There were several components for headway analysis in this study.

Overall, headway increased as a result of lateral and longitudinal warnings and the addition of navigation support. It is not clearly evident if the increase in headway is mainly an outcome of ADAS or functions effect or both. The assumption that their effect is somehow additive is erroneous, linear and over-simplistic. It is important that their effect was not counteractive if, for example, an increase was evident with the addition of ADAS but vanishes with the addition of navigation support. Therefore, hypothesis 1 was accepted with limitations and so accepted for rural, peri-urban roads and motorways. The second hypothesis was also accepted.

An increase in headway within safe limits is not as positive an outcome for traffic efficiency as a decrease in driving with dangerous distance-keeping behavior is. The latter outcome positively affects traffic efficiency by minimizing the possibilities of road accidents that disrupt the stability of the network and create problems in traffic flow.

For a macroscopic investigation of the effect to traffic efficiency, their impact should be further studied with the application of simulation models with different penetration rates for the NDs and bundle of functions (i.e. equipped vs. non-equipped cars).

A step even further would be to conduct a very large scale study with one common site for all Europe that would probably solve direct effects evaluation but would create sequences of others. The management of fleets across Europe would be a very high maintenance endeavor, but the gains and benefits to all members engaged would be enormous.

Navigation support alone was not found to affect traffic efficiency, at least as a primary indicator. It appears to hold tertiary significance. Indirect effects – not possible to be investigated in this study – might exist. The choice of other peri-urban roads or rural roads, for example, because of navigation support, could potentially affect efficiency because of redistribution of traffic volume to smaller roads and subsequent avoidance of traffic jams and delays (route choice).

The development of in-vehicle technology is driven by several factors [VAN 06]. On the one hand, public authorities and road operators recognize that driver support systems offer possibilities to alleviate the problems on their roads. On the other hand, car industries and suppliers consider the systems an important product innovation and a competitive advantage.

In spite of their potential, the take-up of driver support systems in the market is very slow. The development of ACC systems started more than 25 years ago, whereas nowadays only a small percentage of (luxury) cars are fitted with this system. The main reasons for this slow take-up are legal barriers, the competitive situation of the automotive sector, the high cost of the technology, the lack of consumer demand, the lack of information about the potential benefits and a clear business case [EUR 10].

The benefits and know-how distilled from the implementation of this methodological framework constitute a valuable and viable approach for other types of field operational tests (FOTs). The on-site data analysis tools developed will be customized and used for analyzing data resulting from similar studies such as the ones performed within the FP7 European project European FOT on Safe, Intelligent,

Sustainable Road Operation (FOTsis). This is a large-scale effort aiming to test the road infrastructure management systems needed for the operation of seven close-to-market cooperative technologies (e.g. intelligent congestion control and dynamic route planning). Their potential and effectiveness will be assessed with full usage on European roads (Spain, Portugal, Germany and Greece).

It is important to mention that a set of sustainable structural indicators could be set for monitoring the effect of functions on traffic and driver efficiency that could infiltrate the development of the next generation of assistive and information systems and would prove useful for strategies within policymaking for ND usage.

A common goal in current research efforts pertains to global effects. Drivers must be protected from hazardous traffic situations, and a harmonized, stable and efficient network enhances the safety of road users. As controlling the vehicle and the traffic environment simultaneously is a hard and often time consuming task, then future hyper-systems should primarily support the driving task.

20.5. Bibliography

[ABD 07] ABDI H., "Bonferroni and Šidák corrections for multiple comparisons", in SALKIND N. J. (eds), *Encyclopedia of Measurement and Statistics*, Sage, Thousand Oaks, CA, 2007.

[ALK 07] ALKIM T., BOOTSMA G., LOOMAN P., The assisted driver – systems that support driving, Publication of the Dutch Ministry of Transport, Public Works and Water Management, 2007.

[BRO 91] BROQUA F., LERNER G., MAURO V. *et al.*,"Cooperative driving: basic concepts and first assessment of intelligent cruise control strategies", *Proceedings of the DRIVE Conference*, Brussels, Belgium, 1991.

[BRO 03] BROOKHUIS K.A., DE WAARD D., FAIRCLOUGH S.H., "Criteria for driver impairment", *Ergonomics*, vol. 46, pp. 433–445, 2003.

[CON 71] CONOVER W. J., *Practical Nonparametric Statistics*, Wiley, New York, 1971.

[EUR 10] EUROPEAN COMMISSION, Intelligent transport systems in action. Action plan and legal framework for the deployment of intelligent transport systems (ITS) in Europe, Directorate-General for Mobility and Transport, 2010.

[GOL 01] GOLIAS J., YANNIS G., ANTONIOU C., "Impact of advanced driver assistance systems on urban traffic network conditions", *European Journal of Transport and Infrastructure Research*, vol. 1, no. 3, pp. 277–289, 2001.

[KER 04] KERNER B., KLENOV S., "Spatial–temporal patterns in heterogeneous traffic flow with a variety of driver behavioural characteristics and vehicle parameters", *Journal of Physics A: Mathematical and General*, vol. 37, pp. 8753–8788, 2004.

[LIU 99] LIU R., TATE J., BODDY R., Simulation modelling on the network effects of EVSC systems. Deliverable 11.3: External vehicle control. Department of the Environment, Transport and the Region, UK, 1999.

[MIN 99] MINDERHOUD M.M., Supported driving: impacts on motorway traffic flow, TRAIL Thesis Series T99/4, Delft University Press, Delft, The Netherlands, 1999.

[MIN 99] MINDERHOUD M.M., BOVY P.H.L., "Impact of AICC design on motorway capacity", *Transportation Research*, vol. 1679, pp. 1–9, 1999.

[MIN 04] MINDERHOUD M.M., ZUURBIER F., "Empirical data on driving behaviour in stop-and-go traffic", *IEEE Intelligent Vehicles (IV) Symposium*, Parma, Italy, pp. 676–681, 2004.

[YAM 04] YAMASHITA T., IZUMI K., KURUMATANI K., "Car navigation with route information sharing for improvement of traffic efficiency", *7th International IEEE Conference, Intelligent Transportation Systems, Proceedings*, pp. 465–470, 3–6 October, 2004.

[VAN 96] VAN AREM B., HOGEMA J.H., SMOULDERS S.A., "The impact of Autonomous Cruise Control on traffic flow", *Proceedings of Third World Congress on Intelligent Transport Systems*, Orlando, FL, USA, 1996.

[VAN 06] VAN AREM B., VAN DRIEL C.J.G., VISSER R., "The impact of Cooperative Adaptive Cruise Control on traffic-flow characteristics", *IEEE Transactions on Intelligent Transportation Systems*, vol. 7, no. 4, pp. 429–436, 2006.

[VAN 05] VAN DER HEIJDEN R.E.C.M., MARCHAU V.A.W.J., "Advanced Driver Assistance Systems: Behavioural implications of some recent developments", *European Journal of Transport and Infrastructure Research*, vol. 5, no. 4, pp. 239–252, 2005.

21

Impact Evaluation of Traffic Performance and Road Safety: A Case Study on an Urban Motorway in France

Intelligent transportation systems are growing in terms of both acceptability and implementation scale worldwide. Ex-ante and ex-post evaluations are being made to decide upon the implementation of ITS-based measures. Impact evaluation methodologies employ various performance indicators and concern traffic performance (speed, level of service), road safety (occurrences, severities) and environmental externalities (emissions, noise). In this paper, we apply the impact evaluation methodologies with regards to traffic performance and road safety. We use a case study of a dynamic lane operating on the A4–A86 motorway junction in the Ile-de-France region. The different components of the trial are described, in particular the operating strategy concerning the dynamic opening and closing of the lane. The assessment method of the various impacts is presented, based on an ex-post assessment comparison. Results of this study indicate that benefits from congestion reduction may induce safety problems.

21.1. Introduction

Intelligent transportation systems are constantly growing in terms of both acceptability [MAR 99] and implementation scale [FRI 12]. Freeway management figures among the most promising fields for future ITS applications. The latter is due, on one hand, to congestion mitigation concerns and, on the other hand, to the fully controlled environment that freeways offer. Nevertheless, a specific application may well serve a specific purpose while being detrimental to another. In that sense, ramp metering may be beneficial to traffic management but not necessarily to road safety. Furthermore, environmental nuisances (i.e. emissions and noise) may not always decrease by safety- or traffic-related intelligent systems. In view of the

Chapter written by Simon COHEN and Zoi CHRISTOFOROU.

above, we believe that future intelligent transportation systems should be *a priori* evaluated under a systemic approach integrating all related aspects and eventual lateral effects.

The A4–A86 junction is often cited as the most congested site across Europe and with a large number of traffic crashes. Various efforts have been made to mitigate congestion in terms of both intensity and duration. Dynamic lane management was implemented in 2005 and, in particular, the hard shoulder has been used as an additional traffic lane during peak hours in order to increase throughput and alleviate congestion. Already in use for decades in the United States, managed lane operations have more recently increased in Europe [KEL 00, SCH 01]. In the present paper, we use the A4–A86 case study to illustrate the potential benefits of such an application while considering the possible effect on road safety performance. Finally, we propose a measure evaluation context integrating multiple aspects.

This paper is organized as follows: first, the A4–A86 is presented and the ITS application of a managed lane operations is described; second, the ITS application is evaluated in terms of traffic management efficiency; third, a safety analysis is being presented; fourth, the evaluation is discussed and general conclusions are drawn while policy guidelines are given.

21.2. The site and the its application

To establish a context for a systemic and *a priori* evaluation, the A4–A86 highway section from an area, a few miles to the east of Paris, was selected. The A4–A86 junction has a length of 2.3 km and includes four lanes per direction (to and from Paris). The A4–A86 junction is a particular site as it is the point where the Ile-de-France Ring Road (Périphérique-A86) coincides with the Autoroute de l'Est (A4) and merging is prevalent; five lanes are reduced to four in each direction.

Until 2005, the 280,000 vehicles that use this highway section on an everyday basis were causing some of the largest traffic jams in France: more than 10 h of congestion per day with tailbacks exceeding 10 km. Structural insufficiency in capacity is held responsible for systematically causing recurrent congestion; non-recurrent congestion is also an issue. The addition of an extra lane has been under discussion for years, but strong budget constraints hold it.

The ITS application consists of the dynamic use of the hard shoulder as a traffic lane during the periods of high demand; i.e. mainly the morning peak hours (Figure 21.1). For this purpose, lane width has been adapted passing from 3.5 to 3.20 or 3.0 m. Drivers are informed of the opening or the closure of the lane by variable message signs in accordance with the layout. In addition, to emphasize the difference between a traffic lane and a hard shoulder when used as an auxiliary lane, and to show drivers the specific nature of the zone, the auxiliary lane has specific road markings and a light-colored surface. Moreover, it has been entirely resurfaced with low noise asphalt in order to reduce traffic noise. Moveable safety barriers are installed on the right side of the auxiliary lane. When this lane is closed, the device pivots around its upstream end until it forms a lane reduction taper, which takes up almost the entirety of the auxiliary lane. These closure devices are installed at several key locations on the section so that drivers can see them whatever their position and are thus dissuaded from using the lane. In each traffic direction, the first device encountered by drivers is 69 m long and the following ones are 25 m long. Safety has been improved by the installation of automatic incident detection cameras. In the event of an incident or accident when the lane is open, stationary vehicles on the auxiliary lane can be detected, leading to the closure of the lane. Additional safety is provided by speed control radars on the motorway in both traffic directions.

The benefit of the auxiliary lane is that it helps with both recurring and non-recurring congestion. The activation and the deactivation of the auxiliary lane operation are being controlled by the traffic control center according to the following criteria:

– occupancy rate (OCC) measured upstream the considered section (opening if OCC $\geq$ 20%; closure if OCC $\leq$ 15%);

– traffic speed in the downstream Nogent tunnel (speed $\leq$ 15 km/h).

Of course other criteria may also apply such as other lane closure due to road works.

Figure 21.1. *A4–A86 dynamic lane (source: DIRIF)*

21.3. Evaluation of the impact on traffic

In the absence of technical problems and on a typical working day, the auxiliary lane operates approximately 5 h in the Paris-bound direction and 4 h in the opposite direction. On Saturdays, it is open on an average of 4 h towards Paris and 3 h 45 min in the opposite direction. On Sundays, both lanes operate 3 h 20 min in both directions. In this section, we evaluate the performance of the dynamic lane on a traffic basis following a before–after analysis.

21.3.1. *Speed and capacity*

The speed-concentration function was calibrated using 6 min measurements. The operating thresholds were determined by calibrating the speed–density diagram. Several models have been tested and the generalized exponential model of (1) was found to better suit the data:

$$u = a \exp(-b\, k^{\alpha}) \qquad [21.1]$$

where u denotes the mean speed (in km/h), k is the concentration (in veh/km) and a, b and α are parameters to be calibrated. In the all tested cases, this model produced the best fitting [COH 04]. Note that the parameter a is the free-flow speed. Results are provided in Table 21.1.

Road section (period)	Parameter estimation		
	a	b	α
To Paris (before)	85.00	$2.6*10^{-6}$	2.38
To Paris (after)	90.25	$2.9*10^{-6}$	2.36
From Paris (before)	104.50	$1.0*10^{-5}$	2.19
From Paris (after)	93.67	$8.1*10^{-7}$	2.60

Table 21.1. *Calibrated values of the fundamental diagram*

The changes in capacity and free-flow speed can be found in Tables 21.2 and 21.3, respectively. As could have been anticipated, the capacity for the operation of four lanes falls due to the narrower lane width. On the contrary, capacity during the managed lane operation increases significantly. Interestingly, free-flow speed increases in the direction from the suburbs to Paris but drops in the opposite direction. Local measurements of average speed on the considered section reveal similar patterns (Table 21.4).

Road section	Period of reference		
	Before	After	After
	four lanes	four lanes	four lanes + dynamic lane
To Paris	8670	7610	9310
From Paris	8820	8185	9725

Table 21.2. *Changes in capacity due to the ITS application*

Road section	Period of reference		
	Before	After	After
	four lanes	four lanes	four lanes + dynamic lane
To Paris	85.0	90.0	90.0
From Paris	104.5	94.0	94.0

Table 21.3. *Changes in free-flow speed (km/h)*

Road section	Period of reference	
	Before	After
	four lanes	four lanes + dynamic lane
To Paris	57.7	71.9
From Paris	72.1	71.0

Table 21.4. *Changes in average speed (km/h)*

21.3.2. *Level of service*

The concept of level of service (LOS) refers to the traffic conditions observed locally on the common trunk section. Research on this area in France has led to the adoption of the following classification with four levels: free flow, free flow to dense traffic, flow at capacity and saturation.

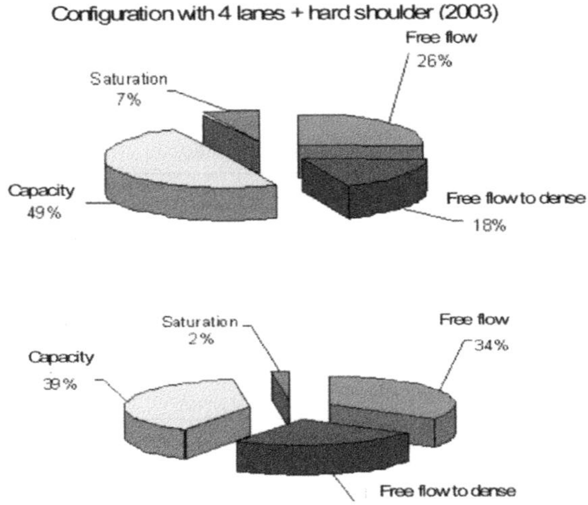

Figure 21.2. *LOS time distribution on weekdays*

The dynamic use of the auxiliary lane leads to a clear LOS improvement on weekdays, as shown in Figure 21.2. Today, we observe an increase in free flow and fluid to dense traffic regimes and a consequent reduction in the congested traffic conditions.

21.3.3. *Congestion*

Traffic mapping allows for the drawing of automatic maps of traffic conditions. Recurrent traffic peaks are identified along with their characteristics such as duration, length and intensity. Four contour levels are defined according to the average speed thresholds: 60, 45, 30 and 15 km/h. Figure 21.3 illustrates, per lane and direction, the speed contours of the traffic peaks between 12 pm and 12 am, both before and after the ITS application.

On the part of the A4 motorway between Bercy and Noisy-le-Grand, there is a clear improvement in traffic conditions upstream the junction and in the evening. This improvement is marked by a significant reduction of approximately 1.5 km in congestion length on the A4 and upstream of the A4–A86 junction. The duration of congestion is reduced by approximately one and a half hours. The improvement in traffic speed in the congestion area can be seen by the disappearance of the slowing zone which is usually observed, with a mean speed of less than 30 km/h.

Lastly, it is important to note that there is no major change to the extent of the congestion downstream the A4–A86 section. This shows that the local improvement of the capacity of A4–A86 does not create any bottlenecks downstream.

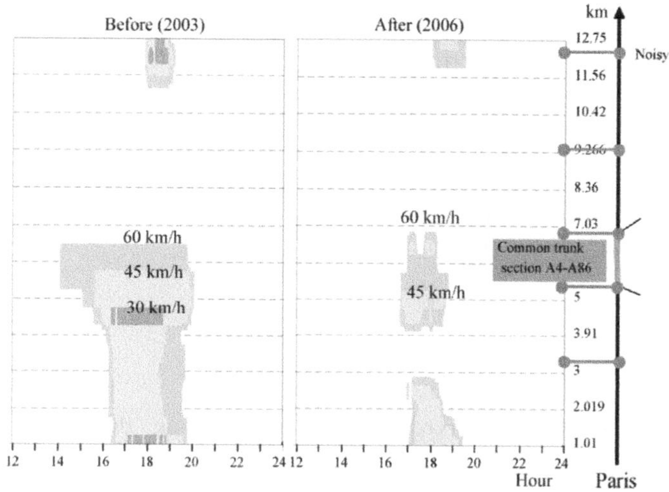

Figure 21.3. *Speed contours during recurrent traffic peaks on the A4*

21.4. Road safety implications

In parallel with the previous analysis, we have undertaken a safety-oriented analysis on the A4–A86 junction [CHR 11]. Using crash and traffic data, we examined the effects of the various traffic parameters collected real-time both at – and prior to – the time of the accident on type of crash. Multivariate probit models were specified on four years of data from the period both before and after the ITS application. We used a disaggregate approach in which the units of analysis were the crashes themselves (rather than aggregations of crashes over time). Accident data were extracted from B.A.A.C. (*Bulletins d'Analyse des Accidents Corporels*) along with the Verbal Proceedings from an INRETS study [ARO 04]. The BAAC files provide a wealth of useful information such as crash-type for all accidents, location and time, lighting conditions and infrastructure characteristics such as road curvature and alignment. Detailed weather data are available on a 30 min basis. We extracted such data directly from the closest meteorological station and for the 30 min interval into which the reported time of the accident occurred. In total,

381 accidents were recorded during the period 2000–2002 and 2006. Traffic data were measurements of volume, speed and density over 6 min intervals. The recorded traffic volume and speed were for the 6 min period ending 6 min before the accident (from the closest downstream detector). This time lag was used to avoid the impact of the crash itself on the traffic variables and as a buffer to compensate for any "inaccuracies" in the exact time of the accident. Such an analysis offers a wide variety of potential benefits; from a methodological standpoint, disaggregation minimizes possible bias [DAV 02], while additional light can be shed on the causal relationship between accidents and several contributing factors such as geometry, traffic conditions and so on.

The general specification for a multivariate probit model of n dependent variables (alternatives) can be expressed as [GRE 03]:

$$Y_i^* = \beta_i X_i + \varepsilon_i , i = 1 \ldots n \quad [21.2]$$

where:

– Y_i^* defines an unobserved variable representing the latent utility (or propensity) for alternative $i = 1,2$;

– X_i is a vector of observed characteristics determining alternative I;

– β_i represents a vector of unknown coefficients to be estimated;

– ε_i represents a vector of error terms that are normally distributed with zero mean and constant variance.

Qualitative results from the multivariate probit model application are illustrated in Figure 21.4. The fundamental diagram depicts the relationship between traffic volume (Q) and speed (V) on a given freeway segment; each crash type (whose probability is traffic-dependent) is related to a particular traffic regime which corresponds to a specific part of the diagram. The following crash types are considered:

– type 2: rear-end involving two vehicles;

– type 3: sideswipe involving two vehicles;

– type 4: rear-end with more than two vehicles;

– type 5: single-vehicle crash.

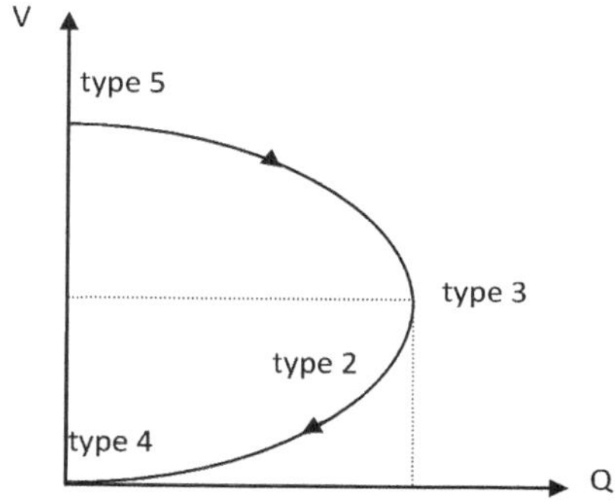

Figure 21.4. *Crash type distribution*

The simultaneous analysis by vehicle involvement and maneuver possibly indicates that aggregation in by-crash-type analyses may lead to erroneous estimations. Rear-end crashes are more probable under congestion, while side-swipes are more probable under "intermediate" density traffic regimes. Similar findings were reported by Golob and Recker [GOL 04]. Finally, as it could have been anticipated, single vehicle crashes mostly occur under free flow.

Using the same dataset, a similar analysis has been undertaken to explore the effect of prevailing traffic conditions on traffic injury severity. Random parameter models of discrete choice were used to explore potential links between crash type severity on one hand and traffic speed and volume on the other [CHR 10]. Empirical results (Table 21.5) indicate that increased traffic volume has a consistently positive effect on severity, while speed has a differential effect depending on flow conditions. While in higher traffic volumes higher speeds aggravate severity outcome, in lower traffic volumes speed does not significantly influence severity in a consistent pattern. This finding indicates that speed-reducing measures should be considered even in rather dense traffic highway segments (that allow however for speed variation among drivers).

21.5. Discussion

The major finding of the traffic analysis performed indicates that the dynamic lane operation has significantly improved the prevailing conditions on the segment considered. In particular, we observe a shift to higher speeds (in Paris towards the suburbs direction), higher capacities (both directions), lower congestion and higher LOS. Free flow and fluid to dense regimes become more frequent while congestion becomes less frequent. Combining the latter with Figure 21.4, we can directly deduce the occurrence of fewer multi-vehicle crashes and of more two-vehicle crashes – either rear-end or sideswipes. Furthermore, according to the second part of the safety analysis, lower traffic volumes generate more severe accidents.

Variable	Coefficient	t-statistics	SD
Constant	2.429	6.48	0.737
Weekend	– 1.680	– 9.62	1.132
Daylight	– 1.544	10.18	2.277
Wet road surface	– 1.064	– 6.39	0.574
No horizontal curvature	– 0.764	5.59	0.092
Adverse weather*experience	– 0.067	– 3.46	0.027
2-wheels	1.805	– 10.79	0.110
Heavy vehicle	– 0.481	1.69	0.030
Average traffic volume per lane and over 6 min (Q6min)	– 0.012	– 6.96	0.000
Average speed for Q6min < 112	0.015	8.59	0.000
Thresholds			
μ1	1.934	13.52	
μ2	4.330	16.65	
Number of observations	893		
Log-likelihood with constant only LLI	– 1207.487		
Log-likelihood at convergence LL(β)	– 590.441		
ρ^2_c =(LLI-LL(b))/LL(c)	0.511		

Table 21.5. *Model estimation results for random parameters ordered probit models: crash severity*

Such detailed impacts upon safety had not been discussed in *a priori* evaluation. To our knowledge, *a priori* evaluations always remain aggregate. Noticeably, regarding the specific site and thanks to the safety measures taken, the absolute number of crash occurrences has not been significantly increased since the dynamic lane operation. Nevertheless, a more in-depth analysis is certainly required in order to investigate whether accident severity has increased, decreased or remained constant.

21.6. Conclusions

In this paper, we performed an empirical analysis of the impacts of an ITS application in the Paris metropolitan region. We chose to highlight the differentiated impacts that occur when considering the traffic and the safety point of view separately. In particular, we demonstrated that traffic performance amelioration does not necessarily imply better safety performance. A similar pattern is drown when considering the environmental footprint of ITS applications varying from ramp metering (see, for example, the emissions from queuing) to dynamic speed smoothing leading at speeds lower to the environmentally optimal speed. Consequently, ex-ante and, of course, ex-post evaluation should be carried out to assess the impact of ITS applications. Otherwise, there is a risk of side-effects having an economic cost exceeding the benefits gained.

21.7. Bibliography

[ARO 04] ARON M., SEIDOWSKY R., Diagnostic d'exploitation et de sécurité du tronc commun autoroutier A4-A86; Analyse statistique des accidents survenus entre 1996 et 2002, SISER-INRETS Report, 2004.

[BAU 04] BAUER K.M., HARWOOD D.G., HUGHES W.E. *et al.*, Safety effects of using narrow lanes and shoulder-use lanes to increase the capacity of urban freeways, *83rd Transportation Research Board Annual Meeting*, Washington D.C., pp. 71–80, 2004.

[CHR 10] CHRISTOFOROU Z., COHEN S., KARLAFTIS M., "Vehicle occupant injury severity on highways: An empirical investigation", *Accident Analysis & Prevention*, vol. 42, pp. 1606–1620, 2010.

[CHR 11] CHRISTOFOROU Z., COHEN S., KARLAFTIS M., "Identifying crash type propensity using real-time traffic data on freeways", *Journal of Safety Research*, vol. 42, pp. 43–50, 2011.

[COH 04] COHEN S., "Using the hard shoulder and narrowing lanes to reduce traffic congestion", *Proceedings of the 12th IEE International Conference Road Transport Information & Control, London*, no. 501, pp. 149–153, 2004.

[DAV 02] DAVIS G.A., "Is the claim that 'variance kills' an ecological fallacy?", *Accident Analysis & Prevention*, vol. 34, pp. 343–346, 2002.

[FRI 12] FRIES R., REISI GAHROOEI M., CHOWDHURY M. *et al.*, "Meeting privacy challenges while advancing intelligent transportation systems", *Transportation Research Part C: Emerging Technologies*, vol. 25, pp. 34–45, 2012.

[GOL 04] GOLOB T., RECKER W., "A method for relating type of crash to traffic flow characteristics on urban freeways", *Transportation Research Part A*, vol. 38, pp. 53–80, 2004.

[GRE 03] GREENE W., *Econometric Analysis*, 5th ed., Prentice-Hall Press, New York, 2003.

[KEL 00] KELLERMANN G., "Experience of using the hard shoulder to improve traffic flows", *Traffic Engineering and Control*, vol. A247, pp. 412–414, 2000.

[MAR 99] MARELL A., WESTIN K., "Intelligent transportation system and traffic safety – drivers perception and acceptance of electronic speed checkers", *Transportation Research Part C: Emerging Technologies*, vol. 7, pp. 111–147, 1999.

[SCH 01] SCHRIJNEN L.M., "Innovations in the better use of motorways in the Netherlands", *Traffic Engineering and Control*, pp. 78–81, 2001.

22

Assessment of the Main New Travel-times Calculation Technologies on Lyon East Ring Road

In recent years, many new technologies for calculating travel time have arisen, but have never been evaluated and compared with each other at the same time. These technologies depend on the type of system used to collect traffic data. The goal of this chapter is to conduct an experiment, which will compare technologies (inductive loops, automatic number plate recognition – ANPR, magnetometer, blue-tooth and three FCD/FMD systems) for calculating travel time. The travel time calculated by ANPR will be the reference of the evaluation of our systems. In fact, ANPR can identify until 90% of the traffic, which is therefore a very good traffic representation. This trial was conducted on Lyon East ring road, in which DIR-CE is the road operator, over a 12 km distance with all kinds of traffic: free flow, heavy traffic and regular daily congestion of 1–2 km at the end of the day. The Lyon East ring road traffic flow is 90,000 vehicles per day with about 10% trucks. The installation of every technology will take place during the month of January with 2 or 3 months of data acquisition after. It is planned to get five measurement of test points and to have four elementary sections. Among these road-sections, one of them does not present crossroads: the accuracy can thus be precisely calculated with no disruption in travel time measurements.

22.1. Introduction

22.1.1. *Background*

DIR-CE (road operator in the Center East of France) deployed several years ago a traffic data collection station fleet, with inductive loops to manage its road network

---

Chapter written by Eric PURSON, Alexis BACELAR, Eric KLEIN, Bruno LEVILLY and Fabrice RECLUS.

through different traffic control centers named “Coraly”, “Hyrondelle”, “Gentiane”, “Osiris” and “Moulins”.

Today, the urban freeways travel times are performed from these with inductive loop stations by interpolation of speed measurements. However, these travel times present, on certain routes and for several periods of the day, significant gaps against real travel times.

It has been decided to develop a system to improve the reliability and quality of the travel-times information to users. Based on various technologies, numerous solutions exist today, but only a few elements allow us to choose the best performance and cost benefit ratio.

22.1.2. *Goals of the experiment*

In this context, DIR-CE can estimate various technical solutions, allowing us to calculate the reliability of the travel time, on the one hand, for the technology of data collection, on the other hand, for calculating estimation methodologies and forecasting real travel times. Their diffusion to the users will be done via the web sites and the road network VMS (variable messages sign). For that purpose, it is planned to appraise in real and known situations, different solutions of travel time calculation from various innovative technologies.

The objective is to estimate the metrological and functional performances of systems and services based on these new technologies of traffic data collection, by comparison of travel times developed by each of them according those obtained from the reference system which is automatic number plate recognition (ANPR), and by comparison to travel time developed from the existing inductive loop stations (see Figure 22.1).

This project of improving reliability of travel time deals with three axes:

– research for an optimal algorithm of travel time development from the spatial aggregation of punctual data that are road flows, speeds and occupation rate supplied by each station;

– evaluation of the system and services based on innovative technologies of data collection from Floating Car Data (FCD), by analyses of travel time measured in exit (historical data);

– research for improving algorithms of predictive travel time reliability with the aim of their diffusion to the users and to the road network operators.

22.2. The trial site

The choice of testing site was operated, according to the profiles of traffic, allowing for a traffic system and sensor evaluation in various and representative situations under typical conditions for these technologies.

Furthermore, it's necessary to have logistic resources for the installation of the systems (presence of energy and communication networks).

Finally, the site must be equipped with traffic data collection stations, for the travel-time calculation from the flow/speed data of this kind of equipment, and to measure the type of traffic.

The chosen experimental platform is a peri-urban freeway from the east of Lyon (ring road N346, 2 × 2 lanes) belonging to the "Coraly" network, South–North direction, between the interchange with A43 motorway at Manissieux until the interchange with A42 motorway.

The coverage area is about 12 km without tunnels.

It is planned to install five points (marked A, B, C, D and E) of vehicle recognition, and to have so a dividing of the section in four elementary segments whose travel times can be independently analyzed.

The implantation of the existing inductive loops stations partly influenced the choice of recognition points.

The localization of these points is as follows:

> The national road N346 is a peri-urban freeway of the "Coraly" network with significant traffic.

This section presents a zone of regular daily congestion about 1–2 km between the kilometer 39 and the kilometer 37 in the South–North direction at the end of the weekdays, caused by a ramp slowing down the trucks.

The capacity of this network is about 3500 vehicles per hour.

A "typical" example of the hourly traffic flows of any vehicles is proposed, where 6 min average speeds are observed during weekdays (except during vacation) on the N346 road (2 lanes).

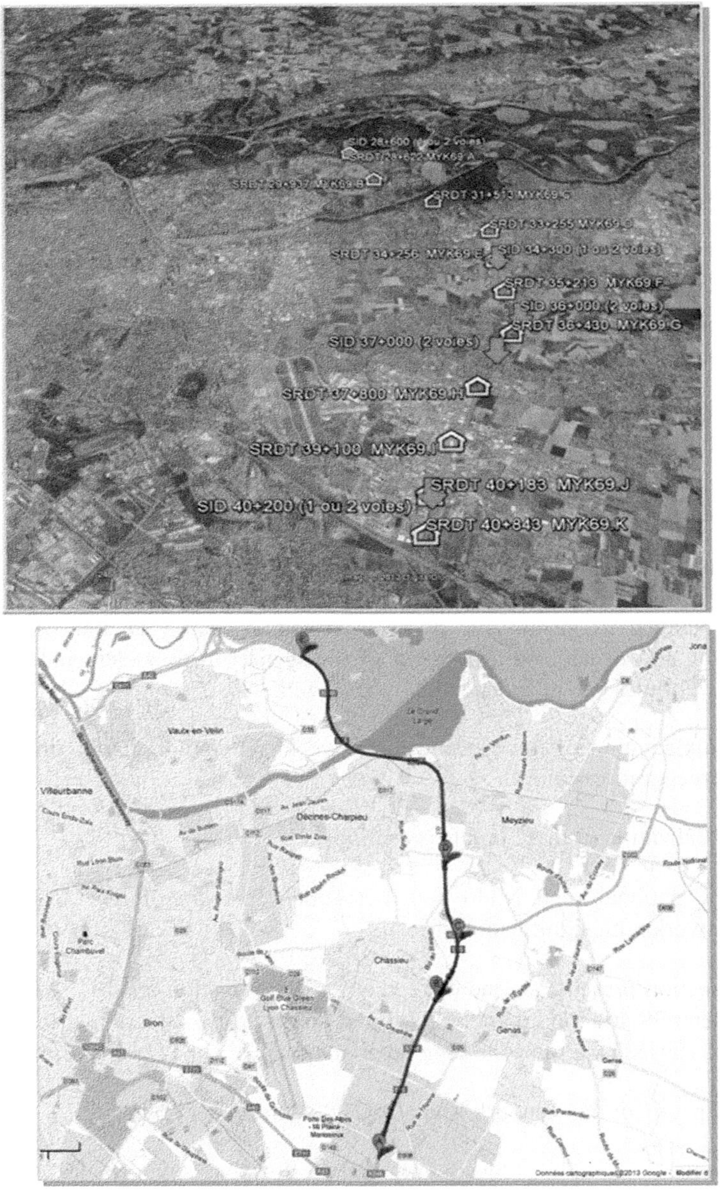

Figure 22.1. *a) Localization of recognition points (green arrow) and inductive loops stations (yellow); b) localization on a map. For a color version of this figure, please see www.iste.co.uk/jacob/traffic.zip*

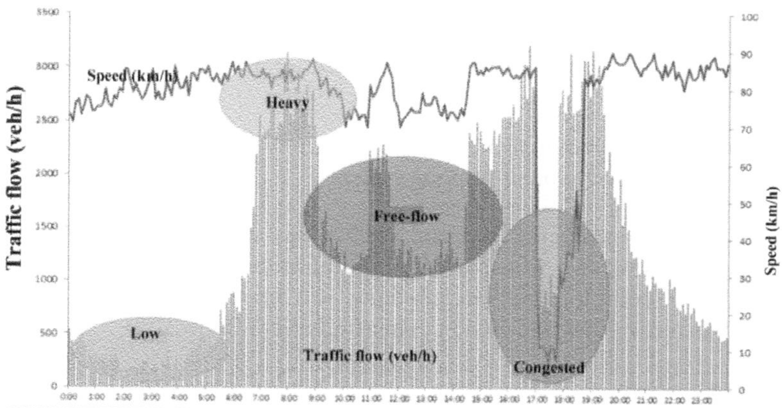

Figure 22.2. *Temporal graph of traffic flows and 6 min average speed of N346 road. For a color version of this figure, please see www.iste.co.uk/jacob/traffic.zip*

Different types of traffic are defined on this road according to traffic volumes and speed limits between 90 and 110 km/h (Table 22.1).

Types of traffic	Traffic flow (veh./h) (2 lanes)	6 min average speed	Indicative time slots
Low	QT < 1000	VT > 50 km/h	9 pm – 6 am
Free-flow	1000 ≤ QT ≤ 2500	VT > 50 km/h	6am – 7am
			9 am – 3 pm
			8 pm – 9 pm
Heavy	QT > 2500	VT > 50 km/h	7 am – 9 am
			3 pm – 5 pm
			6 pm – 8 pm
Congested	∀ QT	VT ≤ 50 km/h	5 pm – 6 pm

Table 22.1. *Definition of types of traffic*

22.3. Assessed technologies

This experiment deals with the estimation of performances of travel-time calculation systems and services from technologies of data collection by floating car data or by a combination of these technologies by data fusion.

Traffic data supplied by these technologies are obtained by identification of these vehicles in a position, then by re-identification of these same vehicles in another position of the analyzed section. Information characterizing the traffic such as speed, traffic flow and travel time are then calculated.

Basically we discern two FCD system approaches:

– radio-assisted communication FCD systems, informing their continual position to an operator;

– beacon-assisted communication FCD systems, which data are collected to equipment-level: these are point-to-point systems.

In this frame, the continual estimated technology will be the Floating Car Data (FCD), implementing the principle of geo-localization by satellite and/or mobile phone, and the "point-to-point" technologies will be Bluetooth/WiFi and vehicle magnetic signature re-identification by magnetometers (Table 22.2).

Holder or anticipated supplier	Technology	Principle
TOMTOM	Floating Car Data (FCD)	Geo-localization by satellite
AUTOROUTE TRAFIC	Floating Car Data (FCD)	Geo-localization by satellite
BLIP Systems	Bluetooth ⊕ WiFi	MAC address detection
STERELA	Bluetooth	MAC address detection
NEAVIA	Bluetooth ⊕ WiFi	MAC address detection
KARRUS ITS	Bluetooth ⊕ WiFi ⊕ Magnetometer	Data fusion

Table 22.2. *Assessed technologies on Lyon East ring road*

22.4. Implemented methodology

22.4.1. *Reference system*

Travel-time analysis requires an accurate reference system that presents a high detection rate, allowing us to extract representative statistic characteristics of the real traffic. For this evaluation, the ANPR system will be the reference system.

Figure 22.3. *"Micropack" SURVISION ANPR*

ANPR are systems equipped with optical character recognition (OCR) cameras, allowing us to identify the vehicle by its license plate. It can thus read every kind of license number, foreign or not. It can also be installed on the shoulder (road side or central reserve), as well as overhanging the lane axis on VMS or road structure.

These devices present a high rate of detection, almost 95% of the traffic, with a variability around 3%, according to the setting-up (shoulder or overhang), the degree of luminosity (twilight, setting sun), the weather conditions (rain, snow, fog) and the traffic conditions (masking of license plates between vehicles or by big size vehicles).

The rate of plates recognized is between 85 and 90% of the detected plates. The OCR errors are mainly not attributable to ANPR, and are mostly due to a damaged or dirty license plate. The rate of recognition, product of the rate of detection by the rate of recognized plates, is thus around 80 to 85% of the traffic.

Every detection of a license plate is timestamped to the second, by the acquisition of a computer system at the moment of reading, and stored in a text file. The installation of ANPR devices is synchronized on the same hourly reference in several points of a road with one license plate reader per lane, thus allowing us to match the same vehicles in each of these positions. According to the re-identification of its license plate, travel time of the mentioned vehicle on the concerned section can be obtained by the difference of passing times.

These devices also allow us to store and save traffic video sequences timestamped to the hundredth of a second.

The license plate matching rate is based on the recognition rate of each entry and exit point of the analyzed section. It also depends on the section characteristics, mainly the number of entry/exit. A closed section (see Figure 22.1(b) – section "BC"), presenting no entry or exit, thus allows us to obtain an optimal representativeness of the road traffic because it maximizes the matching rate.

22.4.2. *Assessment methods*

To obtain similar quantities, the period of individual data aggregation chosen as processing is the same as the one used by the estimated system. The optimal reference period for the analysis of the measured travel time in exit is 5 or 6 min. Other multiple periods or submultiples of these can also be analyzed. The statistical data extracted (average, centiles, etc.) are also the same that those supplied by the evaluated systems. These are compared during the analysis of data collections *a posteriori* in batch mode.

The sampling times are selected to cover the different traffic flows of the trial site (see Figure 22.2). These schemes will be defined with respect to 6 min aggregated data flow/speed provided by traffic data collection stations in the experimental area.

The collection involves sampling at least 30 reference aggregated travel-time sequences for each section studied.

In order that the samples are representative of the observed travel time, the aggregation period will be chosen so that its size reaches the desired minimum value. It must be at least 80% of the average 6 min flow, provided by the nearest data station of the vehicle recognition point. This average rate is considered in relation to the rates observed over the entire duration analysis that represents 30 travel-times sequences.

In addition, to be able to introduce travel time reliability indicators, the number of individual data from which is established the statistical datum on every temporal aggregation group will be taken into account. An example of temporal evolution graph of median travel times up to the 90th centile is represented below.

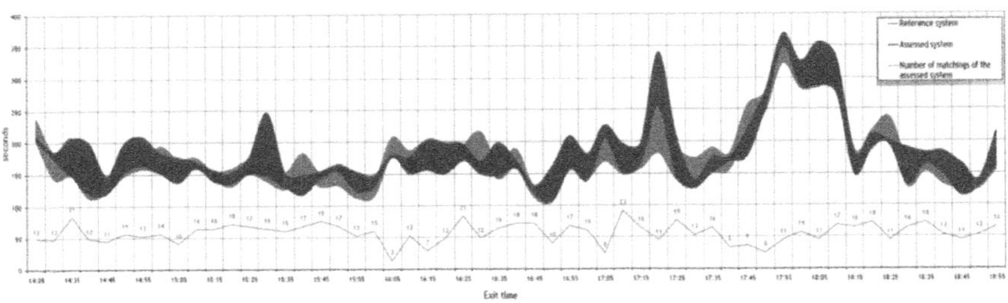

Figure 22.4. *Example of travel time graph and matching rate. For a color version of this figure, please see www.iste.co.uk/jacob/traffic.zip*

This kind of representation is of significant interest to be able to compare at the same time the temporal evolution of the evaluated system travel time as well as the scattering of both systems.

It is therefore possible to build a "reliability indicator" providing a clear picture of performances for each section analyzed. The definition of this indicator makes it possible to take into account both the difference between the two distributions dispersion (the dispersion is defined here as the difference between the time of the 90th percentile and the median), and the absolute differences between the reference travel time and those provided by the evaluated system (the graph position relative to each other).

This indicator is defined as follows:

$$IND_{TT} = \frac{min\left\{(TT_{90\%})_{ANPR};(TT_{90\%})_{A.S.}\right\} - max\left\{(TT_{50\%})_{ANPR};(TT_{50\%})_{A.S.}\right\}}{max\left\{(TT_{90\%} - TT_{50\%})_{ANPR};(TT_{90\%} - TT_{50\%})_{A.S.}\right\}} \quad [22.1]$$

A.S. = Assessed System

This calculation provides a measure of the "degree of coverage" of the two distributions, based on their maximum dispersion. Thus, for a given aggregate range, when the two distributions are identical (perfect superimposition of the two curves), this ratio reaches its maximum value equal to "1". On the contrary, the lower the distribution overlap area (which, in some cases, may be due to a large gap of dispersions), the more the value of this indicator tends to the value "0", but remains positive. A negative value of this indicator means that there is no overlap area of the two distributions. Thus, the larger the travel time gap provided by the two systems, the more this indicator takes a strongly negative value, reflecting a less good estimation.

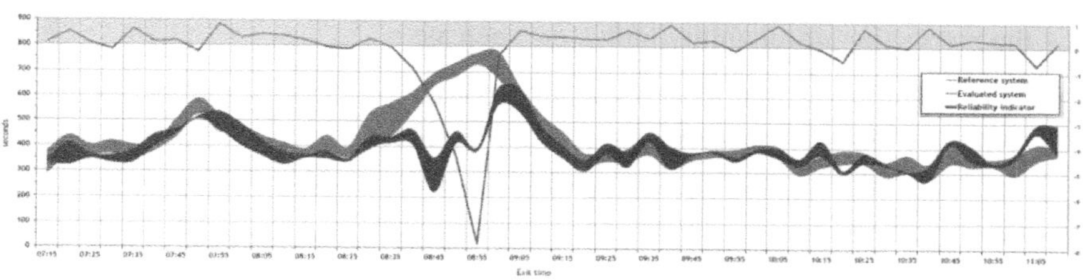

Figure 22.5. *Example of travel time graph and reliability indicator. For a color version of this figure, please see www.iste.co.uk/jacob/traffic.zip*

Overall reliability by section is defined as the ratio of the duration the indicator is positive (shown by the colored stripe on the graph) by the total measurement time. For this example, the travel time overall reliability on this route is 69%.

Other measurement accuracy and reliability indicators can be produced under different traffic flows such as:

– the difference between travel time provided by different systems and those given by the reference devices (provided that their distribution follows a normal distribution), quantified by:

- calculating the relative errors of individual data and mean relative errors of temporally aggregated data,

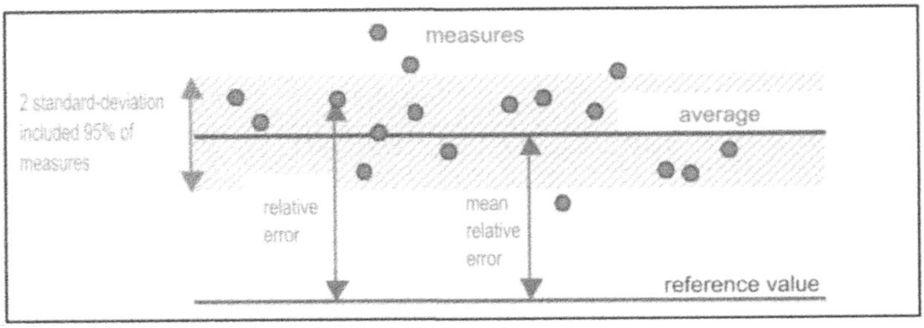

Figure 22.6. *Definition of relative and mean relative errors of data*

- the reference value is statistically of the same kind that the one provided by the system being evaluated (mean, median, etc.),

- calculating uncertainties intervals achieved with a confidence level of p recommended by the CERTU [CER 02] guide at 95%, meaning that 95% of the values of the studied population belong to this interval;

– the production of reliability indicators such as:

- the rate of detection of each system according to the total number of vehicles on each point of vehicle recognition,

- the matching rate relative to the total number of vehicles on a closed section, without entry nor exit,

- the matching rate between two non-consecutive identification points (e.g. AC, AE, etc.),

- the rate of outliers relative to the total number of measurements (a system can be accurate but displays a high rate of outliers, which consequently degrades the performance of the sensor);

– the available data rate for each technology and system throughout the study period;

– the time intervals for renewal and refreshment of the information provided by the different systems, in order to study the ability of devices and systems to develop travel time in real time.

22.5. Innovative administrative procedure

An innovative administrative procedure takes place in this project. In fact, the road operator decided to spend € 10,000 of the cost associated with the evaluation of each sensor technology. Thereby they wrote an administrative procedure adapted for the call for tender in this way. The principle is to assign two scores with the same weighting to each candidate: a technical mark and a financial score. Technical mark ranges from 1 to 10 and is awarded based on the nature of the data and functionality offered. The financial rating is assigned based on the cost of providing, it means 0 if the offer is free and −10 if the offer costs € 10,000 or over. The candidate is retained if its final rating is positive.

22.6. Conclusion

The implementation of the various technical equipment and architectures will take place in the second fortnight of January. Then, an adjustment/validation of the installation will last the week after. At the end of February–beginning of March, the data collection phase will begin with some available treatments mid-March and a duration of 2 or 3 months. The first results will be presented at the TRA conference.

This experiment will allow us to build a large travel-time database developed by various innovative technologies. The results of these evaluations will help road operators to have a better view among the numerous existing and emerging solutions to get travel time, and to make their choice both from the point of view of the required performances as costs to be invested.

22.7. Acknowledgments

The authors deeply thank CERTU for its support with this study and for its interest in the related projects. Special thanks to the DIR-CE for its help in the preparation of the various measure campaigns.

22.8. Bibliography

[KLE 12] KLEIN E., MURUGNEUX E., BACELAR A., "Accurate travel times in a complex urban context", *Proceedings of 9th ITS European Congress*, Dublin, Ireland, 2012.

[OPT 12] OPTIMO D'LYON, Optimiser la mobilité durable en ville, réunion de lancement, présentation, p. 105, 2011.

[CER 02] CERTU, Méthodologie d'évaluation des nouveaux capteurs de trafic routier, technical guide, collection "dossiers", p. 74, 2002.

[MAR 06] MARGULICI J.D., YANG S., TAN C.-W. *et al.*, *Evaluation of Wireless Traffic Sensors, California Center for Innovative Transportation*, Berkeley, California, p. 115, 2006.

[NFP 97] NFP 99-300 STANDARD, Unités de mesure et de traitement, p. 22, 1997.

[NFP 01] NFP 99-330 STANDARD, Nature, exactitude des données de trafic routier – Essais, p. 30, 2001.

23

Rail Externalities: Assessing the Social Cost of Rail Congestion

In the rail sector, there are few academic papers dealing with congestion. Most papers focus on the scarcity of train paths or, following Petersen (1974), on the relationship between scarcity and speed. However, we could suppose that in the rail sector, high traffic density during peak-hours generates an additional delay on other users, so an external cost. In this context, track access charges may reflect the correct marginal social cost of adding a train. The aim of this chapter is to use French infrastructure manager data to assess the relationship between traffic and delay. According to the line and its features (allowed speed, number of tracks, signaling, etc.), the econometrical analysis shows a positive econometric relationship between traffic and unreliability rate: an additional train on the line increases the probability of all trains being late. These results allow computing a marginal cost of congestion, which can be used for pricing and cost benefit analysis of capacity investments. This study may support the implementation of a congestion charge in the French rail network.

23.1. Introduction

In 2012, the French rail regulator (Autorité de Régulation des Activités Ferroviaires, ARAF) pointed out congestion as a crucial issue for the French railway system. In order to improve the utilization of the network, the infrastructure manager should develop track access charges and provide efficient economic incentives (Statement 2011-002 on the 2nd February 2011).

Rail capacity is one of the greatest issues in the French network. Demand for rail transport grows in metropolitan areas and the expansion of rail capacity faces a

---

Chapter written by María PÉREZ HERRERO, Julien BRUNEL and Gregoire MARLOT.

range of obstacles and financial challenges. In the rail sector, capacity shortage has been traditionally studied when demand exceeds the available capacity (scarcity), but we could expect that there is another form of capacity shortage (congestion) during peak-hours as in the road sector.

Economic literature is quite extensive about road congestion; it relies either on a static speed/flow relationship, which can be observed through road statistics, or on a dynamic peak-load approach (bottleneck model, etc.). Nevertheless, literature about congestion is less extensive in sectors where traffic is scheduled in advance. In the air sector, following the seminal papers of [LEV 69] and Carlin and Park [CAR 70], some papers have estimated the marginal congestion cost of adding a new slot. In the rail sector, very few academic papers deal with congestion.

In this context, SNCF Réseau has undertaken several studies in order to define and measure congestion on the national rail network. The present study aims to investigate the relationship between traffic density, reliability rate and average delay for a given rail track. This econometrical analysis has been realized by the economic consulting firm Microeconomix [MIC 12].

The structure of this chapter is as follows: section 23.2 briefly reviews the academic literature dealing with the economics of congestion in transport, and particularly in rail transport. It presents the intuition behind this paper: high traffic density may generate a higher probability of delays. This intuition is empirically confirmed by Gibson *et al.* [GIB 02]. Section 23.3 provides a detailed description of the methodological framework. Section 23.4 describes the data set used in the present paper. Section 23.5 exposes the results of our analysis. Finally, section 23.6 offers concluding remarks.

23.2. Related literature

Many networks suffer from peak-load demand problems. In general, congestion refers to the existence of limited capacity networks whose demand varies periodically. The literature widely deals with congestion in the road sector, from classic contributions such as Pigou [PIG 20], Walters [WAL 61], Vickrey [VIC 63] to more recent works from Arnott *et al.* [ARN 93], Chu [CHU 95] or Verhoef [VER 01]. When car users decide to make an additional trip, they impose additional costs on themselves, on the infrastructure provider and on other users. From an economic perspective, congestion is basically a standard externality problem. Academic literature shows that peak/off-peak pricing is an efficient solution to tackle congestion, and allows users to internalize the external costs generated and reallocate the traffic demand during the day [VIC 63]. The literature also shows that, given certain circumstances, congestion pricing covers the construction costs of highways [MOH 62, ARN 93, HAU 98].

As Quinet [QUI 97] pointed out, congestion does not appear only on roads but also in other transportation modes, even where traffic is scheduled in advance. By contrast with road sector, congestion has received less attention in these sectors. A sizeable amount of literature has nevertheless studied airport congestion that occurs in the neighborhood of large airports due to runways or traffic control saturation.

In a seminal paper, Carlin and Park [CAR 70] estimated the marginal cost of delays in New York's LaGuardia airport. The congestion cost is defined as the additional delay imposed on the following planes in the queue during the busy period. One of the main results is that landing fees are inefficient, because the congestion costs are significantly higher than what is paid by air carriers. This paper explores the possibility of imposing a congestion toll. A contemporary paper of Levine [LEV 69] also advocates more differentiated fees depending on the time of the day in order to reflect congestion during peak hours.

A substantial amount of literature followed these seminal articles. Some of them proposed to assess empirically the cost of congestion. This is notably the case of Morrison and Winston [MOR 89]. This paper intended to econometrically estimate the relationship between airport activity and arrival and departure delays using US data. It states that an increasing level of activity causes an increase in average delays. In other words, when capacity is used to its fullest, an additional slot increases the probability of delays due to a reduction in the ability to recover from an incident. Another interesting contribution to this literature is given by Nombela Merchan and De Rus [NOM 06], which proposes a desegregated analysis of airport delays in Madrid Barajas.

Few academic papers consider congestion in rail transport. Some notable exceptions are the High Level Group on infrastructure charging [NAS 99] and papers by Quinet [QUI 03] and Nash and Matthews [NAS 05]. These papers specify the case of pricing railway congestion from a theoretical point of view.

From an empirical point of view, there are even less papers concerning rail congestion in the economic literature. However, in the operational research literature, some papers have studied rail congestion. These papers have estimated the delays generated by the operational constraints of a railroad network: delays for meets with opposing rail traffic on single-track lines, and for following and overtaking slower rail traffic moving in the same direction, for example. This sizeable literature employs both analytical and simulation-based methods, and study delays and capacity assessment in railroad line networks with specific configurations following Frank [FRA 66] or Petersen [PET 74].

A second type of delays is originated by an incident (failure of the rolling stock, failure of the infrastructure, inadequate behavior of the crew, etc.). An incident generates delays to the following trains, and, given the complexity of the network, a lot of trains can be affected, even on different sections of the network. These delays are obviously unexpected. They increase as capacity utilization rises, because heavy traffic reduces the network manager's ability to resolve the incident, and the delay is transmitted to more trains, with a snowball effect. This idea is quite familiar in airport economics, as stated previously. It is also intuited by the papers of Quinet [QUI 03] and Nash and Matthews [NAS 05] for rail transport.

These delays can be measured with an adequate monitoring system. Very few papers have studied this phenomenon in the economic literature. For instance, it has been empirically studied by the British rail network [GIB 02]. In this chapter, a regression analysis confirms the existence of a relationship between capacity utilization and delays. In this study, an exponential form was chosen to estimate for the relationship between capacity utilization (C_{it}) and reactionary delay (D_{it}) across the network. This relationship is given by:

$$D_{it}=A_i exp\,(\beta C_{it})$$

where A_i is a route section specific constant and β is a route specific constant.

The regression analysis was performed for 24 strategic routes on the British network using 1998 and 1999 financial year data on reactionary delay per train and a measure of capacity utilization based on the winter 1998 timetable.

The results of the regression show that β is statistically significant for 20 out of the 24 routes. This means that there exists a positive relationship between capacity and reactionary delays. This relationship justifies the congestion charge implemented since 2001 by Network Rail. An additional path increases the probability of delays and, therefore, its monetary cost in a performance regime framework. Recently, [HAI 13] have proposed an alternative methodology for the British infrastructure manager which concludes that performance is as much to do with how capacity is used rather than just how much. In other words, the heterogeneity of the timetable is an important factor.

In what follows, we propose to investigate the existence of this relationship between traffic and delays in the case of the French network, using an internal dataset of the French rail infrastructure manager.

23.3. The model and the econometric strategy

The present section proposes a mathematical framework in order to empirically estimate the marginal congestion cost in railways. This mathematical framework enables to isolate the marginal effect of a train on the total delays.

In this section, we notice R_i^* the deviation between the real time and the scheduled time of a train for a given traffic density Q_i. The train can be on time ($R_i^* = 0$), early ($R_i^* < 0$) or late ($R_i^* > 0$).

We define the variable R_i representing the delay of train. One can therefore notice:

$$R_i = \begin{cases} 0 & if R_i^* \leq 0 \\ R_i if R_i^* > 0 \end{cases}$$

The expected delay of train for a given traffic density is:

$$E(R_i) = p(R_i^* \leq 0) \cdot E(R_i|R_i^* \leq 0) + p(R_i^* > 0) \cdot E(R_i|R_i^* > 0)$$

As the expected delay is null when the train is on time or early ($p(R_i^* \leq 0) \cdot E(R_i|R_i^* \leq 0) = 0$), this equation can be written:

$$E(R_i) = p(R_i^* > 0) \cdot E(R_i|R_i^* > 0) \qquad [23.1]$$

This equation indicates that the expected delay of a train for a given traffic density is equal to the product of the expected delay of delayed trains and the number of trains delayed.

The total amount delays of trains for a given traffic is, by definition, the expected delay of train multiplied by the number of trains, i.e. $Q_i \cdot E(R_i)$. Therefore, it follows that the marginal delay imposed by an additional train is the derivative of the total amount of delay function with respect to the level of traffic.

It can also be written as:

$$\frac{\partial[Q_i.E(R_i)]}{\partial Q_i} = Q_i\left[\frac{\partial E(R_i)}{\partial Q_i}\right] + E(R_i) \qquad [23.2]$$

In this equation, the right-hand term is the expected delay of the additional train given the traffic density: this is a direct effect, internalized by the train. The direct effect is equal to the expected delays for a given traffic density. This term, expressed by equation [23.1], can be directly computed from the data set.

The left-hand term of equation [23.2] represents the marginal delay imposed by the additional train on the following trains. It is an indirect effect which corresponds to the pure externality effect of congestion. The indirect effect, cannot be computed directly and needs and econometrical analysis in order to be estimated.

Using equation [23.1], the indirect effect can be rewritten as:

$$\begin{aligned} \frac{\partial E(R_i/Q_i)}{\partial Q_i} &= \frac{\partial}{\partial Q_i}[p(R_i^* > 0) \cdot E(R_i|R_i^* > 0)] \\ &= \frac{\partial p(R_i^*>0)}{\partial Q_i} E(R_i|R_i^* > 0) + (R_i^* > 0) \cdot \frac{\partial E(R_i|R_i^*>0)}{\partial Q_i} \end{aligned} \quad [23.3]$$

In this expression, the expected delay $E(R_i|R_i^* > 0)$ and the probability of being late $p(R_i^* > 0)$ are known from the data set. Nevertheless, the two derivatives should be estimated.

The first term ($\frac{\partial p(R_i^*>0)}{\partial Q_i}$) describes the marginal effect of an additional train on the probability of being late. The second one ($\frac{\partial E(R_i|R_i^*>0)}{\partial Q_i}$) represents the marginal effect of an additional train on the expected delay.

In what follows, we propose to estimate the first effect using a probit model. For each train, the dependent variable equals 1 if the train is delayed ($R_i \; if \; R_i^* > 0$) and 0 if not. The marginal effect on the probability of being late is directly calculated using the results from the probit regression for each level of traffic.

The second effect is estimated using a linear regression:

$$y_{ij} = \alpha_j + \beta x_{ij} + \varepsilon_{ij}$$

where y_{ij} is the delay for train i in the measuring point j, x_{ij} is the traffic density associated at each observation and ε_{ij} is the error term. In this expression, α_j is supposed to be specific for each measuring point. Both the probit and the linear models regressions include fixed effects in order to consider the delay heterogeneity associated to the different measuring points, direction and train use (passenger, freight, etc.).

23.4. The data set

In this empirical research, we use data from an internal database of SNCF Réseau in order to estimate the previous parameters. This internal database records traffic information in the French network, and notably the delays at each measuring point. The data provided by this database allow us to know precisely the performance (reliability rate and delay) of each line at each level of traffic.

The data is recorded by an automatic system which detects the train circulation and registers the traffic details concerning the train. These automatic measuring points are associated to the measuring points which are utilized for the construction of the schedule. The system allows obtaining, for each train which crosses a measuring point, the data presented in the following table.

	Description
Internal circulation number	Specific and unique number associated at each train
Circulation number	Number associated to a specific stopping pattern
Date/Hour	Date and real hour when the train crosses the measuring point
Week day	–
Timetable type	Determines the kind of stop: Origin, Passage, Arrival or Departure (for a stop in a train station) or Terminus
Time deviation	It is the deviation between the real time and the scheduled time (delay)
Statistical category	Informs about the train activity (HSL, regional activity, national activity, freight, etc.) and if the train is loaded or empty

Table 23.1. *Summary variables*

However, railway lines have different characteristics. They have diverse uses (passenger trains or freight trains), different traffic densities (lines with heavy traffic or lines with low traffic) and varied levels of performance. For that reason, we have subdivided the French network in several groups of lines with similar characteristics.

In this classification, the network is divided in four categories depending on uses (freight, regional, national) and speed levels (high speed lines or not):

– high speed lines: routes with a speed higher than 250 kph;

– intercity lines: routes between population centers mainly used by freight and passenger long distance trains;

– regional lines: routes between suburbs, towns and cities, without special speed requirements, and mainly used by regional and commuters trains;

– only freight lines: freight specific routes with no mixed traffic, and generally low traffic density.

At the same time, these categories are subdivided in subcategories depending on the traffic density (trains per weekday per route): high, medium or low traffic density.

The traffic is highly concentrated around several nodes of the networks. For example, we can observe lines with 15 trains per hour during the peak-hours period in some regional railway lines near Paris. By contrast, some local lines can only have one train per hour during the peak-hours. The varied traffic lines density emphasizes that congestion would not emerge with the same intensity in the entire network.

In the present study, we have focused our analysis on 42 lines of the French railway network, with three measuring points for each line. The lines belong to these different groups of lines presented above. The dataset includes 6.4 million trains (i.e. 6.4 million observations). These lines have been assembled in nine subgroups using the strategic segmentation. The dataset used in this research contains all train circulations in these lines during 2011.

23.5. Results

For this analysis, the variable traffic has to be defined. For each observation (each train recorded), we have obtained a level of traffic which equals the number of train scheduled in the same line and direction during the previous hour. Then, an econometrical analysis is pursued to measure the additional delay (in minutes) in a railway route due to an increase of one traffic unit (the marginal delay). As mentioned above, an additional train is likely to be delayed and to impose an additional delay on the next trains. The consequences of an additional train (direct and indirect effect) have been considered separately in our analysis, in order to assess the effect that an additional train generates on other trains. The indirect effect is the pure externality from an economist point of view whereas the indirect effect is internalized by the additional train.

Some of the parameters are directly computed using the data set. Some others are estimated with the econometric analysis, as described above. Two econometrical regressions are conducted in order the estimate the marginal cost of congestion (indirect effect) in minutes: the probit model which estimates the marginal effect of traffic on the probability of being late, and the linear model which estimates the marginal effect of an additional train on the expected delay.

The results of the econometric analysis are presented Table 23.2. The regressions have been estimated separately for the nine groups of lines. Table 23.2 presents the results of the two regressions. The first column represents the average marginal effect of an additional train on the probability of being late. It is the parameter

($\frac{\partial p(R_i^*>0)}{\partial Q_i}$) in equation [23.3]. The second column represents the marginal effect of an additional train on the expected delay. It corresponds to the parameter ($\frac{\partial E(R_i|R_i^*>0)}{\partial Q_i}$) in equation [23.3].

Strategic classification	Type of line	Probit $\frac{\partial p(R_i^* > 0)}{\partial Q_i}$	Linear regression $\frac{\partial E(R_i \mid R_i^* > 0)}{\partial Q_i}$
G1			
	High speed	0.0096***	0.020**
G2		(0.0024)	(0.017)
	Intercity lines	0.020***	0.49**
G3		(0.00042)	(0.12)
	Intercity/Regional lines	0.013***	0.10**
G4		(0.000054)	(0.018)
	Intercity lines high traffic density	0.022***	0.67**
G5		(0.00024)	(0.073)
	Intercity lines low traffic density	0.018***	0.67
G6		(0.00057)	(0.30)
	Intercity lines medium traffic density	0.010***	0.19
G7		(0.0011)	(0.14)
	Regional lines high traffic density	0.025***	0.14**
G8		(0.00024)	(0.024)
	Regional lines low traffic density	0.056***	0.67**
G9		(0.0064)	(0.31)
	Regional lines medium traffic density	-0.025***	1.05
		(0.0024)	(1.10)

Standard error in parentheses. *p < 0.10, **p < 0.05, ***p < 0.001.

Table 23.2. *Regressions results*

These results can be interpreted as follows: for high speed lines, an additional train increases the probability of being late by 0.96 points and increases the expected delay by 0.20 min for the following trains. Moreover, these results show that for certain types of lines, the congestion is not statistically significant. It is the case of intercity lines. It not surprising since this group corresponds to low traffic group of lines.

Once these two regressions have been estimated, it is possible to compute the average direct effect, given by equation [23.1], and the marginal effect by group, in accordance with equation [23.3]. The results of these computations are presented in Table 23.3. This table can be interpreted as follows: for G4, an extra train generates 0.68 min of delay on the forthcoming trains.

Strategic Classification	Marginal Effect
G1	0.19
G2	0.47
G3	0.13
G4	0.68
G5	0.59
G6	0.18
G7	0.19
G8	0.67
G9	0.30

Table 23.3. *Congestion marginal cost*

In order to check the robustness of these results, some tests have been realized. A first test is realized in order to verify the existence of the relationship with another definition of delay. The previous results considered a train delayed if delay was superior to zero. Nevertheless, the data show that many trains have in fact little delays (less than 5 min). A small delay associated with a train could be a measure error in some points, so we decided to test our results using a different delay definition. Two tests have been done considering only delays superior to 3 and 5 min, respectively. In both cases, even if the absolute value of the direct effect is different, the estimated relationships are significant and the hierarchy between lines does not change.

Strategic Classification	Marginal Effect (>3)	Marginal Effect (>5)
G1	0.22	0.20
G2	0.37	0.27
G3	0.12	0.10
G4	0.62	0.53
G5	0.52	0.46
G6	0.076	0.051
G7	0.14	0.082
G8	0.51	0.56
G9	-0.018	0.09

Table 23.4. *Robustness test*

Until now we have considered that marginal effects are homogeneous between measuring points or lines in the same group. Some regressions analyses have been also conducted for several specific points. The test shows that there exist some differences between measuring points and lines. In some measuring points the congestion effects are higher than in others sections of the network, but the effect remains significant from a statistical point of view.

These results therefore provide strong evidence of our intuitive idea: an additional train increases the probability of late trains. It means that there is a form of unexpected congestion in the railways. The direct effect is internalized by the supplementary train, but the indirect effect generates an external cost on other users.

23.6. Conclusions

This chapter investigates one form of rail congestion. It shows that the economics of congestion in rail transport is smaller in comparison with road transport. As traffic is scheduled by advance, one can expect scarcity rather than congestion in rail transport. Nevertheless, congestion can appear when traffic is scheduled in advance: when an airport is highly used, an additional slot increases the probability of delays due to a reduction in the ability to recover from an incident.

We propose an economic analysis of congestion in rail transport and present an econometrical analysis of delays on the French rail network, which establishes a relationship between delays and heavy traffic in several points of the French network. This chapter confirms the existence of an external congestion cost on rail transport.

If delays increase the travelling time for passengers, it has a negative impact on social welfare. From an economic perspective, this phenomenon can be understood as a standard externality problem. High traffic density during peak-hours generates an external cost on other users. Track access charges can send to train operators the correct signal of the marginal social cost of adding a train. This pricing rule would allocate demand in an efficient way during the periods of the day. A congestion fee would permit to internalize the external cost imposed by the additional train on others when rail operators decide the number of paths supplied. It would reflect the external negative effects that an additional train generates on travelling time for passengers.

This study may support the implementation of a congestion charge and the improvement of cost benefit analysis methods. Nevertheless, the observed reliability rates in this study depend on the features of the line, the trade-off between capacity and resilience for the design of the train paths, and on the way train paths are allocated between different trains. In order to determine the optimal level of congestion and the optimal capacity policies, including congestion pricing, it is necessary to consider and analyse all these choices as a whole.

23.7. Acknowledgments

We would like to thank SNCF Réseau, which financed and supervised this study and G. de Muizon, E. Frot and A. Charpin from Microeconomix for their contributions.

23.8. Bibliography

[ARN 93] ARNOTT R., DE PALMA A., LINDSEY R., "A structural model of peak-period congestion: A traffic bottleneck with elastic demand", *American Economic Review*, vol. 83, no. 1, pp. 161–179, 1993.

[BRU 02] BRUECKNER J.K., "Airport congestion when carriers have market power", *American Economic Review*, vol. 92, no. 5, pp. 1357–1375, 2002.

[CAR 70] CARLIN A., PARK R.E., "Marginal cost pricing of airport runway capacity", *American Economic Review*, vol. 60, no. 3, pp. 310–319, 1970.

[CHU 95] CHU X., "Endogenous trip scheduling: The Henderson approach reformulated and compared with the Vickrey approach", *Journal of Urban Economics*, vol. 37, pp. 324–343, 1995.

[DOL 02] DOLL C., Transport User Cost and Benefit Case Studies, UNITE (UNIfication of accounts and marginal costs for Transport Efficiency) Deliverable 7, University of Leeds, Karlsruhe, November 2002, 2002.

[FRA 66] FRANK O., "Two-way traffic in a single line of railway", *Operations Research*, vol. 14, pp. 801–811, 1966.

[GIB 02] GIBSON S., COPPER G., BALL B., "Developments in transport policy: the evolution of capacity charges in the UK network", *Journal of Transport Economics and Policy*, vol. 32, no. 2, pp. 341–354, 2002.

[HAU 98] HAU, TIMOTHY D., "Congestion pricing and road investment", in BUTTON KENNETH J., VERHOEF ERIK T. (eds), *Road Pricing, Traffic Congestion and the Environment: Issues of Efficiency and Social Feasibility*, Edward Elgar, Cheltenham, UK, 1998.

[HAI 13] HAITH J., JOHNSON D., NASH C., Timetable heterogeneity, reactionary delay and the calculation of the congestion charge in Britain, Working Paper University of Leeds 2013.

[JOH 08] JOHNSON D., NASH C., "Charging for scarce rail capacity in Britain: a case study", *Review of Network Economics*, vol. 7, no. 1, March 2008, 2008.

[LEV 69] LEVINE M.E., "Landing fees and the airport congestion problem", *Journal of Law and Economics*, vol. 12, pp. 79–108, 1969.

[MIC 12] MICROECONOMIX, Etude économétrique de l'effet de la congestion ferroviaire sur la régularité des trains à l'attention de RFF, Internal report, 2012.

[MOH 62] MOHRING H., HARWITZ M., *Highway Benefits: An Analytical Framework*, Northwestern University Press, Evanston, Illinois, 1962.

[MOR 89] MORRISON S., WINSTON C., "Enhancing performance of the deregulated air transportation system", *Booking Paper on Economics Affairs, Microeconomics*, pp. 61–123, 1989.

[MOR 07] MORRISON S., WINSTON C., "Another look at airport congestion pricing", *American Economic Review*, vol. 97, no. 5, pp. 1970–1977, 2007.

[NAS 99] NASH C., Calculating transport congestion and scarcity costs, Final report of the expert advisors to the High level group on infrastructure charging, Working group 2, 1999.

[NAS 05] NASH C., MATTHEWS B., Rail infrastructure charges – the issue of scarcity, Working paper, University of Leeds, 2005.

[NOM 06] NOMBELA MERCHÁN G., DE RUS G., "Analisis economico de la congestion en los aeropuertos europeos", in DE RUS G. (ed.), *La politica de transporte europea: el papel del analisis economico*, Fundacion BBVA, 2006.

[PET 74] PETERSEN E.R., "Over the road transit time for a single track railway", *Transportation Science*, vol. 8, pp. 65–74, 1974.

[PIG 20] PIGOU A.C., *The Economics of Welfare*, MacMillan, London, 1920.

[QUI 97] QUINET E., "Full social cost of transportation in Europe", in GREENE D.L. *et al.* (eds), *The Full Costs and Benefits of Transportation: Contribution to Theory, Methods and Measurement*, Springer, 1997.

[QUI 03] QUINET E., "Short term adjustment in rail activity", *Transport Policy*, vol. 10, no. 1, pp. 73–79, 2003.

[VER 01] VERHOEF E.T., "An integrated dynamic model of road traffic congestion based on simple car-following theory: exploring hypercongestion" *Journal of Urban Economics*, vol. 49, pp. 505–5, 2001.

[VIC 63] VICKREY W.S., "Pricing in urban and suburban transport", *American Economic Review*, vol. 53, no. 2, pp. 452–65, 1963.

List of Authors

Dolores ADAMSKI
Chambre de Commerce et d'Industrie de Grenoble
France

Diego ALBESANO
Camera di Commercio di Torino
Turin
Italy

Thomas ALBRECHT
CSC Deutschland GmbH
Dresden
Germany

Konstantinos N. ANDROUTSOPOULOS
Transportation Systems and Logistics Laboratory, Department of Management Science and Technology
Athens University of Economics and Business
Greece

Alexis BACELAR
Centre d'Etudes et d'expertise sur les Risques, l'Environnement, la Mobilité et l'Aménagement (Cerema)
Territoires et Villes
Bron
France

Evangelos BEKIARIS
CERTH – HIT
Thessaloniki
Greece

Sylvain BELLOCHE
Centre d'Etudes et d'expertise sur les Risques, l'Environnement, la Mobilité et l'Aménagement (Cerema)
Direction territoriale Centre-Est
Bron
France

Martin BÖHM
Federal Agency for Technological Measures Ltd
Austria Tech
Vienna
Austria

Patrick BONNEL
Laboratoire d'Économie des Transports
University of Lyon/ENTPE
France

James BROGAN
University of New Mexico
Albuquerque
New Mexico
USA

Julien BRUNEL
SNCF Réseau
Paris
France

Christine BUISSON
Laboratoire Ingénierie de la Circulation Transports (LICIT)
IFSTTAR and ENTPE
Lyon
France

Jan BURGMEIJER
TNO
Delft
The Netherlands

Jorge CABRERA DELGADO
Laboratoire d'Économie des Transports
University of Lyon/ENTPE
France

Marko ČELAN
Faculty of Civil Engineering, Transportation Engineering and Architecture
University of Maribor
Slovenia

Yusen CHEN
TU Delft
The Netherlands

Zoi CHRISTOFOROU
Ecole des Ponts-Paris Tech
Marne-la-Vallée
France

Simon COHEN
Génie des Réseaux de Transport Terrestres et Informatique Avancée (GRETTIA)
IFSTTAR
Marne-la-Vallée
France

Ian CORNWELL
Mott MacDonald Ltd.
Croydon
UK

Winnie DAAMEN
Department of Transport & Planning
Delft University of Technology
The Netherlands

Meena DASIGI
Network Rail
London
UK

Jean Daniel DEMOND
Conseil général de l'Isère
Grenoble
France

Martijn DE KIEVIT
TNO
Delft
The Netherlands

Janiek DE KRUIJFF
TNO
Delft
The Netherlands

Aurélien DURET
Centre d'Etudes et d'expertise sur les Risques, l'Environnement, la Mobilité et l'Aménagement (Cerema)
Direction territoriale Centre-Est
Bron
France

Anastasia FEOFILOVA
Rostov State Civil Engineering University
Rostov-on-Don
Russia

Jörg FREUDENSTEIN
Albrecht Consult GmbH
Aachen
Germany

Evangelia GAITANIDOU
CERTH – HIT
Thessaloniki
Greece

Bernard JACOB
IFFSTAR
Marne-la Vallée
France

Eric KLEIN
Centre d'Etudes et d'expertise sur les Risques, l'Environnement, la Mobilité et l'Aménagement (Cerema)
Direction territoriale Est
Metz
France

Mitja KLEMENČIČ
Faculty of Civil Engineering, Transportation Engineering and Architecture
University of Maribor
Slovenia

Gerdien KLUNDER
TNO
Delft
The Netherlands

Jean-Bernard KOVARIK
Ministry of Infrastructure, Transport and the Sea
Paris La Défense
France

Olivier LATOUILLE
Conseil général de l'Isère
Grenoble
France

Marjan LEP
Faculty of Civil Engineering, Transportation Engineering and Architecture
University of Maribor
Slovenia

Gaele LESTEVEN
CRIA/UMR Géographie-Cités
University Paris 1 Panthéon-Sorbonne
France

Bruno LEVILLY
Direction Interdépartementale des Routes Centre Est (DIR-CE)
Genas
France

Jean-Christophe MAISONOBE
Conseil général de l'Isère
Grenoble
France

Kerry MALONE
TNO
Delft
The Netherlands

Florian MARCZAK
Laboratoire Ingénierie de la Circulation Transports (LICIT)
IFSTTAR and ENTPE
Lyon
France

Giannicola MARENGO
Provincia di Torino
Turin
Italy

Gregoire MARLOT
SNCF Réseau
Paris
France

Gerhard MENZEL
Federal Agency for Technological Measures Ltd
Austria Tech
Vienna
Austria

Isabela MOCANU
Austrian Institute of Technology (AIT)
Vienna
Austria

Enda MURPHY
University College Dublin
Ireland

Philippe NITSCHE
Austrian Institute of Technology (AIT)
Vienna
Austria

María PEREZ HERRERO
SNCF Réseau
Paris
FRANCE

Charlotte PIERREFEU
Centre d'Etudes et d'expertise sur les Risques, l'Environnement, la Mobilité et l'Aménagement (Cerema)
Direction territoriale Centre-Est
Bron
France

Eric PURSON
Centre d'Etudes et d'expertise sur les Risques, l'Environnement, la Mobilité et l'Aménagement (Cerema)
Direction territoriale Est
Metz
France

Fabrice RECLUS
CETE de Lyon
Bron
France

Isabel RIBEIRO
University of Porto
Faculty of Engineering
Porto
Portugal

Vlasta RODOŠEK
Faculty of Civil Engineering, Transportation Engineering and Architecture
University of Maribor
Slovenia

Caroline SORAND
Laboratoire Ingénierie de la Circulation Transports (LICIT)
IFSTTAR and ENTPE
Lyon
France

Rafiqul TAREFDER
University of New Mexico
Albuquerque
New Mexico
USA

Henny TER HUERNE
University of Twente
Enschede
The Netherlands

Aikaterini TOULIOU
CERTH – HIT
Thessaloniki
Greece

Cecília VALE
University of Porto
Faculty of Engineering
Porto
Portugal

Ernst Jan VAN ARK
TNO
Delft
The Netherlands

Eric VAN BERKUM
University of Twente
Enschede
The Netherlands

Diana VONK NOORDEGRAAF
TNO
Delft
The Netherlands

George YANNIS
School of Civil Engineering
National Technical University of Athens
Greece

Konstantinos G. ZOGRAFOS
School of Management
Department of Management Science
Lancaster University
UK

Vladimir ZYRYANOV
Rostov State Civil Engineering University
Rostov-on-Don
Russia

Katharina ZWICK
Federal Agency for Technological Measures Ltd
Austria Tech
Vienna
Austria

Index

Printed and bound by CPI Group (UK) Ltd, Croydon, CR0 4YY
15/09/2026
14962356-0004